Integral equations—
a reference text

P. P. Zabreyko, A. I. Koshelev, M. A. Krasnosel'skii, S. G. Mikhlin. L. S. Rakovshchik, and V. Ya. Stet'senko

Integral equations—a reference text

Translated and edited by

T. O. Shaposhnikova
Department of Mathematics
Leningrad University

R. S. Anderssen
Computer Centre
Australian National University

S. G. Mikhlin
Department of Mathematics
Leningrad University

NOORDHOFF INTERNATIONAL PUBLISHING, LEYDEN

ISBN: 90 286 0393 x

Library of Congress Catalog Card Number: 73-84403

Original title

'Integral'nye uravneneya'

published in 1968 in Moscow by the Nauka Press

Set-in-type in England
Printed in the Netherlands

CONTENTS

THE AUTHORS' PREFACE

The title 'Integral equations' covers many things which have very little connection with each other. However, they are united by the following important feature. In most cases, the equations involve an unknown function operated on by a bounded and often compact operator defined on some Banach space.

The aim of the book is to list the main results concerning integral equations.

The classical Fredholm theory and Hilbert-Schmidt theory are presented in Chapters II and III. The preceding Chapter I contains a description of the most important types of integral equations which can be solved in 'closed' form. Chapter IV is an important addition to Chapters II and III, as it contains the theory of integral equations with non-negative kernels. The development of this theory is mainly due to M. G. Krein. The content of the first four chapters is fairly elementary.

It is well known that the Fredholm theory has been generalized for equations with compact operators. Chapter V is devoted to this generalization.

In Chapter VI one-dimensional (i.e. with one dependent variable) singular integral equations are considered. The last type of equations differ from that considered in the preceding chapters in that singular integral operators are not compact but only bounded in the usual functional spaces.

Because the above mentioned types of integral equations involve lots of applications, Chapter VII is devoted to the application of integral equations of mathematical physics. Chapter VIII contains the theory of integral equations of convolution type. Included among them are such important types of integral equations as Wiener-Hopf and dual. In Chapter IX the main results of the theory for multi-dimensional singular integral equations are presented. Bounded operators, which are not compact, are considered in Chapters VIII, IV, as well as in Chapter VI.

All the above mentioned types of integral equations are only concerned with linear equations. The last chapter, Chapter X, is devoted to non-linear integral equations. In that chapter, some results about the compactness for non-linear operators are presented. Questions of the existence and uniqueness of, as well as the continuity and bifurcation of solutions, are considered for equations with such operators.

Chapters I–III, VI, IX were written by S. G. Mikhlin. Chapters IV, V, X were written jointly by P. P. Zabreyko, M. A. Krasnosel'skii and V. Ya. Stet'senko. Chapter VII was written by A. I. Koshelev and Chapter VIII by L. S. Rakovshchik.

25 May, 1966 The authors

THE TRANSLATORS' AND EDITORS' PREFACE

The translators and editors would like to take this opportunity to record their thanks to people and institutions who have helped in some way or other with the preparation of this book. In particular, we wish to record our thanks to Professor J. R. M. Radok, The Horace Lamb Institute of Oceanography, South Australia; Mrs. Stephanie Venema, Noordhoff International Publishing, The Netherlands; Tony Jakeman and Greg Woodford, Computer Centre, Australian National University. Special thanks are also due to Judy McHugh (nee Knight) of the Computer Centre, Australian National University, for typing the manuscript.

Chapter I

GENERAL INTRODUCTION

Integral equations are sometimes defined as equations which involve the integration of an unknown function. This definition is not apt as it is too wide: it fails to specify what additional operations can be performed on the unknown as well as the integral. Consequently, we could have a situation where certain differential equations take the place of integral equations. For example, if the operation of 'lim' is allowed, then

$$a(x) \lim_{h \to 0} \int_{x}^{x+h} \frac{u(t+h) - 2u(t) + u(t-h)}{h^3} \, dt + b(x)\, u(x) = f(x)$$

is actually the second order differential equation

$$a(x)\, u''(x) + b(x)\, u(x) = f(x).$$

However, it is difficult to specify which operations on the unknown are permissible in order to give a general definition for integral equations. In particular, it is impossible to eliminate the operation of 'lim', which is crucial for the definition of a singular integral (see Chapters VI and IX). This concept is needed, if we are to consider the subject of singular integral equations.

We shall not give a general definition for integral equations. In fact, we limit consideration to some of the most important classes of integral equations. They are summarized in the next two sections.

We shall assume that all functions below are measurable, that independent variables are real except when otherwise stated, and that known and unknown functions can be real or complex.

§1 Fredholm and Volterra equations

1.1 *Fredholm equations*

These represent one of the most important classes of linear integral equations. We shall distinguish between *Fredholm equations of the first and second kinds*.

Equations of the second kind are the most interesting and important in applications. The simplest formulation for such equations is

$$u(x) - \int_a^b K(x, t)\, u(t)\, dt = f(x). \tag{1.1}$$

Here, the unknown function $u(x)$ is a function of the real variable x which, along with t, is defined on the interval $[a, b]$. This is a general condition which will apply to all classes of integral equations considered in this book. The interval $[a, b]$ may be finite or infinite. The functions $K(x, t)$ and $f(x)$ are known and defined almost everywhere on $a \leqslant x, t \leqslant b$ and $a \leqslant x \leqslant b$, respectively. $K(x, t)$ is called the *kernel* of (1.1), $f(x)$ the *right hand side* (or, more briefly, *r.h.s.*). We shall assume that, for Fredholm equations, $K(x, t)$ satisfies

$$\int_a^b \int_a^b |K(x, t)|^2\, dx\, dt < \infty, \tag{1.2}$$

and the r.h.s. satisfies

$$\int_a^b |f(x)|^2\, dx < \infty. \tag{1.3}$$

Frequently, it is necessary to consider Fredholm equations of a more general form. Let Ω be a measurable set in a space with an arbitrary number of variables, x and t points in this set, and μ a non-negative measure defined on Ω. The following equation

$$u(x) - \int_\Omega K(x, t)\, u(t)\, d\mu(t) = f(x) \tag{1.4}$$

is also a *Fredholm equation*, if the Kernel $K(x, t)$ and the r.h.s. $f(x)$ satisfy

$$\int_\Omega \int_\Omega |K(x, t)|^2\, d\mu(x)\, d\mu(t) < \infty \tag{1.5}$$

and

$$\int_{\Omega} |f(x)|^2 \, d\mu(x) < \infty \tag{1.6}$$

Kernels $K(x, t)$ satisfying (1.5) are called *Fredholm kernels*.

When solving (1.1), we shall always assume that $u(x)$ is square summable in (a, b), and hence, belongs to the functional space $L_2(a, b)$. By analogy, we shall assume that the solution of (1.4) belongs to the space $L_2(\mu, \Omega)$ of functions which are square summable in Ω with respect to the measure μ. Inequalities (1.3) and (1.6) imply that the r.h.s. also belongs to the same space.

Let $K(x, t)$ satisfy (1.5). The expression

$$\int_{\Omega} K(x, t) \, u(t) \, d\mu(t) \tag{1.7}$$

defines an integral operator in the space $L_2(\mu, \Omega)$. It is called a *Fredholm operator*. It maps $L_2(\mu, \Omega)$ into itself, and is bounded. Denoting this operator by $\boldsymbol{K}$, we obtain

$$\|\boldsymbol{K}\| \leqslant \left[\int_{\Omega}\int_{\Omega} |K(x, t)|^2 \, d\mu(x) \, d\mu(t) \right]^{1/2} \tag{1.8}$$

If $\mu(x)$ is a Lebesgue measure, i.e. $d\mu(x) = dx$, then $\boldsymbol{K}$ acts in the space $L_2(\Omega)$ and

$$\|\boldsymbol{K}\| \leqslant \left[\int_{\Omega}\int_{\Omega} |K(x, t)|^2 \, dx \, dt \right]^{1/2} \tag{1.8$_1$}$$

The most frequently occurring cases arise when Ω is a region or surface in Euclidean space, and $d\mu$ is a corresponding element of volume or area, respectively. In particular, Ω may be a curve, and $d\mu$ an element of length on this curve.

If $f(x) \equiv 0$ (or, more precisely, if $f(x)$ vanishes almost everywhere in Ω), then (1.4) is said to be *homogeneous*, otherwise (i.e. if $f(x)$ doesn't vanish on a set of positive measure) *non-homogeneous*.

We shall consider some examples of Fredholm equations. In all cases, it is assumed that $f(x)$ satisfies either (1.3) or (1.6).

Example 1.1. Let the real variables x and t be defined on the finite interval $[a, b]$. If $K(x, t)$ is bounded on the square $a \leqslant x \leqslant b$, $a \leqslant t \leqslant b$, then (1.1) defines a second kind Fredholm equation.

Example 1.2. Equation (1.1) is also a Fredholm in the case when the interval $[a, b]$ is finite and the kernel has the form

$$K(x, t) = \frac{A(x, t)}{|x - t|^{\alpha}}, \qquad 0 \leqslant \alpha < \tfrac{1}{2}, \qquad \alpha = \text{const},$$

where $A(x, t)$ is a bounded function. In fact, let $|A(x, t)| \leqslant C$, $C = \text{const}$. Then

$$\int_a^b \int_a^b |K(x, t)|^2 \, dx \, dt \leqslant C^2 \int_a^b \int_a^b \frac{dx \, dt}{|x - t|^{2\alpha}} = C^2 \int_a^b dx \int_a^b \frac{dt}{|x - t|^{2\alpha}}.$$

Further

$$\int_a^b \frac{dt}{|x - t|^{2\alpha}} = \int_a^x \frac{dt}{(x - t)^{2\alpha}} + \int_x^b \frac{dt}{(t - x)^{2\alpha}} = -\left.\frac{(x - t)^{1 - 2\alpha}}{1 - 2\alpha}\right|_a^x + \left.\frac{(t - x)^{1 - 2\alpha}}{1 - 2\alpha}\right|_x^b$$

$$= \frac{(x - a)^{1 - 2\alpha} + (b - x)^{1 - 2\alpha}}{1 - 2\alpha}.$$

The last expression is a continuous function of x, and hence, is bounded on $a \leqslant x \leqslant b$. Let C_1 be its upper bound. Then

$$\int_a^b \int_a^b |K(x, t)|^2 \, dx \, dt \leqslant C^2 C_1 (b - a),$$

and thus, (1.2) is satisfied.

Example 1.3. Let Ω be an m-dimensional surface in Euclidean space of $(m + 1)$-dimensions, x and t be points of Ω, $d\Omega$ denote an elementary square on Ω, and r denote the distance between x and t. Suppose that $K(x, t)$ has the form

$$K(x, t) = \frac{A(x, t)}{r^{\alpha}}; \qquad 0 \leqslant \alpha < \frac{m}{2}, \qquad \alpha = \text{const}, \tag{$*$}$$

where $A(x, t)$ is a bounded function. Then (1.4) with $d\mu(t) = d_t\Omega$ is a Fredholm equation. The subscript t in $d_t\Omega$ implies that integration is performed with respect to t while x is fixed.

Example 1.4. Equation (1.4) is also a Fredholm, when Ω is a region in m-dimensional Euclidean space with $d\mu(t) = dt$, where dt is an element of volume, and the kernel has the form (*) with $A(x, t)$ bounded.

Example 1.5. The equation

$$u(x) - \int_1^\infty e^{-xt}\, u(t)\, dt = f(x)$$

is a Fredholm, since

$$\int_1^\infty \int_1^\infty |K(x, t)|^2\, dx\, dt = \int_1^\infty dt \int_1^\infty e^{-2xt}\, dt = \frac{1}{2}\int_1^\infty \frac{e^{-2x}}{x}\, dx < \infty.$$

Example 1.6. The equation

$$u(x) - \int_0^\infty e^{-xt}\, u(t)\, dt = f(x)$$

is not a Fredholm. In fact,

$$\int_0^\infty \int_0^\infty e^{-2xt}\, dx\, dt = \frac{1}{2}\int_0^\infty \frac{dx}{x} = \infty.$$

Fredholm equations of the first kind differ from the second kind in that the isolated term '$u(x)$' in (1.1) and (1.4) is absent. The simplest form for it is

$$\int_a^b K(x, t)\, u(t)\, dt = f(x), \tag{1.9}$$

where $K(x, t)$ and $f(x)$ satisfy (1.2) and (1.3), respectively. In general, Fredholm equations of the first kind have the form

$$\int K(x, t)\, u(t)\, d\mu(t) = f(x). \tag{1.10}$$

In this case, inequalities (1.5) and (1.6) must be satisfied.

1.2 *Equations with a weak singularity*

They have the form (1.4) with Ω a bounded region in m-dimensional Euclidean space, or a bounded surface in $(m + 1)$-dimensional Euclidean space (the last case can be easily reduced to the first by introducing a parameterization of the surface), and $d\mu(t) = dt$ in the first case and $d\mu(t) = d_t\Omega$ in the second. The kernel has the form

$$K(x, t) = \frac{A(x, t)}{r^\alpha}, \qquad 0 < \alpha < m, \qquad \alpha = \text{const}, \tag{1.11}$$

with $A(x, t)$ a bounded function, and r the distance between the points x and t.

If $\alpha < m/2$, then equations with a weak singularity are also of Fredholm type. If $K(x, t)$ is a kernel with a weak singularity, then the operator (1.7) is called an *integral operator with a weak singularity*.

1.3 *Volterra equations*

For functions with one independent variable, *Volterra equations of the second kind* have the form

$$u(x) - \int_a^x K(x, t)\, u(t)\, dt = f(x), \qquad a \leqslant x \leqslant b. \tag{1.12}$$

The situation where $a = -\infty$ and/or $b = +\infty$ are not excluded. Volterra equations for functions with more than one independent variable can be defined in the following way.

Let $x_1, x_2, \ldots, x_m$ and $t_1, t_2, \ldots, t_m$ be the Cartesian components of points x and t which lie in m-dimensional Euclidean space. We set

$$\rho = \left\{ \sum_{k=1}^{m-1} (x_k - t_k)^2 \right\}^{1/2}$$

and denote by Q_x the set of points t bounded by the cone $\rho = c(x_m - t_m)$ and the plane $t_m = a$, where a and c are constants with $c > 0$. We now define

$$u(x) - \int_{Q_x} K(x, t)\, u(t)\, dt = f(x) \tag{1.13}$$

as a Volterra equation of the second kind.

Here, we shall not formulate conditions which could be imposed on the kernel and the r.h.s. of Volterra equations. This will be done in Chapter II, where methods for the solution of integral equations are discussed.

Volterra equations of the first kind are characterized by the absence of the isolated term $u(x)$ in (1.13). Volterra equations of the first kind for functions with one independent variable have the form

$$\int_a^x K(x, t)\, u(t)\, dt = f(x). \tag{1.14}$$

A particular case of this equation is the *Abel equation*

$$\int_a^x \frac{u(t)}{(x - t)^\alpha} = f(x), \qquad 0 < \alpha < 1, \qquad \alpha = \text{const.} \tag{1.15}$$

For functions with more than one independent variable, Volterra equations of the first kind take the form

$$\int_{Q_x} K(x, t)\, u(t)\, dt = f(x). \tag{1.16}$$

Under certain circumstances, we can consider Volterra equations to be particular cases of Fredholm equations. For functions of one independent variable, let $f(x)$ satisfy (1.3) and the kernel

$$\int_a^b \left\{ \int_a^x |K(x, t)|^2 \, dt \right\} dx < \infty. \tag{1.17}$$

Thus, $K(x, t)$ is defined for $a \leqslant t \leqslant x$. We complete its definition by requiring that

$$K(x, t) = 0, \qquad x < t \leqslant b.$$

Consequently, (1.12) can be rewritten as (1.1) with the kernel satisfying (1.2).

§2 Other classes of integral equations

2.1 *Equations with convolution kernels*

Integral equations with convolution kernels have the form

$$au(x) - \int_{E_m} K(x - t) \, u(t) \, dt = f(x), \tag{1.18}$$

where a is a constant, E_m is m-dimensional Euclidean space, x and t are points of E_m, and the kernel $K(x - t)$ depends only on the difference between the vectors x and t, i.e. on the differences $x_k - t_k$ between the Cartesian components of x and t. If $m = 1$, then (1.18) becomes

$$au(x) - \int_{-\infty}^{+\infty} K(x - t) \, u(t) \, dt = f(x), \tag{1.19}$$

where x, t are real variables.

Sometimes, 'almost convolution' equations are considered. They have the form

$$a(x) \, u(x) - \int_{E_m} K(x, x - t) \, u(t) \, dt = f(x). \tag{1.20}$$

Here, the dependence on the first argument in the kernel $K(x, x - t)$ is unimportant in some sense. This will be discussed in detail in Chapter VIII.

Volterra equations with convolution kernels are also of some interest. When $m = 1$, they take the form

$$au(x) - \int_{-\infty}^{x} K(x - t)\, u(t)\, dt = f(x) \tag{1.21}$$

or

$$au(x) - \int_{0}^{x} K(x - t)\, u(t)\, dt = f(x). \tag{1.21$_1$}$$

Equation (1.21) is a particular case of (1.19), and can be obtained from the latter if we put $K(z) = 0$, $z < 0$.

2.2 *The Wiener-Hopf equations*

The Wiener-Hopf equation has a convolution kernel and is defined on the half line $[0, \infty)$:

$$au(x) - \int_{0}^{\infty} K(x - t)\, u(t)\, dt = f(x), \qquad a = \text{const.} \tag{1.22}$$

It is possible to construct equations of such a type in multidimensional space. The integration along the half-line $[0, \infty)$ is replaced by integration over the interior of a cone with vertex at the origin. The case when the cone is a half-space is regularly investigated. In this situation, we deal with Weiner-Hopf equations of the form

$$au(x) - \int_{x_m > 0} K(x - t)\, u(t)\, dt = f(x), \qquad a = \text{const.} \tag{1.23}$$

2.3 *Dual equations*

For functions of one independent variable, dual equations take the form

$$au(x) - \int_{-\infty}^{+\infty} K_1(x - t)\, u(t)\, dt = f(x); \qquad x > 0, \tag{1.24}$$

$$bu(x) - \int_{-\infty}^{+\infty} K_2(x - t)\, u(t)\, dt = f(x); \qquad x < 0, \tag{1.24}$$

where a and b are constants.

2.4 *Integral transforms*

Any one of the known integral transforms can be regarded as an integral equation operating on the function with respect to which the transform is defined. We give some examples. Following the notation developed above, we denote the function on which the transform operates by u, and the result by f.

a) *The Fourier cosine-transform*

$$\sqrt{\left(\frac{2}{\pi}\right)} \int_0^\infty u(t) \cos tx\, dt = f(x). \tag{1.25}$$

b) *The Fourier sine-transform*

$$\sqrt{\left(\frac{2}{\pi}\right)} \int_0^\infty u(t) \sin tx\, dt = f(x). \tag{1.26}$$

c) *The complex Fourier transform*

$$\frac{1}{\sqrt{(2\pi)}} \int_{-\infty}^{+\infty} u(t) e^{-itx}\, dt = f(x). \tag{1.27}$$

d) *The multi-dimensional complex Fourier transform*

$$\frac{1}{(2\pi)^{m/2}} \int_{E_m} u(t) e^{-i(x,t)}\, dt = f(x), \tag{1.28}$$

where

$$(x, t) = \sum_{k=1}^{m} x_k t_k$$

denotes the inner product between vectors x and t.

e) *The Laplace transform*

$$\int_0^\infty e^{-tx}\, u(t)\, dt = f(x). \tag{1.29}$$

f) *The Mellin transform*

$$\int_0^\infty t^{x-1}\, u(t)\, dt = f(x). \tag{1.30}$$

g) *The Hankel transform*

$$\int_0^\infty \sqrt{(tx)}\, J_\nu(tx)\, u(t)\, dt = f(x), \tag{1.31}$$

where J_ν is the Bessel function of the first kind of order ν.

h) *The Kantorovich-Lebedev transform*

$$\frac{1}{2}\int_0^\infty t \operatorname{sh} \pi t e^{\pi t/2}\, H_{it}^{(2)}(kx)\, u(t)\, dt = f(x), \tag{1.32}$$

where $H_{it}^{(2)}$ is the Hankel function of the second kind with the purely imaginary order 'it'.

2.5 *Singular integral equations*

Let Γ be a finite or countable number of simple closed or open arcs iying in the complex plane. The one-dimensional singular integral equation has the form

$$a(t)\, u(t) - \frac{b(t)}{\pi i}\int_\Gamma \frac{u(\zeta)}{\zeta - t}\, d\zeta + \int_\Gamma K(t, \zeta)\, u(\zeta)\, d\zeta = f(t), \tag{1.33}$$

where ζ and t are complex points on Γ, $a(t)$, $b(t)$, $f(t)$ are given functions defined on Γ, and $K(t, \zeta)$ is a Fredholm kernel. If the sum of the arc lengths generating Γ is finite, then it is possible to let $K(t, \zeta)$ be a kernel with a weak singularity. The divergent integral in (1.33) is interpreted in the sense of its Cauchy principal value

$$\int_{\Gamma} \frac{u(\zeta)}{\zeta - t}\, d\zeta = \lim_{\epsilon \to 0} \int_{\Gamma \cap (|\zeta - t| > \varepsilon)} \frac{u(\zeta)}{\zeta - t}\, d\zeta.$$

In the simplest multi-dimensional case, when the range of integration is the m-dimensional Euclidean space E_m, the multi-dimensional singular integral equation takes the form

$$a(x)\, u(x) + \int_{E_m} \frac{f(x, \theta)}{|x - t|^m}\, u(t)\, dt + \int_{E_m} K(x, t)\, u(t)\, dt = f(t), \tag{1.34}$$

where $\theta = (t - x)/|t - x|$, and $K(x, t)$ is a Fredholm kernel. The divergent integral in (1.34) is again interpreted in the sense of its Cauchy principal value

$$\int_{E_m} \frac{f(x, \theta)}{|x - t|^m}\, u(t)\, dt = \lim_{\varepsilon \to 0} \int_{|x - t| > \varepsilon} \frac{f(x, \theta)}{|x - t|^m}\, u(t)\, dt.$$

The space E_m can be replaced by other m-dimensional varieties. A particularly interesting case arises when an m-dimensional surface in a Euclidean space of more than m-dimensions is examined. The singular integral equation can now be rewritten as

$$a(x)\, u(x) + \int_{S} \frac{f(x, \theta)}{|x - t|^m}\, u(t)\, d_t S + \int_{S} K(x, t)\, u(t)\, d_t S = f(x). \tag{1.34$_1$}$$

In Chapter IV, more general singular equations will be investigated.

2.6 Non-linear integral equations

Below we cite some important classes of non-linear integral equations.

The simplest of them is Hammerstein's

$$u(x) - \int_{\Omega} K(x, t) F(t, u(t)) \, dt = 0, \tag{1.35}$$

where $K(x, t)$ is a Fredholm kernel. A more general class is Urison's

$$u(x) - \int_{\Omega} K(x, t, u(t)) \, dt = 0, \tag{1.36}$$

where $K(x, t, u)$ is usually required to be continuous for $x, t \in \bar{\Omega}$, $|u| \leqslant a$, where a is a sufficiently large positive constant, and $\bar{\Omega}$ is the closure of some bounded set Ω.

Let $u(x)$ and $v(x)$ be continuous functions defined on Ω which is assumed to be closed and bounded. In addition, let α and β be vectors with non-negative integer components

$$\alpha = (\alpha_0, \alpha_1, \ldots, \alpha_\rho), \qquad \beta = (\beta_0, \beta_1, \ldots, \beta_\rho),$$

and $K_{\alpha,\beta}(x, t^{(1)}, t^{(2)}, \ldots, t^{(\rho)})$ be a continuous function of its arguments, each of which varies over Ω. We call the following expression an *integro-power term*

$$L_{\alpha,\beta}(u, v) = \int_{\Omega} \ldots \int_{\Omega} K_{\alpha,\beta}(x, t^{(1)}, t^{(2)}, \ldots, t^{(\rho)}) u^{\alpha_0}(x) \, u^{\alpha_1}(t^{(1)}) \ldots u^{\alpha_\rho}(t^{(\rho)})$$
$$\times \, v^{\beta_0}(x) \, v^{\beta_1}(t^{(1)}) \ldots v^{\beta_\rho}(t^{(\rho)}) \, dt^{(1)} \, dt^{(2)} \ldots dt^{(\rho)}. \tag{1.37}$$

The series

$$L(u, v) = \sum_{\alpha,\beta} L_{\alpha,\beta}(u, v), \tag{1.38}$$

in which ρ is fixed and the summation is taken over all possible vectors α and β, is called an *integro-power series*. The equation

$$L(u, v) = 0, \tag{1.39}$$

in which u is unknown and v is known is called *Liapounov's equation*.

Non-linear one-dimensional singular integral equations can be represented in the following way

$$F(t, u(t), \boldsymbol{S}(u), \boldsymbol{T}_1(u), \boldsymbol{T}_2(u), \ldots, \boldsymbol{T}_k(u)) = 0 \tag{1.40}$$

For brevity, we have written

$$\boldsymbol{S}(u) = \frac{1}{\pi i} \int_\Gamma \frac{u(\zeta)}{\zeta - t} \, d\zeta,$$

$$T_j(u) = \int_\Gamma K_j(t, \zeta) \, u(\zeta) \, d\zeta \qquad (j = 1, 2, \ldots, k),$$

where $K_j(t, \zeta)$ are Fredholm kernels.

Multi-dimensional singular integral equations can have the form

$$F(x, u(x), \boldsymbol{\sigma}_1(u), \boldsymbol{\sigma}_2(u), \ldots, \boldsymbol{\sigma}_n(u), \boldsymbol{T}_1(u), \boldsymbol{T}_2(u), \ldots, \boldsymbol{T}_k(u)) = 0, \tag{1.41}$$

where

$$\boldsymbol{\sigma}_j(u) = \int_S \frac{f_j(x, \theta)}{|x - t|^m} \, u(t) \, dt \qquad (j = 1, 2, \ldots, n).$$

and, as above, $\boldsymbol{T}_j$ are Fredholm operators, i.e. integral operators with Fredholm kernels. More general non-linear singular integral equations could also be defined.

§3 Some inversion formulas

If $\boldsymbol{A}$ is an operator, and the equation

$$\boldsymbol{A}u = f \tag{1.42}$$

has a unique solution

$$u = \boldsymbol{A}^{-1} f, \tag{1.43}$$

then (1.43) is called an *inversion formula* for (1.42). In this section, we shall give some examples of integral equations for which comparatively simple inversion formulas are known.

§3 *Some inversion formulas*

3.1 *The inversion of integral transforms*

Following the order in §2.4, we list the inverse transforms corresponding to the integral transforms cited there {see, for example, [26] and [27]}

$$\text{a)}\quad u(x) = \sqrt{\left(\frac{2}{\pi}\right)} \int_0^{\infty} f(t) \cos tx \, dt. \tag{1.44}$$

$$\text{b)}\quad u(x) = \sqrt{\left(\frac{2}{\pi}\right)} \int_0^{\infty} f(x) \sin tx \, dt. \tag{1.45}$$

$$\text{c)}\quad u(x) = \frac{1}{\sqrt{(2\pi)}} \int_{-\infty}^{+\infty} e^{itx} f(x) \, dt. \tag{1.46}$$

$$\text{d)}\quad u(x) = \frac{1}{(2\pi)^{m/2}} \int_{E_m} e^{i(t,x)} f(t) \, dt. \tag{1.47}$$

$$\text{e)}\quad u(x) = \frac{1}{2\pi i} \int_{k-i\infty}^{k+i\infty} e^{tx} f(t) \, dt. \tag{1.48}$$

In (1.48), k is a positive constant. This inverse transform is valid for all $f(t)$ which are regular in the complex half-plane $\operatorname{Re} t > k', k' < k$, and have in this half-plane the asymptotic estimate $O(|t|^{-\alpha})$, $\alpha > 0$, for large $|t|$. This ensures that the r.h.s. of (1.48) equals zero when $x < 0$.

$$\text{f)}\quad u(x) = \frac{1}{2\pi i} \int_{c-i\infty}^{c+i\infty} f(t)\, x^{-t} \, dt. \tag{1.49}$$

The inverse transform (1.49) is valid for all $f(t)$ for which

$$\int_0^{\infty} x^{k-1} \, |u(x)| \, dx$$

converges for some $k > 0$. It is necessary to require that $c > k$ {see [26], [27]}.

g) $$u(x) = \int_0^\infty f(t)\sqrt{(tx)}\, J_\nu(tx)\, dt. \tag{1.50}$$

h) $$u(x) = \int_0^\infty e^{\pi x/2}\, H_{ix}^{(2)}(kt)\, f(t)\, \frac{dt}{t}. \tag{1.51}$$

A general inverse transform {see [26]} from which many of the above follow as particular cases can be defined. Consider

$$\int_0^\infty K(x, t)\, u(t)\, dt = f(t) \tag{1.52}$$

and let

$$\boldsymbol{K}(x) = \int_0^\infty t^{x-1}\, K(t)\, dt$$

be the Mellin transform of the kernel $K(x)$. If there exists a function $H(x)$ such that its Mellin transform

$$\boldsymbol{H}(x) = \int_0^\infty t^{x-1}\, H(t)\, dt$$

satisfies

$$\boldsymbol{K}(x)\, \boldsymbol{H}(1 - x) = 1, \tag{1.53}$$

then the inverse transform corresponding to (1.52) is given by

$$u(x) = \int_0^\infty H(x, t)\, f(t)\, dt. \tag{1.54}$$

3.2 *Inversion fomulas for equations with convolution kernels*

It is easy to investigate (1.19), if use is made of the fact that the product of two complex Fourier transformations corresponds to the convolution of these functions. In order to clarify this, we assume that $f(x) \in L_2(-\infty, +\infty)$, and seek the solution of (1.19) which is contained in $L_2(-\infty, +\infty)$. Initially, we take the Fourier transform of (1.19). Assuming that the convolution theorem is applicable, we obtain

$$[a - \sqrt{(2\pi)}\,\tilde{K}(x)]\,\tilde{u}(x) = \tilde{f}(x),$$

where '$\tilde{u}$' denotes the Fourier transform of u. If $[a - \sqrt{(2\pi)}\,\tilde{K}(x)]^{-1}$ is bounded, then

$$\tilde{u}(x) = \frac{\tilde{f}(x)}{a - \sqrt{(2\pi)}\,\tilde{K}(x)},$$

and we obtain the following inversion formula

$$u(x) = \frac{1}{\sqrt{2\pi}} \int_{-\infty}^{+\infty} \frac{e^{itx}\,\tilde{f}(t)}{a - \sqrt{2\pi}\,\tilde{K}(t)}\,dt. \tag{1.55}$$

If $[a - \sqrt{(2\pi)}\,\tilde{K}(x)]^{-1}$ is not bounded {this will be the case, for example, when $\tilde{K}(x)$ is continuous and $a - \sqrt{(2\pi)}\,\tilde{K}(x)$ vanishes at some real x}, then (1.19) has no solution for arbitrary $f(x)$. Nevertheless, the inversion formula (1.55) remains effective if the following additional conditions are satisfied:

1) $a - \sqrt{(2\pi)}\,\tilde{K}(t)$ vanishes only on a set of measure zero,
2) $\tilde{f}(t)[a - \sqrt{(2\pi)}\,\tilde{K}(t)]^{-1}$ is square summable with respect to t for $-\infty < t < +\infty$.

Example 3.1. Consider the equation

$$au(x) - \int_{-\infty}^{+\infty} e^{-(x-t)^2}\,u(t)\,dt = f(x). \tag{1.56}$$

In this case, $K(x) = e^{-x^2}$ and

$$\tilde{K}(x) = \frac{1}{\sqrt{(2\pi)}} \int_{-\infty}^{+\infty} e^{-t^2 - itx}\,dt.$$

The evaluation of this last integral is straightforward. We have

$$\int_{-\infty}^{+\infty} e^{-t^2-itx}\,dx = e^{-x^2/4}\int_{-\infty}^{+\infty} e^{-(t+ix/2)^2}\,dt.$$

Applying Cauchy's theorem to the integral

$$\int_{\Pi} e^{-z^2}\,dz,$$

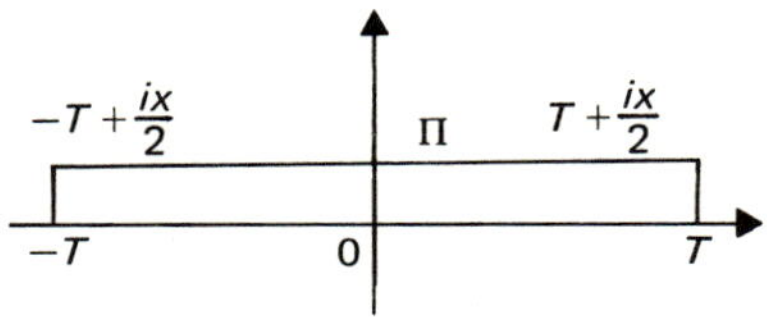

Figure 1

where Π is a rectangle in the complex z-plane (see Fig. 1); and then letting $T \to \infty$, we obtain

$$\int_{-\infty}^{+\infty} e^{-(t+ix/2)^2}\,dt = \int_{-\infty}^{+\infty} e^{-t^2}\,dt = \sqrt{\pi},$$

and hence, $\tilde{K}(x) = [1/\sqrt{(2)}]e^{-x^2/4}$. Taking the Fourier transform of (1.56), we obtain

$$\left(a - \frac{1}{\sqrt{2}}\,e^{-x^2/4}\right)\tilde{u}(x) = \tilde{f}(x).$$

If $a < 0$ or $a > (1/\sqrt{2})$, then the solution of (1.56) exists, is unique and is defined by

$$u(x) = \frac{1}{\sqrt{(2\pi)}}\int_{-\infty}^{+\infty} \frac{\tilde{f}(t)}{(a - e^{-t^2/4}/\sqrt{2})}\,e^{itx}\,dt. \tag{1.57}$$

Consider the case $a = 0$. Here, the solution is unique, since $\tilde{K}(x) = [1/\sqrt{2}]e^{-x^2/4}$ does not vanish. It exists, if $f(x)$ is such that $e^{x^2/4}\tilde{f}(x) \in L_2(-\infty, +\infty)$, and it is defined by

$$u(x) = -\frac{1}{\sqrt{\pi}} \int_{-\infty}^{+\infty} e^{t^2/4}\, \tilde{f}(t)\, e^{ixt}\, dt. \tag{1.58}$$

Example 3.2. {see [30]} The equation

$$\int_0^{\infty} \frac{v(\tau)}{\tau + \xi}\, d\tau = g(\xi) \tag{1.59}$$

can be transformed into (1.19). In fact, setting

$$\xi = e^x,\ \tau = e^t,\ e^{t/2}v(e^t) = u(t),\ e^{x/2}g(e^x) = f(x),$$

we obtain

$$\int_{-\infty}^{+\infty} \frac{u(t)\, dt}{2 \operatorname{ch} \dfrac{x - t}{2}} = f(x). \tag{1.60}$$

It follows from (1.55), that its solution is

$$u(x) = \frac{1}{\sqrt{(2\pi^3)}} \int_{-\infty}^{+\infty} \tilde{f}(t) \operatorname{ch} \pi t\, e^{itx}\, dt.$$

The transformation back to the original variables is not difficult.

For more general equations of the form (1.18), which satisfy the condition $[a - (2\pi)^{m/2}\, \tilde{K}(t)]^{-1}$ is bounded, we obtain the inversion formula

$$u(x) = \frac{1}{(2\pi)^{m/2}} \int_{E_m} \frac{\tilde{f}(t) e^{i(t,x)}}{a - (2\pi)^{m/2}\, \tilde{K}(t)}\, dt. \tag{1.61}$$

The formula (1.57) is also an inversion formula for the Volterra equation

(1.21) with the obvious difference that in this case

$$\tilde{K}(x) = \frac{1}{\sqrt{(2\pi)}} \int_0^{\infty} K(t) e^{-ixt} \, dt.$$

3.3 *Volterra's equations with one independent variable and a convolution kernel*

Taking Laplace transforms of (1.21_1), we obtain

$$a\tilde{u}(\xi) - \int_0^{\infty} e^{-\xi x} \left\{ \int_0^x K(x-t)\, u(t)\, dt \right\} dx = \tilde{f}(\xi),$$

where '$\tilde{u}$' denotes the Laplace transform of u. Changing the order of integration in the double integral, and then setting $x - t = z$, we generate the algebraic equation

$$[a - \tilde{K}(\xi)]\, \tilde{u}(\xi) = \tilde{f}(\xi). \tag{1.62}$$

Suppose that, in the right hand half of the complex plane, there is a straight line parallel to the imaginary axis on which $[a - \tilde{K}(\xi)]$ is non-zero. If the equation of this line is $\operatorname{Im} \xi = k$, then

$$u(x) = \frac{1}{2\pi i} \int_{k-i\infty}^{k+i\infty} \frac{\tilde{f}(\xi)}{a - \tilde{K}(\xi)} e^{\xi t} \, d\xi \tag{1.63}$$

yields the solution of (1.21_1). Of course, we assume that

$$\frac{\tilde{f}(\xi)}{a - \tilde{K}(\xi)}$$

is such that (1.48) is applicable.

3.4 *The Abel equation*

An inversion formula for (1.15) is easily obtained in the following manner. We replace x by s, multiply both sides of (1.15) by $(x - s)^{\alpha - 1}\, ds$ and integrate with respect to s from a to x. After some elementary manipulations and the use of

$$\int_0^1 \frac{dy}{y^\alpha(1-y)^{1-\alpha}} = \frac{\pi}{\sin \alpha\pi},$$

we obtain the inversion formula

$$u(x) = \frac{\sin \alpha\pi}{\pi} \frac{d}{dx} \int_a^x \frac{f(t)\,dt}{(x-t)^{1-\alpha}}. \tag{1.64}$$

It is applicable, if, for example, $f(x)$ is continuously differentiable for $x \geqslant a$.

We can easily reduce

$$\int_a^x \frac{u(t)\,dt}{\sqrt{(x^2-t^2)}} = f(x) \tag{1.65}$$

to the Abel form. It is sufficient to put $x^2 = x_1$, $t^2 = t_1$. We can also solve it using the above procedure: replace x by s, multiply both sides by $2s(x^2 - s^2)^{-1/2}\,ds$, and integrate with respect to s from a to x. This readily yields

$$u(x) = \frac{2}{\pi} \frac{d}{dx} \int_a^x \frac{sf(s)\,ds}{\sqrt{(x^2-s^2)}}. \tag{1.66}$$

In an analogous way, we can also solve certain equations with more than one variable.

Consider {see [18], [20]}

$$\iint_{\sigma_0} \frac{u(x,y)\,dx\,dy}{\sqrt{[(y_0-y^2)-(x_0-x)^2]}} = f(x_0, y_0), \tag{1.67}$$

where σ_0 is a right angled isosceles triangle with vertex (x_0, y_0) and base on the x-axis. Its solution is given by

$$u(x_0, y_0) = \frac{1}{2\pi^2}\left(\frac{\partial^2 g}{\partial x_0^2} - \frac{\partial^2 g}{\partial y_0^2}\right), \tag{1.68}$$

where

$$g(x_0, y_0) = \iint_{\sigma_0} \frac{f(x, y)\, dx\, dy}{\sqrt{[(y_0 - y)^2 - (x_0 - x)^2]}}.$$

3.5 *Integral equations with kernels defined by hypergeometric functions* {see [1]}

Consider the first kind equations

$$\int_0^a K(x, t)\, u(t)\, dt = f(x), \tag{1.69}$$

where a is a constant, $0 < a \leqslant \infty$, and

$$K(x, t) = (x^2 + t^2)^{-p}\, F\left(\frac{p}{2}, \frac{p+1}{2}, \frac{q}{2}; \frac{4x^2t^2}{(x^2 + t^2)^2}\right). \tag{1.70}$$

The parameters p and q satisfy the inequalities

$$0 < 2p < q < 2p + 2. \tag{1.71}$$

The solution of (1.69) has the form

$$u(x) = \frac{x^{q-2}}{\Gamma(1 + p - q/2)} \frac{d}{dx} \int_x^a \frac{g(t) t\, dt}{(t^2 - x^2)^{q/2 - p}}$$

with

$$g(t) = \frac{2\Gamma(p) \sin{(p - q/2)\pi}}{\pi\Gamma(q/2)}\, t^{1-2p} \frac{d}{dt} \int_0^t \frac{s^{q-1} f(s)\, ds}{(t^2 - s^2)^{q/2 - p}}. \tag{1.72}$$

If $a = \infty$ and $f(t)$ is differentiable, then the solution can be rewritten as either

$$u(x) = \frac{\Gamma(p)\,\Gamma(q-p)\sin(p-q/2)\pi}{\pi\Gamma(q/2)\,\Gamma(1+q/2)}\frac{d}{dt}\int_0^\infty \frac{(xt)^q f'(t)}{(x^2+t^2)^{q-p}}$$

$$\times F\left(\frac{q-p}{2}, \frac{q-p+1}{2}, \frac{q}{2}+1; \frac{4x^2t^2}{(x^2+t^2)^2}\right) dt, \quad (1.73)$$

or

$$u(x) = \frac{\Gamma(p)\,\Gamma(q-p-2)\sin(p-q/2)\pi}{\pi\Gamma(q/2)\,\Gamma(q/2-1)} x^{q-2}$$

$$\times \frac{d}{dx}\int_0^\infty \frac{d}{dt}[t^{q-2} f(t)]\Big\{(x^2+t^2)^{p+2-q}$$

$$\times F\left(\frac{q-p-2}{2}, \frac{q-p-1}{2}, \frac{q}{2}-1; \frac{4x^2t^2}{(x^2+t^2)^2} - \frac{1}{t^{2q-2p-4}}\right)\Big\} dt. \quad (1.74)$$

This last formula is valid when (1.71) and the following inequalities

$$q > p+2, \qquad q \geqslant 3, \quad (1.75)$$

are satisfied simultaneously.
Particular cases are:

1) $p = 1, q = 3, K(x, t) = \dfrac{1}{2xt} \ln \left|\dfrac{x+t}{x-t}\right|.$

Setting

$$\frac{u(t)}{2t} = U(t), \qquad xf(x) = F(x),$$

in the resulting equation, we obtain

$$\int_0^a U(t) \ln \left|\frac{x+t}{x-t}\right| dt = F(x), \quad (1.76)$$

which has the solution

$$U(x) = -\frac{2}{\pi^2}\frac{d}{dx}\int_x^a \frac{dt}{\sqrt{(t^2 - x^2)}}\frac{d}{dt}\int_0^t \frac{sF(s)\,ds}{\sqrt{(t^2 + s^2)}}. \qquad (1.77)$$

If $a = \infty$, $p = 1$ and $q = 3$, we obtain

$$\int_0^\infty \frac{1}{2xt}\ln\left|\frac{x + t}{x - t}\right| u(t)\,dt = f(x). \qquad (1.78)$$

It follows from (1.72) that it has the solution

$$u(x) = \frac{x}{\pi^2}\frac{d}{dx}\int_0^\infty \frac{d}{dt}\,[2tf(t)]\ln\left|1 - \frac{x^2}{t^2}\right| dt. \qquad (1.79)$$

2) $0 < p < 1, q = p + 1, K(x, t) = |x^2 - t^2|^{-p}$.

The resulting equation

$$\int_0^a \frac{u(t)}{|x^2 - t^2|^p}\,dt = f(x) \qquad (1.80)$$

has the solution

$$u(x) = -\frac{2x^{p-1}\,\Gamma(p)\cos\dfrac{p\pi}{2}}{\pi\left[\Gamma\left(\dfrac{1+p}{2}\right)\right]^2}\frac{d}{dx}\int_x^a \frac{t^{2-2p}\,dt}{(t^2 - x^2)^{(1-p)/2}}\int_0^t \frac{f(s)s^p\,ds}{(t^2 - s^2)^{(1-p)/2}}.$$

3) $p = \frac{1}{2}$, $q = 2$, $K(x, t) = \dfrac{2}{\pi}\dfrac{1}{x + t}K\left(\dfrac{2\sqrt{(xt)}}{x + t}\right)$, where K denotes the following elliptic integral of the first kind

$$K(k) = \int_0^1 \frac{dx}{\sqrt{[(1 - x^2)(1 - k^2x^2)]}}.$$

The resulting equation is

$$\frac{2}{\pi}\int_0^a \frac{1}{x+t} K\left(\frac{2\sqrt{(xt)}}{x+t}\right) u(t)\, dt = f(x). \tag{1.81}$$

It follows from (1.72) that it has the solution

$$u(x) = -\frac{2}{\pi}\frac{d}{dx}\int_0^a \frac{t\, dt}{\sqrt{(t^2 - x^2)}}\frac{d}{dt}\int_0^t \frac{sf(s)\, ds}{\sqrt{(t^2 - s^2)}}. \tag{1.82}$$

Chapter II

THE FREDHOLM THEORY

Note. A more detailed treatment of the Fredholm theory can be found in many books. Among them, we cite the following: [8], [15], [22], [23], [25], [28]. Though a little old fashioned, [8] is a rich source of information.

§1 Basic concepts and the Fredholm theorems

1.1 *Basic concepts*

For simplicity, we shall assume that the measure μ of (1.4) is Lebesgue, and denote an element of volume by dx. In this way, (1.4) becomes

$$u(x) - \int_\Omega K(x, t)\, u(t)\, dt = f(x). \tag{2.1}$$

The inequalities (1.5) and (1.6) are replaced by

$$\int_\Omega \int_\Omega |K(x, t)|^2\, dx\, dt < \infty \tag{2.2}$$

and

$$\int_\Omega |f(x)|^2\, dx < \infty. \tag{2.3}$$

Below, the integral on the left hand side of (2.2) will play a fundamental role. Therefore, we denote it by B_K^2:

$$\int_\Omega \int_\Omega |K(x, t)|^2\, dx\, dt = B_K^2. \tag{2.4}$$

It is pertinent to consider not only the single equation (2.1), but also the following family

$$u(x) - \lambda \int_{\Omega} K(x, t)\, u(t)\, dt = f(x) \tag{2.5}$$

with parameter λ which can be either real or complex.

Let $K(x, t)$ be a Fredholm kernel. The expression

$$\int_{\Omega} K(x, t)\, u(t)\, dt \tag{2.6}$$

represents an operator which maps $u(x)$ into

$$v(x) = \int_{\Omega} K(x, t)\, u(t)\, dt.$$

It is not difficult to prove that this operator maps $L_2(\Omega)$ into itself. We shall call this operator (2.6) *a Fredholm operator*, denote it by $\boldsymbol{K}u$, and write

$$(\boldsymbol{K}u)(x) = \int_{\Omega} K(x, t)\, u(t)\, dt. \tag{2.7}$$

In $L_2(\Omega)$, $\boldsymbol{K}$ is bounded

$$\|\boldsymbol{K}\| \leqslant B_K, \tag{2.8}$$

with B_K defined by (2.4).

In addition, we define the *identity operator* $\boldsymbol{I}$ by

$$\boldsymbol{I}u = u.$$

Consequently, (2.5) can be rewritten as

$$(\boldsymbol{I} - \lambda \boldsymbol{K})\, u = f. \tag{2.5$_1$}$$

For arbitrary λ, the homogeneous equation

$$u(x) - \lambda \int_{\Omega} K(x, t)\, u(t)\, dt = 0 \tag{2.9}$$

has as a solution the zero function {viz. $u(x) \equiv 0$} which is often called *the trivial solution*. For a given $K(x, t)$, and hence $\boldsymbol{K}$, λ is a *regular point* of $K(x, t)$ and $\boldsymbol{K}$, if (2.9) has only the trivial solution. It is called *a characteristic point* (*value*) of $K(x, t)$ and $\boldsymbol{K}$, if (2.9) has a non-trivial solution. With respect to a given characteristic value λ, such solutions define the *eigenfunctions* of $K(x, t)$, and hence, $\boldsymbol{K}$.

The inverse of a characteristic value is called an *eigenvalue*.

Corresponding to a given λ, let $u_1(x), u_2(x), \ldots, u_k(x)$ be eigenfunctions of $K(x, t)$, and let $C_1, \ldots, C_k$ be arbitrary constants, then

$$u_0(x) = C_1 u_1(x) + C_2 u_2(x) + \ldots + C_k u_k(x).$$

(if it is non-trivial) is also an eigenfunction of $K(x, t)$ with respect to this λ. The number of linear independent eigenfunctions corresponding to a given λ is called the *rank* of λ. If the rank of λ equals one, then λ is called *simple*.

Example 1.1. Let Ω be the interval $[0, 2\pi]$. For $K(x, t) = \sin(x - t)$, determine the regular and characteristic values for λ. We consider the homogeneous equation

$$u(x) - \lambda \int_0^{2\pi} \sin(x - t)\, u(t)\, dt = 0 \tag{2.10}$$

and solve it in the following way. We rewrite (2.10) as

$$u(x) - \lambda(a \sin x - b \cos x) = 0, \tag{2.10_1}$$

with

$$a = \int_0^{2\pi} u(t) \cos t\, dt, \qquad b = \int_0^{2\pi} u(t) \sin t\, dt.$$

Multiplying (2.10_1) by $\cos x$ and integrating, and then by $\sin x$ and integrating, we obtain the following algebraic system for a and b:

$$a + \pi\lambda b = 0, \qquad b - \pi\lambda a = 0.$$

Its determinant is $1 + \lambda^2\pi^2$, which vanishes for $\lambda = \pm i/\pi$. If $\lambda \neq \pm i/\pi$ then it is necessary that $a = b = 0$, and it follows from (2.10_1) that $u(x) \equiv 0$.

Consequently, all $\lambda \neq \pm i/\pi$ are regular. When $\lambda = \pm i\pi$, $b = \pm ia$ with a arbitrary. We obtain from (2.10_1), the non-trivial solutions of (2.10)

$$u(x) = ce^{\mp ix},$$

with c an arbitrary constant. Thus, the kernel $\sin(x - t)$, $0 < x, t < 2\pi$, has two simple characteristic values i/π and $-i/\pi$ and the corresponding eigenfunctions are e^{-ix} and e^{ix}.

The kernels $K(x, t)$ and

$$K^*(x, t) = \overline{K(t, x)} \tag{2.11}$$

are *conjugate*. So are the operators (2.7) and $\boldsymbol{K}^*$, where

$$(\boldsymbol{K}^*u)(x) = \int_\Omega K^*(x, t)\, u(t)\, dt = \int_\Omega \overline{K(t, x)}\, u(t)\, dt, \tag{2.12}$$

as well as the equations (2.1) and

$$u(t) - \int_\Omega \boldsymbol{K}^*(x, t)\, u(t)\, dt = g(x),$$

where $g(x)$ is an arbitrary function from $L_2(\Omega)$. Corresponding to (2.5), the conjugate equation is

$$u(x) - \lambda \int_\Omega K^*(x, t)\, u(t)\, dt = g(x). \tag{2.13}$$

Obviously, $\boldsymbol{K}^*$ is a Fredholm operator. In fact, conjugate operators $\boldsymbol{K}$ and $\boldsymbol{K}^*$ satisfy the identity

$$(\boldsymbol{K}u, v) = (u, \boldsymbol{K}^*v), \qquad u, v \in L_2(\Omega), \tag{2.14}$$

where the round brackets denote the scalar product in $L_2(\Omega)$: if $\phi(x)$ and $\psi(x)$ are contained in $L_2(\Omega)$, then

$$(\varphi, \psi) = \int_\Omega \varphi(x)\, \overline{\psi(x)}\, dx.$$

The kernels $K(x, t)$ and $K(t, x)$ are called *transposed* or *adjoint*. So are the equations (2.5) and

$$u(x) - \lambda \int_{\Omega} K(t, x)\, u(t)\, dt = h(x). \tag{2.15}$$

If $h(x) = \overline{g(x)}$ and $\psi(x)$ is the solution of the conjugate equation (2.12), then $\overline{\psi(x)}$ is the solution of the adjoint equation (2.15), and conversely. In particular, this correspondence between the solutions of conjugate and adjoint equations holds when the equations are homogeneous.

If $\boldsymbol{K}$ and $\boldsymbol{L}$ are Fredholm operators with kernels $K(x, t)$ and $L(x, t)$, respectively, then the product $\boldsymbol{KL}$ is also a Fredholm operator. The kernel of this operator is

$$\int_{\Omega} K(x, s)\, L(s, t)\, ds$$

with

$$B_{KL} \leqslant B_K B_L, \tag{2.16}$$

where the values B_K etc. are defined by (2.4). In particular, positive integer powers of a Fredholm operator $\boldsymbol{K}$ also define Fredholm operators. If we denote the kernel of the operator $\boldsymbol{K}^n$ (the n-th power of $\boldsymbol{K}$) by $K_n(x, t)$, then

$$\begin{aligned} K_1(x, t) &= K(x, t), \\ K_{p+q}(x, t) &= \int_{\Omega} K_p(x, s)\, K_q(s, t)\, ds, \\ K_n(x, t) &= \int_{\Omega} \dots \int_{\Omega} K(x, s_1)\, K(s_1, s_2) \dots K(s_{n-1}, t)\, ds_1\, ds_2 \dots ds_{n-1}. \end{aligned} \tag{2.17}$$

For the sake of brevity, we write

$$B_n^2 = B_{K^n}^2 = \int_{\Omega}\int_{\Omega} |K_n(x, t)|^2\, dx\, dt.$$

Then

$$B_n \leqslant B_K^n. \tag{2.18}$$

The kernel $K_n(x, t)$ is called the *n-th iterated kernel* with respect to $K(x, t)$.

1.2 *The basic theorems*

The following four *Fredholm theorems* are valid for Fredholm equations of the form (2.5).

THEOREM 1.1. *The set of eigenvalues of a Fredholm equation is either finite or denumerable. In the latter case, successive eigenvalues increase without limit.*

Theorem 1.1 can also be formulated in the following manner: if $K(x, t)$ is a Fredholm kernel, then, in any finite region of the complex λ-plane, there exists no more than a finite number of eigenvalues of $K(x, t)$.

THEOREM 1.2. *If the value of λ is regular, then the conjugate equations* (2.5) *and* (2.13) *have one and only one solution for arbitrary $f(x)$ and $g(x)$ contained in $L_2(\Omega)$.*

THEOREM 1.3. *If λ is a characteristic value then the conjugate equations*

$$u(x) - \lambda \int_{\Omega} K(x, t)\, u(t)\, dt = 0 \tag{2.19}$$

and

$$v(x) - \bar{\lambda} \int_{\Omega} K^*(x, t)\, v(t)\, dt = 0 \tag{2.20}$$

have the same number of linearly independent solutions.

Theorem 1.3 can be reformulated in the following manner: if λ is a characteristic value of the kernel $K(x, t)$, then $\bar{\lambda}$ is a characteristic value of the conjugate kernel $K^*(x, t)$. The number of linearly independent eigenfunctions is the same for both characteristic values.

THEOREM 1.4. *In order that the non-homogeneous equation* (2.1) *should have a solution, it is necessary and sufficient that its r.h.s.* $f(x)$ *be orthogonal to any solution* $v(x)$ *of the homogeneous conjugate equation* (2.20)*:*

$$(f, v) = 0. \tag{2.21}$$

If the value of λ is regular, then Theorem 1.4 is trivial for, in this case, (2.20) only has the trivial solution $v \equiv 0$ {Theorem 1.2}, and thus, condition (2.21) is satisfied by any $f(x)$. Simultaneously, on the basis of Theorem 1.2, (2.1) always has a solution. Theorem 1.4 plays an important role when λ is a characteristic value.

An important extension of Theorem 1.1 is the following theorem due to Schur [8].

THEOREM 1.5. *If the kernel* $K(x, t)$ *satisfies the inequality* (2.2) *and* λ_n $(n = 1, 2, \ldots)$ *denote its characteristic values, then*

$$\sum_{n=1}^{\infty} \frac{1}{|\lambda_n|^2} \leqslant B_K^2 = \int_\Omega \int_\Omega |K(x, t)|^2 \, dx \, dt. \tag{2.22}$$

If the kernel $K(x, t)$ *is symmetric, i.e.* $K(x, t) = K^*(x, t)$, *then equality holds in* (2.22). (An explanation of this is given in Chapter III.)

We shall make one additional point concerning Theorem 1.1. There exist kernels which have no characteristic values: for example, the Volterra kernels. A complete classification for such kernels is given by the following theorem of Lalesko:

THEOREM 1.6. *Let* $K(x, t)$ *be a Fredholm kernel and* $K_n(x, t)$ *be the corresponding iterated kernels. The kernel* $K(x, t)$ *has no characteristic values if and only if*

$$A_n = \int_\Omega K_n(x, x) \, dx = 0, \qquad n = 3, 4, 5, \ldots \tag{2.23}$$

We note that the numbers A_n are called the *traces* of $K(x, t)$. Lalesko established his theorem for bounded operators. A general proof has been given by Krachkovsky [10].

In some situations, it is possible to show that the solution of a Fredholm equation is not only square summable but also continuous, or at least bounded. The following two theorems describe the various possibilities (see [20]).

THEOREM 1.7. *Let Ω be a bounded region, and the kernel $K(x, t)$ satisfy not only the inequality* (2.2) *but also the inequality*

$$\int_{\Omega} |K(x, t)|^2 \, dt \leqslant A = \text{const.} \tag{2.24}$$

If the r.h.s. of (2.5) *is bounded, then any solution of* (2.5) *is also bounded.*

We shall say that the kernel $K(x, t)$ is *globally continuous with respect to x in Ω*, if $K(x, t)$ satisfies the following condition: given $\varepsilon > 0$, there exists a $\delta > 0$ such that for any $x_1, x_2 \in \Omega$, with $|x_1 - x_2| < \delta$, the following inequality holds

$$\int_{\Omega} |K(x_1, t) - K(x_2, t)| \, dt < \varepsilon.$$

THEOREM 1.8. *Let Ω be a closed bounded region, $K(x, t)$ satisfy* (2.24) *and be globally continuous with respect to $x \in \Omega$. If the r.h.s. of* (2.5) *is continuous in Ω, then any solution of* (2.5) *is continuous in Ω.*

§2 The solution of Fredholm equations: The method of successive approximation

Below, attention is restricted to exact methods. Approximate methods for the solution of integral equations are surveyed in [14], Chapter IV.

2.1 *The construction of approximations: The Neumann series*

We restrict attention to (2.5). It follows from Theorem 1.2 that (2.5) is solvable and has a unique solution for arbitrary $f(x)$, if λ is sufficiently small, or more exactly, if

$$|\lambda| < |\lambda_1|, \tag{2.25}$$

where λ_1 is the characteristic value of $K(x, t)$ with smallest magnitude. In this case, the solution can be constructed using the *method of successive approximations*. Usually, the r.h.s. of (2.1) is chosen as the initial approximation, i.e. $u_0(x) \equiv f(x)$. If the $(n - 1)$-th approximation, $u_{n-1}(x)$, has been constructed, then the n-th approximation, $u_n(x)$, is determined in the following way: (2.5) is rearranged to give

$$u(x) = f(x) + \lambda \int_\Omega K(x, t)\, u(t)\, dt,$$

then $u(t)$ is replaced by $u_{n-1}(t)$ under the integral sign, and $u(t)$ by $u_n(t)$ on the left hand side to yield

$$u_n(x) = f(x) + \lambda \int_\Omega K(x, t)\, u_{n-1}(t)\, dt. \tag{2.26}$$

Setting $n = 1, 2, \ldots$, we obtain for the first and second approximations

$$u_1(x) = f(x) + \lambda \int_\Omega K(x, t)\, u_0(t)\, dt = f(x) + \lambda \int_\Omega K(x, t)\, f(t)\, dt,$$

$$u_2(x) = f(x) + \lambda \int_\Omega K(x, t)\, u_1(t)\, dt$$

$$= f(x) + \lambda \int_\Omega K(x, t)\, f(t)\, dt + \lambda^2 \int_\Omega K_2(x, t)\, f(t)\, dt.$$

In terms of iterated kernels, the general formula for $u_n(x)$ is

$$u_n(x) = f(x) + \sum_{j=1}^{n} \lambda^j \int_\Omega K_j(x, t)\, f(t)\, dt. \tag{2.26$_1$}$$

If $n \to \infty$ and the condition (2.25) holds, then $u_n(x)$ converges to the exact solution $u(x)$ of (2.5). This solution has the following series representation

$$u(x) = f(x) + \sum_{j=1}^{\infty} \lambda^j \int_\Omega K_j(x, t)\, f(t)\, dt, \tag{2.27}$$

which is known as a *Neumann series*. If, for a given λ, the Neumann series converges for arbitrary $f(x)$ satisfying (2.3), then $|\lambda| < |\lambda_1|$.

In general cases, when the kernel only satisfies (2.2), we can only affirm that the Neumann series converges in the mean in Ω. However, if the kernel satisfies not only (2.2) but also (2.24), then under the condition (2.25), the Neumann series converges uniformly in Ω.

If the kernel has no characteristic values, then the Neumann series converges for all λ. It is trivial when any of the iterated kernels are identically zero: for, in this case, all subsequent iterated kernels vanish, and the series (2.27) reduces to a finite sum. There is another non-trivial and more interesting case: if the integral equation is of Volterra type with a kernel satisfying (2.2), then the Neumann series converges for arbitrary λ. The condition (2.25) is comparatively difficult to verify. It is possible to give a simpler condition: in order that the Neumann series converge, it is sufficient that

$$|\lambda| < \frac{1}{B_K} = \left\{ \int_\Omega \int_\Omega |K(x, t)|^2 \, dx \, dt \right\}^{-1/2}. \tag{2.28}$$

In this case, the rate of convergence of the Neumann series is greater than that of a geometric progression with multiplier $|\lambda| B_K$.

In particular, if Ω has finite measure (for instance, if Ω is a bounded region) and the kernel is bounded $|K(x, t)| \leqslant M =$ const., then, in order that the Neumann series converges, it is sufficient that

$$|\lambda| < \frac{1}{M \cdot \text{mes } \Omega}. \tag{2.29}$$

Under these conditions, the rate of convergence of the uniformly convergent Neumann series is greater than that of a geometric progression with multiplier $|\lambda| M \cdot$ measure (Ω).

2.2 *The resolvent kernel*

It is assumed that (2.25) is satisfied. We can interchange the order of summation and integration in (2.27). This yields a new form for the solution of (2.5), viz.

$$u(x) = f(x) + \lambda \int_\Omega \Gamma(x, t; \lambda) f(t) \, dt, \tag{2.30}$$

where

$$\Gamma(x, t; \lambda) = \sum_{j=1}^{\infty} \lambda^{j-1} K_j(x, t). \tag{2.31}$$

The function $\Gamma(x, t; \lambda)$, which allows the solution of (2.5) to be represented in the form (2.30) is called the *resolvent of the kernel* $K(x, t)$, or more briefly, the *resolvent kernel*. The series (2.31) is a power series representation for $\Gamma(x, t; \lambda)$, and is valid for sufficiently small λ. Below (in §4.5), we shall give representations for the resolvent which are valid for all regular λ.

In particular, we shall consider Volterra equations of the second kind (equations (1.12) and (1.13)). If x and t are points of a bounded region, then under rather general conditions successive approximations converge for arbitrary values of λ (and hence, for arbitrary λ, the resolvent can be represented by the series (2.31)). We cite two independent classes of conditions:

1) the kernel and the r.h.s. are square summable;
2) the kernel is bounded and the r.h.s. is summable.

In some cases, first kind Volterra equations can be reduced to second kind, which can then be solved using successive approximations. Let x and t be real variables, and in the first kind Volterra equation

$$\int_a^x K(x, t)\, u(t)\, dt = f(x)$$

let the kernel be continuous and have a continuous partial derivative with respect to x. If a summable solution exists, then it follows that the r.h.s. $f(x)$ has a continuous first derivative. Consequently, differentiation with respect to x yields

$$K(x, x)\, u(x) + \int_a^x \frac{\partial K(x, t)}{\partial x}\, u(t)\, dt = f'(x).$$

If the function $K(x, x)$ is non-zero everywhere, then we obtain the following second kind Volterra equation

$$u(x) + \int_a^x \frac{1}{K(x, x)} \frac{\partial K(x, t)}{\partial x}\, u(t)\, dt = f'(x)/K(x, x),$$

which can be solved by the method of successive approximation.

In the case when $K(x, x) \equiv 0$, we again obtain a first kind Volterra equation. The above procedure can be reapplied to try to reduce this new equation to one of second kind.

§3 The solution of Fredholm equations: Degenerate equations and the general case

3.1 *Equations with degenerate kernels*

A Fredholm kernel $K(x, t)$ is called *degenerate*, if it can be expressed as the finite sum

$$K(x, t) = \sum_{k=1}^{n} a_k(x)\, b_k(t). \tag{2.32}$$

The functions a_k, as well as b_k, are assumed to be linearly independent, otherwise n can be reduced.

An equation with a degenerate kernel is called *degenerate*.

An equation with a degenerate kernel can be readily solved by reducing it to a system of algebraic equations. By making use of (2.32), we can rewrite (2.5) in the form

$$u(x) - \lambda \sum_{k=1}^{n} c_k a_k(x) = f(x), \tag{2.33}$$

where the c_k are constants defined by

$$c_k = \int_{\Omega} u(t)\, b_k(t)\, dt. \tag{2.34}$$

It follows from (2.33) that $u(x)$ is determined once the c_k are known. Multiplying both sides of (2.33) by $b_j(x)$ and integrating over Ω, we obtain the following system of algebraic equations for the determination of the c_k

$$c_j - \lambda \sum_{k=1}^{n} \alpha_{jk} c_k = f_j, \qquad j = 1, 2, \ldots, n, \tag{2.35}$$

where

$$\alpha_{jk} = \int_{\Omega} a_k(x)\, b_j(x)\, dx, \qquad f_j = \int_{\Omega} f(x)\, b_j(x)\, dx. \tag{2.36}$$

The determinant of (2.35) is a polynomial in λ of degree not greater than n. It is not identically equal to zero (it equals unity when $\lambda = 0$), and hence has no more than n distinct roots which are the characteristic values of the kernel (2.32). All other values of λ are regular for this kernel.

Once the solution of (2.35) has been found and substituted in (2.32), the following solution for a Fredholm equation with the above degenerate kernel is obtained

$$u(x) = f(x) + \sum_{k=1}^{n} c_k a_k(x).$$

Example 3.1. Consider the equation

$$u(x) - \int_0^\pi \sin(x + t)\, u(t)\, dt = f(x), \tag{2.37}$$

where Ω is the interval $[0, \pi]$, and x and t are real variables contained in the same interval. Clearly, (2.37) is degenerate. Its kernel can be expressed in the form

$$\sin(x + t) = \sin x \cos t + \cos x \sin t.$$

A comparison with (2.32) yields

$$n = 2, \quad a_1(x) = \sin x, \quad a_2(x) = \cos x, \quad b_1(t) = \cos t, \quad b_2(t) = \sin t.$$

It follows from (2.36) that

$$\alpha_{11} = \int_0^\pi \sin x \cos x\, dx = 0,$$

$$\alpha_{12} = \int_0^\pi \sin^2 x\, dx = \frac{\pi}{2},$$

$$\alpha_{21} = \int_0^\pi \cos^2 x\, dx = \frac{\pi}{2},$$

$$\alpha_{22} = \int_0^\pi \cos x \sin x \, dx = 0,$$

$$f_1 = \int_0^\pi f(x) \cos x \, dx, \qquad f_2 = \int_0^\pi f(x) \sin x \, dx.$$

Consequently, in this case, the system (2.35) has the form

$$\begin{aligned} c_1 - \frac{\lambda\pi}{2} c_2 &= f_1, \\ -\frac{\lambda\pi}{2} c_1 + c_2 &= f_2. \end{aligned} \tag{2.38}$$

Its determinant

$$\begin{vmatrix} 1 & -\lambda\pi/2 \\ -\lambda\pi/2 & 1 \end{vmatrix} = 1 - \lambda^2\pi^2/4$$

has the roots

$$\lambda_1 = -2/\pi, \qquad \lambda_2 = 2/\pi, \tag{2.39}$$

which are the characteristic values of (2.37). The remaining values of λ are regular. If λ is a regular value, then the determinant of the system (2.38) is non-zero, the coefficients c_1 and c_2 are uniquely defined, and the unique solution of (2.37) is given by

$$u(x) = f(x) + c_1 \sin x + c_2 \cos x.$$

It is not difficult to find the eigenfunctions of (2.37). It is only necessary to solve the homogeneous system corresponding to (2.38), in which λ has been replaced by one of its characteristic values. When $\lambda = \lambda_1 = -2/\pi$, we obtain the system

$$c_1 + c_2 = 0.$$

Hence, $c_2 = -c_1$, and the corresponding eigenfunction is

$$u_1(x) = c_1(\sin x - \cos x).$$

The constant c_1 can be determined using the condition

$$\|u_1\|^2 = \int_0^\pi |u_1^2(x)|\, dx = 1,$$

which yields $c_1 = 1/\sqrt{\pi}$, and hence,

$$u_1(x) = \frac{1}{\sqrt{\pi}}(\sin x - \cos x).$$

Analogously, when $\lambda = \lambda_2 = 2/\pi$, we obtain $c_1 - c_2 = 0$ and

$$u_2(x) = \frac{1}{\sqrt{\pi}}(\sin x + \cos x).$$

If, in the non-homogeneous equation (2.37), $\lambda = \lambda_1 = -2/\pi$, then the system (2.38) takes the form

$$c_1 + c_2 = f_1, \qquad c_1 + c_2 = f_2.$$

In order to solve this equation, it is necessary and sufficient that $f_1 = f_2$. This implies the orthogonality of the functions $f(x)$ and $u_1(x)$. In the same way, if $\lambda = \lambda_2 = 2/\pi$, then (2.37) is solvable if and only if $f_1 = -f_2$, i.e. when $f(x)$ and $u_2(x)$ are orthogonal. These are the orthogonality conditions of Theorem 1.4 of §1. Here, the kernel $K(x, t)$ coincides with its conjugate.

3.2 *The general case*

In the general case, when the kernel is not degenerate and the parameter λ is not small, the solution of (2.5) can be constructed in the following manner. We express $K(x, t)$ as a double sum of the form

$$K(x, t) = \sum_{k,n=1}^{\infty} A_{kn}\varphi_k(x)\,\psi_n(t), \tag{2.40}$$

where the A_{kn} are constants, and the ϕ_k and ψ_n are contained in $L_2(\Omega)$. The equality (2.40) is interpreted with the following sense

$$\lim_{N\to\infty} \int_\Omega\int_\Omega \left|K(x, t) - \sum_{k,n=1}^{N} A_{kn}\varphi_k(x)\,\psi_n(t)\right|^2 dx\, dt = 0. \tag{2.41}$$

The expansion (2.40) can be constructed in the following way. Let $\phi_k(x)$ $(k = 1, 2, 3, \ldots)$ and $\psi_n(x)$, $(n = 1, 2, \ldots)$ denote complete and orthogonal systems in $L_2(\Omega)$ (the case $\psi_k(x) = \phi_k(x)$ is allowed). Then the system

$$\varphi_k(x)\,\psi_k(t), \qquad k, n = 1, 2, \ldots, \tag{2.42}$$

is orthogonal and complete on the Cartesian product $\Omega \times \Omega$, (2.40) is the Fourier series of $K(x, t)$ with respect to the system (2.42), and (2.41) is a consequence of the completeness of the system (2.42). In this case, the coefficients A_{kn} are defined by the standard formula for Fourier coefficients

$$A_{kn} = \int_\Omega\int_\Omega K(x, t)\,\overline{\varphi_k(x)}\;\overline{\psi_n(t)}\,dx\,dt.$$

We shall assume that the expansion (2.40) has been constructed, and consider (2.5) for the values of λ for which its modulus is bounded above by a fixed integer R:

$$|\lambda| \leqslant R.$$

We introduce the notation

$$K''(x, t) = \sum_{k,n=1}^{N} A_{kn}\varphi_k(x)\,\psi_k(t),$$

$$K'(x, t) = K(x, t) - K''(x, t).$$

It follows from (2.41) that a suitably large N can be chosen so that

$$B_{K'}^2 = \int_\Omega\int_\Omega |K'(x, t)|^2\,dx\,dt \leqslant \frac{1}{4R^2}. \tag{2.43}$$

We rewrite (2.5) in the form

$$u(x) - \lambda\int_\Omega K'(x, t)\,u(t)\,dt = f(x) + \lambda\int_\Omega K''(x, t)\,u(t)\,dt. \tag{2.44}$$

Temporarily considering the r.h.s. of (2.44) as known, we can apply the method of successive approximation to it; or, equivalently, apply (2.30).

Let us put

$$\Gamma'(x, t; \lambda) = \sum_{n=1}^{\infty} \lambda^{n-1} K_n'(x, t), \tag{2.45}$$

where $K_n'(x, t)$ are kernels iterated with respect to the kernel $K'(x, t)$. Using (2.30), we obtain

$$u(x) - \lambda \int_{\Omega} \sum_{k,n=1}^{N} A_{kn}\varphi_k(x, \lambda)\, \psi_n(t)\, u(t)\, dt = f(x, \lambda), \tag{2.46}$$

where

$$\varphi_k(x, \lambda) = \varphi_k(x) + \lambda \int_{\Omega} \Gamma'(x, \tau; \lambda)\, \varphi_k(\tau)\, d\tau,$$

$$f(x, \lambda) = f(x) + \lambda \int_{\Omega} \Gamma'(x, \tau; \lambda)\, f(\tau)\, d\tau.$$

The kernel of (2.46) is degenerate. Solving this equation in the manner we have described above, we obtain the solution of (2.5).

We can also obtain an equation with a degenerate kernel in other ways. We introduce the new unknown $v(t)$:

$$u(x) - \lambda \int_{\Omega} K'(x, t)\, u(t)\, dt = v(t).$$

Solving for it in terms of $u(x)$ using (2.30), we obtain

$$u(x) = v(x) + \lambda \int_{\Omega} \Gamma'(x, t; \lambda)\, v(t)\, dt. \tag{2.47}$$

Substituting this in (2.44), we find a new integral equation, the kernel of which is degenerate

$$v(x) - \lambda \int_{\Omega} \sum_{k,n=1}^{N} A_{kn}\varphi_k(x)\, \psi_n(t; \lambda)\, v(t)\, dt = f(x), \tag{2.48}$$

where

$$\psi_n(t; \lambda) = \psi_n(t) + \lambda \int_{\Omega} \Gamma'(\tau, t; \lambda)\, \psi_n(\tau)\, d\tau.$$

Once (2.48) has been solved, we obtain the solution of the given equation (2.5) by using (2.47).

§4 The Fredholm resolvent

4.1 *The Fredholm resolvent*

The function $\Gamma(x, t; \lambda)$ which is defined almost everywhere in Ω for all regular values of the parameter λ, and which yields in conjunction with (2.30) the solution of (2.5), is called *The Fredholm Resolvent* (or, more simply, *resolvent*). In §2, the resolvent has been constructed for sufficiently small values of λ. The trick described in §3 allows the resolvent to be constructed for all regular values of λ, and to be given an explicit expression which is valid, for fixed R, inside any circle $|\lambda| \leqslant R$ in the complex λ-plane.

Assume that the following kernel decomposition

$$K(x, t) = K'(x, t) + K''(x, t),$$

where the kernel $K'(x, t)$ satisfies the inequality (2.43), and the kernel $K''(x, t)$ is degenerate, has been given. We shall write the degenerate kernel in the form (2.32) and put

$$a_k(x, \lambda) = a_k(x) + \lambda \int_{\Omega} \Gamma'(x, t; \lambda)\, a_k(t)\, dt,$$

$$b_k(x, \lambda) = b_k(x) + \lambda \int_{\Omega} \Gamma'(t, x; \lambda)\, b_k(t)\, dt,$$

where $\Gamma'(x, t; \lambda)$ is the resolvent of the kernel $K'(x, t)$ and is defined by (2.45).

We introduce the notation

$$D_R(\lambda) = \begin{vmatrix} 1 - \lambda\alpha_{11}(\lambda) & -\lambda\alpha_{12}(\lambda) & \dots & -\lambda\alpha_{1n}(\lambda) \\ -\lambda\alpha_{21}(\lambda) & 1 - \lambda\alpha_{22}(\lambda) & \dots & -\lambda\alpha_{2n}(\lambda) \\ \dots & \dots & \dots & \dots \\ -\lambda\alpha_{n1}(\lambda) & -\lambda\alpha_{n2}(\lambda) & \dots & 1 - \lambda\alpha_{nn}(\lambda) \end{vmatrix},$$

where

$$\alpha_{jk}(\lambda) = \int_{\Omega} a_k(x)\, b_j(x, \lambda)\, dx,$$

and denote by $\Delta_{jk}(x)$ the cofactor of the (k, j)-element of the determinant $D_R(\lambda)$. Making use of this notation, we obtain the following representation for the resolvent

$$\Gamma(x, t; \lambda) = \Gamma'(x, t; \lambda) + \frac{1}{D_R(\lambda)} \sum_{j,k=1}^{n} \Delta_{jk}(\lambda)\, a_j(x, \lambda)\, b_k(t, \lambda). \tag{2.49}$$

4.2 *Properties of the resolvent*

We list the important properties of the resolvent.

1) The resolvent is unique.

2) If the kernel satisfies inequality (2.2), then, for any regular λ, the resolvent satisfies the same inequality. If the set Ω is bounded and closed, and the kernel is continuous in Ω, then the resolvent is also continuous in Ω for any regular λ.

3) The resolvent is a meromorphic function of λ on the complex λ-plane.

4) The poles of the resolvent coincide with the characteristic values of λ.

5) For small absolute magnitudes of λ, the resolvent has the representation (2.31).

6) The resolvent satisfies the integral equations

$$\left.\begin{aligned} \Gamma(x, t; \lambda) &= K(x, t) + \lambda \int_{\Omega} K(x, \tau)\, \Gamma(\tau, t; \lambda)\, d\tau, \\ \Gamma(x, t; \lambda) &= K(x, t) + \lambda \int_{\Omega} K(\tau, t)\, \Gamma(x, \tau; \lambda)\, d\tau. \end{aligned}\right\} \tag{2.50}$$

7) If the resolvent $\Gamma(x, t; \lambda)$ is the kernel of a new integral equation with parameter μ, then the resolvent of the new equation is $\Gamma(x, t; \lambda + \mu)$.

§5 The solution of Fredholm equations: The Fredholm series

We shall now consider another method for the construction of exact solutions of Fredholm equations. This method was proposed by Fredholm.

§5 *Fredholm equations: the Fredholm series*

5.1 *The Fredholm series. Fredholm determinants and minors*

Being a meromorphic function, the resolvent can be represented as the quotient of two entire functions of λ. The poles of the resolvent, which are the characteristic values of the given kernel, are independent of x and t. Consequently, we can ensure that the denominator in the above mentioned representation for the resolvent depends only on λ. Thus, the resolvent can have the form

$$\Gamma(x, t; \lambda) = \frac{D(x, t; \lambda)}{D(\lambda)}, \tag{2.51}$$

where $D(x, t; \lambda)$ and $D(\lambda)$ are entire functions of λ. If it is possible to construct such functions, then the resolvent will be known explicitly, and therefore, the solution of the given integral equation can be found using (2.30).

For the numerator and the denominator in (2.51), Fredholm proposed the following series representations, which are now called *Fredholm series*,

$$D(x, t; \lambda) = \sum_{n=0}^{\infty} \frac{(-1)^n}{n!} B_n(x, t)\lambda^n, \tag{2.52}$$

$$D(\lambda) = \sum_{n=0}^{\infty} \frac{(-1)^n}{n!} c_n\lambda^n, \tag{2.53}$$

where

$$c_0 = 1, \qquad B_0(x, t) = K(x, t), \tag{2.54}$$

$$c_n = \int_\Omega B_{n-1}(x, x)\, dx, \qquad n > 0, \tag{2.55}$$

$$B_n(x, t) = \int_\Omega \cdots \int_\Omega K\begin{pmatrix} x, t_1, t_2, \ldots, t_n \\ t, t_1, t_2, \ldots, t_n \end{pmatrix} dt, dt_2 \ldots dt_n, \tag{2.56}$$

with

$$K\begin{pmatrix} x_1, x_2, \ldots, x_p \\ t_1, t_2, \ldots, t_p \end{pmatrix} = \begin{vmatrix} K(x_1, t_1) & K(x_1, t_2) & \ldots & K(x_1, t_p) \\ K(x_2, t_1) & K(x_2, t_2) & \ldots & K(x_2, t_p) \\ \ldots & \ldots & \ldots & \ldots \\ K(x_p, t_1) & K(x_p, t_2) & \ldots & K(x_p, t_p) \end{vmatrix}. \tag{2.57}$$

If the kernel is not continuous, then the integral

$$\int_{\Omega} K(x, x)\, dx$$

defining the coefficient c_1 may not have sense. This difficulty can be overcome in the following way: it is sufficient to set $K(x, x)$ equal to an arbitrary function of x which is summable on Ω. For instance, if we put $K(x, x) = 0$, then $c_1 = 0$.

If the kernel $K(x, t)$ and the region Ω are bounded, then the convergence of the Fredholm series for all finite values of λ follows by virtue of a certain theorem of Hadamard concerning the evaluation of determinants (see, for example, [8], [15], [23]). Carleman ([29]) has proved that the Fredholm series also represent entire functions of λ when Ω is a bounded region and the kernel satisfies inequality (2.2). Another proof, valid for unbounded Ω, is given in [19].

The formula (2.56) is inconvenient for it requires the multiple integration of high order determinants. A more convenient formula is

$$B_n(x, t) = c_n K(x, t) - n \int_{\Omega} K(x, \tau)\, B_{n-1}(\tau, t)\, d\tau. \tag{2.58}$$

Together with (2.54) and (2.55), it allows the coefficients $B_n(x, t)$ and c_n to be calculated recursively.

It is worth mentioning the formula

$$\frac{D'(\lambda)}{D(\lambda)} = -\sum_{n=1}^{\infty} A_n \lambda^{n}, \tag{2.59}$$

where

$$A_n = \int_{\Omega} K_n(x, x)\, dx, \qquad n = 1, 2, \ldots$$

is the trace of $K(x, t)$ introduced above. The series (2.59) converges when (2.25) is satisfied.

The function $D(\lambda)$ is called the *Fredholm determinant*, while $D(t, t; \lambda)$ is called the *first minor of Fredholm*.

Fredholm's minor of order p is defined by

$$D\begin{pmatrix} x_1, x_2, \ldots, x_p \\ t_1, t_2, \ldots, t_p \end{pmatrix}\lambda\Bigg) = K\begin{pmatrix} x_1, x_2, \ldots, x_p \\ t_1, t_2, \ldots, t_p \end{pmatrix}$$
$$+ \sum_{n=1}^{\infty} \frac{(-1)^n}{n!} \lambda^n \int_{\Omega} \cdots \int_{\Omega} K\begin{pmatrix} x_1, x_2, \ldots, x_p, \tau_1, \ldots, \tau_n \\ t_1, t_2, \ldots, t_p, \tau_1, \ldots, \tau_n \end{pmatrix} d\tau_1 \ldots d\tau_n. \tag{2.60}$$

Clearly, $D(x, t; \lambda) = D(^x_t \mid \lambda)$.

For any p and $x_k, t_k \in \Omega$, the series (2.60) is an entire function of λ. This statement can be proved when the kernel is bounded and measurable on a bounded region Ω. These are the kernels we shall consider below.

5.2 *The representation of the eigenfunctions of a kernel in terms of the minors of Fredholm*

We can express eigenfunctions of a given kernel in terms of the minors of Fredholm. Let λ_0 be a characteristic value of the kernel $K(x, t)$, and let all minors of order less than p be zero when $\lambda = \lambda_0$, while the minor of order p is non-zero.

Let fixed points $x_1^0, x_2^0, \ldots, x_p^0, t_1^0, t_2^0, \ldots, t_p^0$ be such that

$$D\left(\begin{matrix} x_1^0, & x_2^0, \ldots, x_p^0 \\ t_1^0, & t_2^0, \ldots, t_p^0 \end{matrix} \;\middle|\; \lambda_0\right) \neq 0. \tag{2.61}$$

Then the functions

$$u_k(x) = D\left(\begin{matrix} x_1^0, \ldots, x_{k-1}^0, & x, & x_{k+1}^0, \ldots, x_p^0 \\ t_1^0, \ldots, t_{k-1}^0, & t_k^0, & t_{k+1}^0, \ldots, t_p^0 \end{matrix} \;\middle|\; \lambda_0\right), \tag{2.62}$$
$$k = 1, 2, \ldots, p,$$

are the linearly independent eigenfunctions of the kernel $K(x, t)$ corresponding to the characteristic value λ_0. In fact, there are no eigenfunctions of the kernel $K(x, t)$ corresponding to characteristic value λ_0 which together with (2.62) form a linearly independent set. This can be formulated briefly as follows: the functions (2.62) form a complete system of eigenfunctions corresponding to the characteristic values λ_0 of the kernel $K(x, t)$.

The functions $v_k(x)$, defined by

$$\overline{v_k(x)} = D\left(\begin{matrix} x_1^0, \ldots, x_{k-1}^0, & x_k^0, & x_{k+1}^0, \ldots, x_p^0 \\ t_1^0, \ldots, t_{k-1}^0, & x, & t_{k+1}^0, \ldots, t_p^0 \end{matrix}\middle| \lambda_0 \right), \tag{2.63}$$

$$k = 1, 2, \ldots, p,$$

form a complete system of eigenfunctions of the conjugate kernel $K^*(x, t) = \overline{K(t, x)}$ corresponding to the characteristic values λ_0 of this kernel.

§6 Equations with a weak singularity

6.1 *Boundedness of the integral operator with a weak singularity*

All four theorems given in §1 of this chapter remain valid for equations with a weak singularity—see the definition in Chapter I, §1.3.

An integral operator with a weak singularity is bounded in $L_2(\Omega)$. If Ω is a region in m-dimensional Euclidean space, the operator is defined by (1.7) with $d\mu = dt$, and the kernel is defined by (1.11), then

$$\|K\| \leqslant \frac{C\omega_m h^{m-\alpha}}{m - \alpha}, \tag{2.64}$$

where

$$\omega_m = \frac{2\pi^{m/2}}{\Gamma(m/2)}$$

is the surface area of the unit sphere in m-dimensional Euclidean space, h is the diameter of the region Ω {i.e. the supremum of all distances between points of Ω} and $C = \sup |A(x, t)|$.

Let $\boldsymbol{K}$ and $\boldsymbol{L}$ be two operators with weak singularities, and their kernels $K(x, t)$ and $L(x, t)$ satisfy

$$|K(x, t)| \leqslant \frac{C_1}{r^\alpha}, \ |L(x, t)| \leqslant \frac{C_2}{r^\beta}, \ C_1, C_2 = \text{const.} \tag{2.65}$$

Then the kernel $M(x, t)$ of the product $\boldsymbol{KL}$ of $\boldsymbol{K}$ and $\boldsymbol{L}$ {the kernel $M(x, t)$ is defined by (2.15)} has the following estimate

$$|M(x,t)| \leqslant \begin{cases} \dfrac{C}{r^{\alpha+\beta-m}}, \alpha + \beta > m, \\ C \ln \dfrac{h}{r} + C, \alpha + \beta = m, \\ C, \alpha + \beta < m. \end{cases} \tag{2.66}$$

6.2 *Iteration of a kernel with a weak singularity*

Let $K(x, t)$ be a kernel with a weak singularity for which (2.65) holds. We denote by $K_n(x, t)$ its iterated kernels and n_0 the smallest integer exceeding $m/2(m - \alpha)$. It is evident that

$$n_0 = \left[\frac{m}{2(m-\alpha)}\right] + 1. \tag{2.67}$$

All iterated kernels with index $n \geqslant n_0$ are of Fredholm type. If $n \geqslant n_1$, where

$$n_1 = \left[\frac{m}{m-\alpha}\right] + 1,$$

then the iterated kernels $K_n(x, t)$ are bounded.

The Theorem of Schur for a kernel $K(x, t)$ with a weak singularity of the form (1.11) is: *if* $\lambda_1, \lambda_2, \ldots$ *are the characteristic values of* $K(x, t)$, *then*

$$\sum_{k=1}^{\infty} \frac{1}{|\lambda_k|^{2n_0}} \leqslant \int_{\Omega}\int_{\Omega} |K_{n_0}(x,t)|^2 \, dx \, dt. \tag{2.68}$$

6.3 *The method of successive approximation*

Equations with a weak singularity can be solved by the method of successive approximation for small values of λ. If Ω is a region in m-dimensional Euclidean space, then the method of successive approximation converges when

$$|\lambda| < \frac{m-\alpha}{C\omega_m h^{m-\alpha}}. \tag{2.69}$$

If Ω is a closed region, and the functions $f(x)$ and $A(x, t) = r^\alpha K(x, t)$ are continuous in Ω, then any solution of an equation with a weak singularity is continuous in Ω.

§7 Systems of integral equations

7.1 *The vector form for systems of integral equations*

We shall consider systems of integral equations of the form

$$u_i(x) - \lambda \sum_{j=1}^{n} \int_{\Omega} K_{ij}(x, t)\, u_j(t)\, d\mu(t) = f_i(x) \tag{2.70}$$

$$(i = 1, 2, \ldots, n).$$

Introducing n-component vectors

$$u(x) = (u_1(x), u_2(x), \ldots, u_n(x)),$$
$$f(x) = (f_1(x), f_2(x), \ldots, f_n(x))$$

and the matrix $K(x, t)$ of order n with elements $K_{ij}(x, t)$ $(i, j = 1, 2, \ldots, n)$, we can rewrite the system (2.70) in terms of a single vector equation which coincides with (2.1).

7.2 *Methods of solution for Fredholm kernels*

If the kernels $K_{ij}(x, t)$ are of Fredholm type or have a weak singularity, and the $f_i \in L_2(\Omega)$, then all the Fredholm theorems remain valid for the system (2.70). Methods of solution presented in §2-5 of this chapter can be applied to systems. In particular, the method of successive approximation converges for small values of λ. If the kernels $K_{ij}(x, t)$ are of Fredholm type, then convergence holds when

$$|\lambda| \leqslant \left\{ \sum_{i,j=1}^{n} B_{ij}^2 \right\}^{-1}, \qquad B_{ij}^2 = \int_{\Omega}\int_{\Omega} |K_{ij}(x, t)|^2\, d\mu(x)\, d\mu(t). \tag{2.71}$$

A more general condition for convergence is available. We denote the matrix of elements B_{ij} by B, the matrix conjugate to B by B^*, and the largest eigenvalue of B^*B by σ^2. The method of successive approximation converges for the system (2.70) if

$$|\lambda| < \sigma^{-1}. \tag{2.72}$$

7.3 *Methods of solution for kernels with a weak singularity*

Let the kernels $K_{ij}(x, t)$ have a weak singularity, viz.

$$|K_{ij}(x, t)| \leqslant \frac{C_{ij}}{r_{ij}^{\alpha}}, \qquad C_{ij} = \text{const}, \qquad 0 \leqslant \alpha_{ij} < m.$$

We denote the matrix of elements $C_{ij}\omega_m h^{m-\alpha_{ij}}/(m - \alpha_{ij})$ by B, and, as usual, its conjugate by B^*. Let σ^2 be the largest eigenvalue of the matrix B^*B. Then (2.72) is again sufficient for the convergence of the method of successive approximations. In particular, if $C_{ij} = C, \alpha_{ij} = \alpha$ then condition (2.72) takes the form

$$|\lambda| < \frac{m - \alpha}{Cn\omega_m h^{m-\alpha}}. \tag{2.73}$$

If the kernels $K_{ij}(x, t)$ are of Volterra type, then the method of successive approximation converges for the system (2.69) for all values of the parameter λ.

§8 The structure of the resolvent in the neighbourhood of a characteristic value

The questions we shall examine in this section were first investigated by Goursat—see [8]. In 1916, Riesz [24] investigated equations more general than those of Fredholm type. Recently, the main results have come from Krachkovsky and Gol'dman [2]–[4], [9], [11]–[13], and Gokhberg, Kreĭn and Marcus [5]–[7], [16]–[17].

8.1 *Orthogonal kernels*

Two Fredholm kernels $K(x, t)$ and $L(x, t)$ are said to be *orthogonal in the region* Ω if

$$\int_\Omega K(x, s)\, L(s, t)\, ds \equiv 0, \qquad \int_\Omega L(x, s)\, K(s, t)\, ds \equiv 0. \tag{2.74}$$

In other words, $K(x, t)$ and $L(x, t)$ are othogonal if $\boldsymbol{KL} = \boldsymbol{LK} = 0$ where $\boldsymbol{K}$ and $\boldsymbol{L}$ are the Fredholm operators with kernels $K(x, t)$ and $L(x, t)$, respectively.

We denote the resolvents of the kernels $K(x, t)$, $L(x, t)$ and $K(x, t) + L(x, t)$ by $\Gamma_{\boldsymbol{K}}(x, t; \lambda)$, $\Gamma_{\boldsymbol{L}}(x, t; \lambda)$, $\Gamma_{\boldsymbol{K}+\boldsymbol{L}}(x, t; \lambda)$, respectively. If the kernels $K(x, t)$ and $L(x, t)$ are orthogonal, then: a) $\Gamma_{\boldsymbol{K}}(x, t; \lambda)$ and $\Gamma_{\boldsymbol{L}}(x, t; \mu)$ are orthogonal for all values of λ and μ; b) $\Gamma_{\boldsymbol{K}+\boldsymbol{L}}(x, t; \lambda) = \Gamma_{\boldsymbol{K}}(x, t; \lambda) + \Gamma_{\boldsymbol{L}}(x, t; \lambda)$.

Let λ_0 be a characteristic value of the kernel $K(x, t)$. Then λ_0 is a pole of the resolvent of this kernel which we denote by $\Gamma(x, t; \lambda)$. Let n be the multiplicity of the pole λ_0. Then in the neighbourhood of the point λ_0, the following decomposition is valid

$$\Gamma(x, t; \lambda) = \sum_{k=1}^{n} \frac{\gamma_n(x, t)}{(\lambda - \lambda_0)^k} + H(x, t; \lambda),$$

where $H(x, t; \lambda)$ is regular for $\lambda = \lambda_0$. Setting, for brevity,

$$\sum_{k=1}^{n} \frac{\gamma_k(x, t)}{(\lambda - \lambda_0)^k} = \gamma(x, t; \lambda), \tag{2.75}$$

we obtain

$$\Gamma(x, t; \lambda) = \gamma(x, t; \lambda) + H(x, t; \lambda). \tag{2.76}$$

It turns out that the kernels $\gamma(x, t; \lambda)$ and $H(x, t; \mu)$ are orthogonal for all values of λ and μ.

8.2 *The principal kernels*

We write

$$\gamma(x, t; 0) = k(x, t), \qquad H(x, t; 0) = H(x, t). \tag{2.77}$$

Then the kernels $k(x, t)$ and $H(x, t)$ are orthogonal and

$$K(x, t) = k(x, t) + H(x, t). \tag{2.78}$$

If we consider the functions $k(x, t)$ and $H(x, t)$ as kernels, then their resolvents will be $\gamma(x, t; \lambda)$ and $H(x, t; \lambda)$, respectively. The kernel $k(x, t)$ is called the *principal kernel* corresponding to the characteristic value λ_0. The principal kernels corresponding to different characteristic values are mutually orthogonal.

All principal kernels are degenerate. Consequently, the principal kernel $k(x, t)$ corresponding to the pole λ_0 of the resolvent has the form

$$k(x, t) = \sum_{j=1}^{l} \varphi_j(x)\, \overline{\psi_j(t)}. \tag{2.79}$$

The functions $\phi_j(x)$, as well as functions $\psi_j(t)$, can be regarded as linearly independent.

8.3 *The canonical kernels*

If the functions $\varphi_j(x)$ are subjected to a linear transformation, the form of (2.79) remains unchanged. We can choose this transformation in such a way that the decomposition (2.79) has a *canonical structure* in the following sense. The functions $\varphi_j(x)$ $(j = 1, 2, \ldots, l)$ can be classified into the groups

$$\begin{array}{l} \varphi_1^{(1)}, \varphi_2^{(1)}, \ldots, \varphi_{p_1}^{(1)}, \\ \varphi_1^{(2)}, \varphi_2^{(2)}, \ldots, \varphi_{p_2}^{(2)}, \\ \cdots\cdots\cdots\cdots \\ \varphi_1^{(q)}, \varphi_2^{(q)}, \ldots, \varphi_{p_q}^{(q)}, \end{array} \tag{2.80}$$

such that

$$\begin{aligned} &\lambda_0 \boldsymbol{K} \varphi_1^{(s)} = \varphi_1^{(s)}, \\ &\lambda_0 \boldsymbol{K} \varphi_2^{(s)} = \varphi_1^{(s)} + \varphi_2^{(s)}, \ldots, \lambda_0 \boldsymbol{K} \varphi_{p_s}^{(s)} = \varphi_{p_s - 1}^{(s)} + \varphi_{p_s}^{(s)}. \end{aligned} \tag{2.81}$$

where the $\varphi_1^{(s)}(x)$ $(s = 1, 2, \ldots, q)$ form a complete system of eigenfunctions for the kernel $K(x, t)$ corresponding to the characteristic value λ_0, and the functions $\varphi_j^{(s)}(x)$ satisfy

$$(I - \lambda \boldsymbol{K})^j \varphi_j^{(s)} = 0 \qquad (j = 1, 2, \ldots, p_s). \tag{2.82}$$

The functions $\varphi_j^{(s)}(x)$ are called the *zero elements* of the kernel $K(x, t)$ corresponding to the characteristic value λ_0.

Using (2.80), the principal kernel can be rewritten in the form

$$k(x, t) = \sum_{s=1}^{q} k^{(s)}(x, t), \tag{2.83}$$

where $k^{(s)}(x, t)$ has the form

$$k^{(s)}(x, t) = \sum_{j=1}^{p_s} \varphi_j^{(s)}(x)\, \overline{\psi_j^{(s)}(t)}. \tag{2.84}$$

The kernels $k^{(s)}(x, t)$ are called *the canonical kernels* corresponding to the characteristic value λ_0. The canonical kernels are mutually orthogonal. The functions $\psi_j^{(s)}$ are the zero-elements of the conjugate kernel $K^*(x, t)$ corresponding to the characteristic value $\bar{\lambda}_0$, and the functions $\psi_1^{(s)}$ ($s = 1, 2, \ldots, q$) form a complete system of eigenfunctions for this kernel corresponding to $\bar{\lambda}_0$.

The resolvent of the canonical kernel $k^{(s)}(x, t)$ has the form

$$\gamma^{(s)}(x, t; \lambda) = \sum_{h=1}^{p_s} \left(\frac{\lambda}{\lambda_0}\right)^{h-1} \left(1 - \frac{\lambda}{\lambda_0}\right)^{-h} \sum_{j=1}^{p_s - h} \varphi_j(x)\, \overline{\psi_{j+h}(t)}. \tag{2.85}$$

It follows from the mutual orthogonality of the canonical kernels that the resolvent of the principal kernel has the form

$$\gamma(x, t; \lambda) = \sum_{s=1}^{q} \gamma^{(s)}(x, t; \lambda). \tag{2.86}$$

Every characteristic value λ_0 is associated with three integers. They are:
1) the rank r of λ_0; i.e. the number of linearly independent eigenfunctions corresponding to λ_0;
2) the multiplicity n of the pole as λ_0 is a pole of the resolvent;
3) the multiplicity ν of λ_0 as a zero of the Fredholm determinant.

These three integers are related to the subscripts introduced in (2.80) in the following way:

$$r = q, \qquad n = \max_s p_s, \qquad \nu = \sum_s p_s.$$

§9 The rate of growth of eigenvalues

The rate of growth of the eigenvalues of kernels of Fredholm type is defined by the Theorem of Schur—see (2.22). Using special assumptions about the

Fredholm kernel, it is possible to obtain more definitive results concerning the growth of eigenvalues. Below, we present some results due to Gel'fond {see the Appendix of [15]} related to this problem.

Let x and t be real variables, Ω be the square $0 \leqslant x, t \leqslant 1$, and the kernel $K(x, t)$ be a function of the two real variables x and $t \in \Omega$. Then the following Theorems hold:

THEOREM 9.1. *In Ω, let the kernel $K(x, t)$ have bounded derivatives with respect to t of order up to and including m. If $\lambda_1, \lambda_2, \ldots$ are the characteristic values of $K(x, t)$, then the following series converges*

$$\sum_{n=1}^{\infty} \frac{1}{|\lambda_n|^{\rho+\varepsilon}}, \qquad \rho = \frac{2}{2m+1}, \tag{2.87}$$

where $\varepsilon > 0$. Here, the Fredholm determinant is an entire function with order less than ρ.

A stronger result can be formulated for kernels analytic with respect to t. It is

THEOREM 9.2. *In a neighbourhood of any point $\tau \in [0, 1]$, let the decomposition*

$$K(x, t) = \sum_{n=0}^{\infty} \varphi_n(x, \tau)(t - \tau)^n, \tag{2.88}$$

and the inequality

$$|\varphi_n(x, \tau)| \leqslant c\rho^n \tag{2.89}$$

hold, where the positive constants c and ρ are independent of x and τ. Then the growth of the characteristic values of the kernel $K(x, t)$ is bounded below by a certain geometric progression.

Finally, we state a Theorem concerning kernels which can be unbounded near the boundary of Ω.

THEOREM 9.3. *Let $K(x, t)$ satisfy the following conditions:*

1) *There exist constants γ and δ, $\gamma > 1$ and $0 < \delta \leqslant \frac{1}{2}$, such that*

$$|K(x, t)| \leqslant \frac{\gamma}{[xt(1-x)(1-t)]^{(1/2)-\delta}}.$$

2) *There exists a constant* $q \geqslant 1$ *such that inside the following crescent*

$$\left| z - \frac{1}{2} \pm \operatorname{ctg} \frac{\pi}{2q} \right| < \frac{1}{2 \sin \dfrac{\pi}{2q}}, \qquad z = x + it,$$

the kernel is a holomorphic function of z.

3) *In the bounded region, the contour of which is defined by*

$$z = \frac{[1 + (1 - 2\varepsilon)e^{i\varphi}]^{1/q}}{[1 + (1 - 2\varepsilon)e^{i\varphi}]^{1/q} + [1 - (1 - 2\varepsilon)e^{i\varphi}]^{1/q}},$$

let the following inequality hold

$$|K(x, t)| < \frac{\gamma}{[x(1 - x)]^{(1/2) - \delta} \varepsilon^{p/q}},$$

where p *is a positive constant and* $0 < \varepsilon \leqslant \frac{1}{2}$. *Then the characteristic values of* $K(x, t)$ *satisfy*

$$|\lambda_n| > ab^{\sqrt{(n)}},$$

where a *and* b *are constants such that* $a > 0, b > 1$.

Chapter III

SYMMETRIC EQUATIONS[1])

§1 Basic properties

1.1 *Symmetric kernels*

In the present chapter, we limit consideration to symmetric equations with Fredholm kernels. Equations of more general type will be considered in Chapter V. We retain all notions, terminology and notation of Chapter II.

The Fredholm kernel $K(x, t)$ is called *symmetric* if it coincides with its conjugate

$$K(x, t) = K^*(x, t) = \overline{K(t, x)}. \tag{3.1}$$

A real kernel is symmetric if

$$K(x, t) = K(t, x). \tag{3.2}$$

Example 1.1. Let Ω be an interval $[a, b]$ of the real axis. The kernels $x + t$, $x^2 + t^2$ and $i(x - t)$ are symmetric, while the kernel $i(x + t)$ is not.

Example 1.2. Let Ω be an m-dimensional bounded region and let $r = |x - t|$. The kernel $K(x, t) = r^{-\lambda}$, $0 \leqslant \lambda < m/2$ is a symmetric Fredholm kernel. The kernel $K(x, t) = \ln (1/r)$ is also symmetric.

If a kernel is symmetric then all its iterated kernels are also symmetric.

If a kernel $K(x, t)$ is symmetric and $\boldsymbol{K}$ is the Fredholm operator corresponding to $K(x, t)$ (see formula (2.7)), then this operator coincides with its conjugate. Operators, characterized by the last property, are called *self-adjoint*. The identity (2.14) takes a simpler form for self-adjoint operators, viz.

$$(\boldsymbol{K}u, v) = (u, \boldsymbol{K}v), \qquad u, v \in L_2(\Omega). \tag{3.3}$$

---

[1]) The detailed treatment of the theory of symmetric equations can be found in [8], [15], [20], [22], [23], [25], [28].

An integral equation with a symmetric kernel is also called symmetric. The Fredholm operator with a symmetric kernel is known as the Hilbert-Schmidt operator.

1.2 *Basic theorems connected with symmetric kernels*

THEOREM 1.1. *A symmetric kernel, which is not identically equal to zero, has at least one characteristic value.*

THEOREM 1.2. *All characteristic values of a symmetric kernel are real.*

THEOREM 1.3. *Eigenfunctions of a symmetric kernel, corresponding to different characteristic values, are orthogonal.*

1.3 *Systems of characteristic values and eigenfunctions*

Several of the linearly independent eigenfunctions can correspond to the same characteristic value. Using a process of orthogonalization (see below), they can be orthogonalized. Further, any eigenfunction can be normalized. It is sufficient to divide it by its norm. It therefore follows that the system of eigenfunctions of a symmetric kernel can be considered to be orthonormal.

It is customary to write the characteristic values and eigenfunctions in the form

$$\lambda_1, \lambda_2, \ldots, \lambda_n, \ldots, \qquad u_1(x), u_2(x), \ldots, u_n(x), \ldots, \tag{3.4}$$

using the following rules:

1) The characteristic values are written in ascending order with respect to their absolute values, viz. $|\lambda_n| \leqslant |\lambda_{n+1}|$.

2) If n is the rank of a characteristic value (see §1.1, Chapter II), then we repeat this value n times. Consequently, one and the same value λ_k can occur in the sequence (3.4) several times. However, there is only one eigenfunction corresponding to each occurrence of λ_k.

3) All eigenfunctions $u_n(x)$ are normalized and orthogonal; that is, the system $\{u_k(x)\}$ is orthonormal.

The systems (3.4) are usually referred to as the system of characteristic values and eigenfunctions of the given symmetric kernel.

1.4 *Orthogonalization*

Let

$$\varphi_1(x), \varphi_2(x), \ldots, \varphi_n(x), \ldots$$

be a finite or denumerable system of functions which are contained in $L_2(\Omega)$ and such that $\varphi_1(x), \varphi_2(x), \ldots, \varphi_n(x)$ are linearly independent for any n. If the number of these functicns is finite, we shall simply assume that they are linearly independent. The orthogonalization process consists of the construction of the orthonormalized system

$$\omega_1(x), \omega_2(x), \ldots, \omega_n(x), \ldots$$

according to the following algorithm:

$$\psi_1(x) = \varphi_1(x), \qquad \omega_1(x) = \frac{\psi_1(x)}{\|\psi_1\|},$$

$$\psi_n(x) = \varphi_n(x) - \sum_{k=1}^{n-1} (\varphi_n, \omega_k)\, \omega_k(x),$$

$$\omega_n(x) = \frac{\|\psi_n\|}{\psi_n(x)}, \qquad n = 2, 3, 4, \ldots$$

Note. For any n, the function ω_n is a linear combination of $\varphi_1, \varphi_2, \ldots, \varphi_n$ and, conversely, the function φ_n is a linear combination of $\omega_1, \omega_2, \ldots, \omega_n$.

§2 The Hilbert-Schmidt series and its properties

2.1 *Hilbert-Schmidt theorem*

Let $\boldsymbol{K}$ be a Hilbert-Schmidt operator. Let the system of characteristic values and eigenfunctions of the kernel $K(x, t)$ of the operator $\boldsymbol{K}$ be given by (3.4). Finally, let $g(x)$ be a function from the range of the operator $\boldsymbol{K}$ such that

$$g(x) = (\boldsymbol{K}h)(x) = \int_\Omega K(x, t)\, h(t)\, dt, \qquad h \in L_2(\Omega). \tag{3.5}$$

Then

$$g(x) = \sum_{n=1}^{\infty} \frac{(h, u_n)}{\lambda_n} u_n(x), \tag{3.6}$$

where the series (3.6) converges in the mean in Ω. If the kernel $K(x, t)$ satisfies the inequality (2.24), then (3.6) converges uniformly in Ω.

The series (3.6) is called the *Hilbert-Schmidt series*. It represents the orthogonal decomposition of the function $g(x)$ with respect to eigenfunctions of the kernel $K(x, t)$. The *Hilbert-Schmidt theorem* can be formulated as follows:

The system of eigenfunctions of a Hilbert-Schmidt operator is complete in the range of this operator.

In particular, if the range of a certain Hilbert-Schmidt operator is dense in $L_2(\Omega)$, then the system of its eigenfunctions is complete in $L_2(\Omega)$.

2.2 *The solution of symmetric integral equations*

The integral equation (2.5) can be easily solved, if its kernel is symmetric and the system (3.4) of its characteristic values and eigenfunctions is known. Namely, if λ is a regular value, then

$$u(x) = f(x) + \sum_{n=1}^{\infty} \frac{(f, u_n)}{\lambda_n - \lambda} u_n(x). \tag{3.7}$$

The formula (3.7) also holds for characteristic values of λ. Let $\lambda = \lambda_{p+1} = \lambda_{p+2} = \ldots = \lambda_q$. Then a necessary and sufficient condition that (2.5) has a solution is {Theorem 1.4, Chapter II}

$$(f, u_n) = 0, \qquad n = p + 1, p + 2, \ldots, q.$$

If these conditions hold, then the coefficients in the series (3.7) with the above mentioned indices take the form $\frac{0}{0}$. Thus, for arbitrary values of these coefficients, (3.7) yields the solution of the symmetric equation (2.5) for the case when λ is a characteristic value.

2.3 *The resolvent of a symmetric kernel*

It is easy to obtain an expression for the resolvent of a symmetric kernel from (3.7):

$$\Gamma(x, t; \lambda) = \sum_{n=1}^{\infty} \frac{u_n(x)\,\overline{u_n(t)}}{\lambda_n - \lambda}. \tag{3.8}$$

For regular values of λ, the series (3.8) converges in the mean with respect to both variables x and t.

2.4 *The bilinear series of a kernel and its iterations*

For $\lambda = 0$, the formula (3.8) becomes a bilinear series for the kernel $K(x, t)$

$$K(x, t) = \sum_{n=1}^{\infty} \frac{u_n(x)\,\overline{u_n(t)}}{\lambda_n}. \tag{3.9}$$

The bilinear series for iterated kernels has the form

$$K_p(x, t) = \sum_{n=1}^{\infty} \frac{u_n(x)\,\overline{u_n(t)}}{\lambda_n^p}. \tag{3.10}$$

From (3.10), it follows that the system of characteristic values and eigenfunctions of the iterated kernel $K(x, t)$ has the form

$$\lambda_1^p, \lambda_2^p, \ldots, \lambda_n^p, \ldots, \qquad u_1(x), u_2(x), \ldots, u_n(x), \ldots \tag{3.11}$$

It also follows from (3.10) that

$$\sum_{n=1}^{\infty} \frac{1}{\lambda_n^{2p}} = \int_\Omega \int_\Omega |K_p(x, t)|^2\,dx\,dt. \tag{3.12}$$

In particular, if $p = 1$, we obtain

$$\sum_{n=1}^{\infty} \frac{1}{\lambda_n^2} = \int_\Omega \int_\Omega |K(x, t)|^2\,dx\,dt. \tag{3.13}$$

This case has been mentioned in §1 of Chapter II.

It follows from (3.9) that the symmetric kernel is degenerate when and only when it has a finite number of characteristic values or, equivalently, a finite number of orthonormalized eigenfunctions.

We comment briefly on the convergence of the bilinear series (3.9) and (3.10). In the general case, these series converge in the mean with respect to both variables. If the kernel $K(x, t)$ satisfies the inequality (2.24), then for $p > 2$ the series (3.10) converges uniformly with respect to both variables, and for $p = 2$, it converges uniformly with respect to one variable for almost any fixed value of the second variable. If all the characteristic values are positive (in this case the kernel $K(x, t)$ is said to be *positive*, see below), the kernel itself is continuous and the set Ω is compact, then the bilinear series (3.9) converges uniformly with respect to both variables (Mercer's Theorem—see [22] or [23]). In particular, if the kernel is continuous and the set Ω is compact, then, for $p = 2$, the series (3.10) converges uniformly with respect to both variables.

§3 The classification of symmetric kernels

A symmetric kernel $K(x, t)$, defined on a region Ω, is called *complete* (sometimes called *closed*) if its system of eigenfunctions is *complete* in $L_2(\Omega)$, otherwise it is called *incomplete*.

A symmetric kernel is called *positive* (or *negative*) if all its characteristic values are positive (or negative). A complete positive kernel is called *positive definite*. A *negative definite* kernel is defined in a similar manner.

An equivalent definition of positive (and negative) kernels is: the symmetric kernel $K(x, t)$, defined on a region Ω, is called *positive* if the following quadratic form

$$(\boldsymbol{K}u, u) = \int_{\Omega}\int_{\Omega} K(x, t)\,\overline{u(x)}\,u(t)\,dx\,dt, \tag{3.14}$$

is non-negative for all $u \in L_2(\Omega)$. The same kernel is called *positive definite* if the quadratic form (3.14) is positive for all $u \in L_2(\Omega)$ which are not identically equal to zero.

The non-symmetric kernel $H(x, t)$, defined on a region Ω, is called *left symmetrisable*, if there exists a positive definite kernel $G(x, t)$ such that the kernel

$$\int_{\Omega} G(x, s)\,H(s, t)\,ds \tag{3.15}$$

is symmetric. The notion of a right symmetrisable kernel is defined in an analogous manner. If a kernel is either left or right symmetrisable, then it has at least one characteristic value and all its characteristic values are real.

§4 Extremal properties of characteristic values and eigenfunctions

Theorems cited below can be obtained as corollaries of the properties of the bilinear series (3.9).

§4 Extremal properties of characteristic values

THEOREM 4.1. *Let λ_1 be the characteristic value with smallest magnitude of a symmetric kernel $K(x, t)$ and let $u_1(x)$ be the corresponding eigenfunction. Then,*

$$\frac{1}{\lambda_1} = \max_{u \neq 0} \frac{|(\boldsymbol{K}u, u)|}{\|u\|^2} = \max_{\|u\|=1} |(\boldsymbol{K}u, u)| \tag{3.16}$$

and the maximum value is attained when $u = u_1$.

THEOREM 4.2. *Let $\lambda_1, \lambda_2, \ldots, \lambda_n$ be the first n characteristic values of a symmetric kernel $K(x, t)$, written in ascending order of absolute value, and let $u_1, u_2, \ldots, u_n$ be the corresponding orthonormal eigenfunctions. Then, the characteristic eigenvalue λ_{n+1}, which is nearest to λ_n in absolute value, satisfies*

$$\frac{1}{|\lambda_{n+1}|} = \max \frac{|(\boldsymbol{K}u, u)|}{\|u\|^2}, \tag{3.17}$$

where the maximum is evaluated over the set of non-zero functions u which are orthogonal to the eigenfunctions $u_1, u_2, \ldots, u_n$:

$$(u, u_j) = 0, \qquad j = 1, 2, \ldots, n. \tag{3.18}$$

The quantity (3.17) *attains its maximum value at the eigenfunction $u_{n+1}(x)$ corresponding to the eigenvalue λ_{n+1}.*

The formula (3.17) is equivalent to

$$\frac{1}{|\lambda_{n+1}|} = \max |(\boldsymbol{K}u, u)|,$$

where the maximum is evaluated over the set of functions satisfying (3.18) and $\|u\| = 1$.

Theorem 4.1 can be sharpened in the following way.

Let H^+ and H^- denote the sets of functions from $L_2(\Omega)$ for which the quadratic form $(\boldsymbol{K}u, u)$ takes only positive and negative values, respectively. Further, let λ_1^+ and λ_1^- denote the smallest (in absolute magnitude) positive and negative characteristic value of the kernel $K(x, t)$, respectively. Then

$$\frac{1}{\lambda_1^+} = \max_{u \in H^+} \frac{(\boldsymbol{K}u, u)}{\|u\|^2}, \qquad \frac{1}{\lambda_1^-} = \min_{u \in H^-} \frac{(\boldsymbol{K}u, u)}{\|u\|^2}. \tag{3.19}$$

Theorem 4.2 can be sharpened in an analogous manner. In particular, if the kernel $K(x, t)$ is positive, then

$$\frac{1}{\lambda_1} = \max_{u \neq 0} \frac{(\boldsymbol{K}u, u)}{\|u\|^2} = \max_{\|u\|=1} (\boldsymbol{K}u, u). \tag{3.20}$$

§5 Schmidt kernels and bilinear series for non-symmetric kernels

Let $L(x, t)$ be a non-symmetric kernel satisfying an inequality of the form (2.2), and let $L^*(x, t)$ be its conjugate. As usual, we shall denote the corresponding Fredholm kernels by $\boldsymbol{L}$ and $\boldsymbol{L}^*$. The operators $\boldsymbol{K}_1 = \boldsymbol{L}^*\boldsymbol{L}$ and $\boldsymbol{K}_2 = \boldsymbol{L}\boldsymbol{L}^*$ are symmetric and positive; and their kernels

$$K_1(x, t) = \int_\Omega L^*(x, s)\, L(s, t)\, ds = \int_\Omega \overline{L(s, x)}\, L(s, t)\, ds \tag{3.21}$$

and

$$K_2(x, t) = \int_\Omega L(x, s)\, L^*(s, t)\, ds = \int_\Omega L(x, s)\, \overline{L(t, s)}\, ds \tag{3.22}$$

are symmetric positive kernels. It can be proved that their characteristic values coincide. The kernels $K_1(x, t)$ and $K_2(x, t)$ are called Schmidt kernels corresponding to the kernel $L(x, t)$.

We shall denote by μ_k $(k = 1, 2, \ldots)$ the characteristic values of the Schmidt kernels, by $u_k(x)$ the orthonormal eigenfunctions of $K_2(x, t)$ and by $v_k(x)$ the orthonormal eigenfunctions of $K_1(x, t)$. We can multiply each of the functions $u_k(x)$ and $v_k(x)$ by an arbitrary value the modulus of which is unity. These values can be chosen so that the following formulas are valid

$$L(x, t) = \sum_{k=1}^{\infty} \frac{u_k(x)\, \overline{v_k(t)}}{\sqrt{\mu_k}}, \qquad L^*(x, t) = \sum_{k=1}^{\infty} \frac{v_k(x)\, \overline{u_k(t)}}{\sqrt{\mu_k}}. \tag{3.23}$$

These formulas can be regarded as bilinear series for the kernels $L(x, t)$ and $L^*(x, t)$. The series (3.23) converge in the mean in Ω with respect to both variables x and t. The following inequality holds

$$\sum_{k=1}^{\infty} \frac{1}{\mu_k} \leqslant B_L^2 = \int_\Omega \int_\Omega |L(x, t)|^2\, dx\, dt. \tag{3.24}$$

§6 The solution of integral equations of the first kind

6.1 *Symmetric equations*

Let $K(x, t)$ be a symmetric kernel defined on a region Ω and let its characteristic values and eigenfunctions be given by (3.4). The solutions of the homogeneous first kind equation

$$\int_{\Omega} K(x, t)\, u(t)\, dt = 0 \tag{3.25}$$

coincide with the functions which are orthogonal to the space spanned by the eigenfunctions of $K(x, t)$. It follows that (3.25) has only a trivial solution if the kernel $K(x, t)$ is complete. If it is incomplete, then (3.25) has a finite or denumerable set of linearly independent solutions. This set will be denumerable if, for instance, the kernel $K(x, t)$ is degenerate.

Now, consider the non-homogeneous first kind equation

$$\int_{\Omega} K(x, t)\, u(t)\, dt = f(x), \tag{3.26}$$

and assume that $f \in L_2(\Omega)$. For (3.26) to have a solution, it is necessary and sufficient that the r.h.s. $f(x)$ possess a series expansion with respect to the eigenfunctions of $K(x, t)$, viz.

$$f(x) = \sum_{k=1}^{\infty} f_k u_k(x), \qquad f_k = (f, u_k),$$

which converges in the mean, and hence, that the series

$$\sum_{k=1}^{\infty} \lambda_k^2 \, |f_k|^2$$

converges. The general solution of (3.26) is then given by

$$u(x) = u_0(x) + \sum_{k=1}^{\infty} \lambda_k f_k u_k(x), \tag{3.27}$$

where $u_0(x)$ is any solution of the homogeneous equation (3.25). If the kernel

$K(x, t)$ is complete, then $u_0(x) \equiv 0$, and the equation (3.26) has the unique solution

$$u(x) = \sum_{k=1}^{\infty} \lambda_k f_k u_k(x) \tag{3.28}$$

if the above mentioned conditions are satisfied.

6.2 *Non-symmetric equations*

Let the non-symmetric kernel $L(x, t)$ possess the bilinear decomposition (3.23). The solutions of the homogeneous first kind equation

$$\int_{\Omega} L(x, t)\, u(t)\, dt = 0 \tag{3.29}$$

are the functions which are orthogonal to the space spanned by $v_k(x)$ $(k = 1, 2, \ldots)$. Hence, it follows that (3.29) only has a trivial solution, if the Schmidt kernel $K_1(x, t)$ is complete. This equation has a finite or denumerable set of linearly independent solutions, if this kernel is incomplete.

The non-homogeneous first kind equation

$$\int_{\Omega} L(x, t)\, u(t)\, dt = f(x), \qquad f \in L_2(\Omega), \tag{3.30}$$

has a solution, if and only if $f(x)$ can be expanded as a series in $u_k(x)$ $(k = 1, 2, \ldots)$, viz.

$$f(x) = \sum_{k=1}^{\infty} f_k u_k(x), \qquad f_k = (f, u_k),$$

which converges in the mean, and hence, if and only if the series

$$\sum_{k=1}^{\infty} \mu_k\, |f_k|^2$$

converges. If these conditions are satisfied, then the general solution of (3.30) is given by

$$u(x) = u_0(x) + \sum_{k=1}^{\infty} \sqrt{(\mu_k)} f_k v_k(x), \tag{3.31}$$

where $u_0(x)$ is any solution of the equation (3.29). If the Schmidt kernel $K_1(x, t)$ is complete and the above mentioned conditions are satisfied, then $u_0(x) \equiv 0$, and (3.30) has the unique solution

$$u(x) = \sum_{k=1}^{\infty} \sqrt{(\mu_k)} f_k v_k(x). \tag{3.32}$$

Chapter IV

INTEGRAL EQUATIONS WITH NON-NEGATIVE KERNELS

§1 Positive eigenvalues

1.1 *The formulation of the problem*

In this section, we shall consider non-negative kernels $K(x, t)$ for $x, t \in \Omega$, where Ω is a compact set in a finite-dimensional space. Only non-essential additional assumptions are required to ensure that these kernels have non-negative eigenfunctions. It is customary to call the corresponding eigenvalues *positive*. Each positive eigenvalue λ is, evidently, non-negative. Below, we shall prove, under sufficiently general conditions, that the non-negative eigenfunction is unique {up to a scalar multiple}. In addition, we shall study the conditions under which a positive eigenvalue is simple. In fact, we say in this chapter that an eigenvalue λ is *simple* if there exists a unique {up to some scalar multiple} eigenfunction which corresponds to it and if the equation

$$\int_{\Omega} K(x, t)\, \psi(t)\, dt = \lambda \psi(x) + \varphi(x)$$

has no solution

Under fairly broad suppositions, the positive eigenvalue majorizes the absolute magnitudes of the other eigenvalues (real or complex).

1.2 *The kernels to be examined*

In this section, we assume that the kernel $K(x, t)$ is non-negative and either continuous or measurable and square integrable:

$$\int_{\Omega}\int_{\Omega} K^2(x, t)\, dx\, dt < \infty. \tag{4.1}$$

If the kernel is square summable then we shall deal with its square summable eigenfunctions. If, in particular, the kernel is continuous, then its square summable eigenfunctions are also continuous. Analogously, all functions which occur below must be regarded as continuous, if $K(x, t)$ is continuous, and square integrable, if (4.1) is satisfied.

The continuous non-negative kernel $K(x, t)$ is called *non-factorable*, if under any division of the set Ω into two non-empty and non-intersecting subsets Ω_1 and Ω_2, there exists $x_0 \in \Omega_1$, $t_0 \in \Omega_2$ such that $K(x_0, t_0) > 0$.

The kernel $K(x, t)$ is non-factorable if and only if for any non-negative continuous function $\varphi(x)$, which is not identically zero, there exists an iteration $K^{(N)}(x, t)$ of the kernel $K(x, t)$

$$K^{(N)}(x, t) = \int_\Omega \int_\Omega \ldots \int_\Omega K(x, s_1)\, K(s_1, s_2) \cdot \ldots \cdot K(s_{N-1}, t)\, ds_1 \ldots ds_{N-1}, \tag{4.2}$$

such that

$$\int_\Omega K^{(N)}(x, t)\, \varphi(t)\, dt > 0 \qquad (x \in \Omega). \tag{4.3}$$

If the kernel $K(x, t)$ is non-factorable, then every iterated kernel $K^{(p)}(x, t)$ is also non-factorable and, conversely, if at least one iterated kernel $K^{(p)}(x, t)$ is non-factorable, then the kernel $K(x, t)$ is non-factorable.

Let $u_0(x)$ be a fixed non-negative function which is positive on some set with positive measure. The kernel $K(x, t)$ is called $u_0(x)$-*upper-bounded*, if for every function $u(x)$ there exist $p = p(u)$ and $a = a(u) > 0$ such that

$$\int_\Omega K^{(p)}(x, t)\, u(t)\, dt \leqslant a u_0(x). \tag{4.4}$$

Analogously, the kernel $K(x, t)$ is called $u_0(x)$-*lower-bounded*, if for every non-negative function $u(x)$, which is positive on a set with non-zero measure,

$$\int_\Omega K^{(q)}(x, t)\, u(t)\, dt \geqslant b \cdot u_0(x), \tag{4.5}$$

where q and b depend on $u(x)$ with $b > 0$.

A continuous kernel $K(x, t)$ is, evidently, u_0-upper-bounded if u_0 is a positive function. If the kernel is non-factorable, then it is u_0-lower-bounded with $u_0(x) \equiv 1$. If $K(x, t)$ satisfies the condition (4.1), then it is u_0-upper-bounded with

$$u_0(x) = \left(\int_\Omega K^2(x, t)\, dt \right)^{1/2} \qquad (x \in \Omega). \tag{4.6}$$

A kernel which is u_0-upper-bounded and u_0-lower-bounded is called *u_0-bounded.*

The Green's functions of some boundary value problems yield important examples of non-negative kernels. The following table contains the simplest examples of Green's functions for boundary value problems of ordinary differential equations. In all these examples, $G(t, x) = G(x, t)$, and therefore, are specified only for $x \leqslant t$.

Differential Operator	Boundary Conditions	Green's Function $G(x, t)$
y''	$y(0) = y(l) = 0$ $y(0) = y'(l) = 0$ $y'(0) = y(l) = 0$	$x\left(1 - \frac{t}{l}\right)$ t $l - s$
$[-f(x)y']'$, $f(x) > 0$, $q(x) = \int_0^x \frac{dx}{f(s)}$	$y(0) = y(l) = 0$ $y(0) = y'(l) = 0$ $y(0) = y(l) + cy'(l)$ $(c > 0)$	$q(x) - \frac{q(x)\, q(t)}{q(l)}$ $q(x)$ $q(x) - \frac{q(x)\, q(t)\, f(l)}{c + q(l)\, f(l)}$
$-xy'' - y'$ $(-xy')' + \frac{n^2}{x} y$	$y(0)$ finite, $y(l) = 0$ $y(0)$ finite, $y(l) = 0$	$-\ln \frac{t}{l}$ $\frac{1}{2n}\left[\left(\frac{x}{t}\right)^n - \frac{1}{l^{2n}}(xt)^n\right]$

All the Green's functions $G(x, t)$ in this table are u_0-bounded kernels with

$$u_0(x) = \int_0^l G(x, t)\, dt. \tag{4.7}$$

1.3 *Existence of a positive eigenfunction*

THEOREM 1.1. *Let $K(x, t)$ be a non-negative kernel and let $\varphi_0(x)$ be a function which is positive on some set with positive measure and satisfies*

$$\int_{\Omega} K^{(p)}(x, t)\, \varphi_0(t)\, dt \geqslant \alpha\varphi_0(x) \qquad (x \in \Omega), \tag{4.8}$$

where $\alpha > 0$ and $K^{(p)}(x, t)$ is an iterated kernel of some order p.

Then the kernel $K(x, t)$ has at least one positive eigenvalue λ_0. This positive eigenvalue satisfies the inequality $\lambda_0 \geqslant \alpha^{1/p}$.

From Theorem 1.1, it follows that u_0-lower-bounded kernels and, in particular, non-factorable kernels have a positive eigenvalue.

Another sufficient condition for the existence of a positive eigenvalue is:

THEOREM 1.2. *Let a non-negative kernel $K(x, t)$ have at least one non-zero (real or complex) eigenvalue.*

Then it has a positive eigenvalue λ_0.

We would like to warn the reader against the notion that every non-negative kernel has a non-negative eigenfunction. For instance, (in the case where $\Omega = [a, b]$), the kernels $K(x, t)$ satisfying the condition $K(x, t) \equiv 0$ for $t \geqslant x$ have no eigenfunctions corresponding to non-zero eigenvalues.

1.4 *The comparison of positive eigenvalues with other eigenvalues*

We shall denote the largest positive eigenvalue of a non-negative kernel by Λ.

THEOREM 1.3. *All (real or complex) eigenvalues λ of a non-negative kernel $K(x, t)$ satisfy the inequality $|\lambda| \leqslant \Lambda$.*

The eigenvalue, of a kernel $K(x, t)$, with the largest absolute value plays an important role in the theory of linear integral equations. On the basis of Theorem 1.2, we can assert that in the case of non-negative kernels there always exists a positive eigenvalue which attains the largest absolute value.

1.5 *The multiplicity of the positive eigenvalue*

THEOREM 1.4. *Let the kernel $K(x, t)$ be non-factorable. Then it has a unique positive and simple eigenvalue Λ, which is greater than the absolute values of all other eigenvalues.*

This theorem is a consequence of the following more general statement.

THEOREM 1.5. *Let the kernel $K(x, t)$ be $u_0(x)$-bounded. Then it has a unique positive and simple eigenvalue Λ, which is greater than the absolute values of all other eigenvalues.*

Theorem 1.5 can be applied, in particular, to the Green's functions of the boundary value problems which were mentioned in §1.2.

1.6 *Stochastic kernels*

A non-negative continuous kernel $K(x, t)$ is called *stochastic* if

$$\int_{\Omega} K(x, t)\, dt \equiv 1 \qquad (x \in \Omega). \tag{4.9}$$

It is evident that the function $\varphi_0(x) \equiv 1 (x \in \Omega)$ is the eigenfunction of a stochastic kernel corresponding to the eigenvalue $\lambda_0 = 1$. The other eigenvalues λ of a stochastic kernel satisfy the inequality $|\lambda| \leqslant 1$. Examples show that stochastic kernels can have eigenvalues λ which are different from 1 and satisfy $|\lambda| = 1$. These eigenvalues are called *permutators*.

THEOREM 1.6
1) *All eigenvalues of a stochastic kernel with absolute value equal to unity are primitive roots of unity.*
2) *The set of eigenfunctions of a stochastic kernel, corresponding to the eigenvalue $\lambda = 1$, has a basis consisting of non-negative functions $\varphi_1(x)$, $\varphi_2(x), \ldots, \varphi_r(x)$ which satisfy the following properties:*

a) *For every function $\varphi_j(x)$ $(j = 1, 2, \ldots, r)$, there exists at least one point at which this function is positive and at which the remaining functions are zero.*

b) *For every point of the region Ω, there exists at least one function of the basis which is positive at this point.*

c) *The set of eigenfunctions of the transposed equation*

$$\lambda\psi(x) = \int_{\Omega} K(t, x)\, \psi(t)\, dt, \tag{4.10}$$

corresponding to the eigenvalue $\lambda = 1$, has a basis consisting of non-negative *functions* $\psi_1(x), \psi_2(x), \ldots, \psi_r(x)$ *which is biorthogonal to the basis* $\varphi_1(x), \varphi_2(x), \ldots, \varphi_r(x)$ *and such that*

$$\psi_i(x)\,\psi_j(x) \equiv 0 \qquad (x \in \Omega, i \neq j, i, j = 1, 2, \ldots, r).$$

1.7 *Notes*

All the statements presented in this section are contained in the general theory (see [3]–[5]) of linear operators which have an invariant cone in a Banach space. Some results from this theory are given in Chapter V, §5.

The validity of all the theorems formulated in this section for kernels $K(x, t)$ which are not square summable, but give rise to linear integral operators

$$\boldsymbol{K}u(x) = \int_\Omega K(x, t)\, u(t)\, dt,$$

which are compact in some function space, follows from the general theory.

§2 Positive solutions of the non-homogeneous equation

2.1 *Existence of a positive solution*

Consider the non-homogeneous equation

$$\mu u(x) = \int_\Omega K(x, t)\, u(t)\, dt + f(x) \tag{4.11}$$

with a non-negative kernel $K(x, t)$ which is either continuous or satisfies the condition (4.1). As in the preceding section, all functions are assumed to be continuous, if $K(x, t)$ is continuous, or square summable, if $K(x, t)$ satisfies the condition (4.1).

THEOREM 2.1. *Let* $\mu > \Lambda$, *where* Λ *is the largest positive eigenvalue of the kernel* $K(x, t)$.

Then, for any non-negative function $f(x)$, *the equation* (4.11) *has the unique non-negative solution* $u^*(x)$ *which can be obtained using the method of* *successive approximations*

$$\mu u_{n+1}(x) = \int_{\Omega} K(x, t)\, u_n(t)\, dt + f(x) \tag{4.12}$$

$(n = 0, 1, 2, \ldots)$

with the initial approximation $u_0(x)$ arbitrary.

If we put $u_0(x) \equiv 0$ then

$$u_n(x) = \sum_{m=0}^{n-1} \frac{\boldsymbol{K}^m f(x)}{\mu^{m+1}}$$

and hence

$$u^*(x) = \sum_{m=0}^{\infty} \frac{\boldsymbol{K}^m f(x)}{\mu^{m+1}},$$

where $\boldsymbol{K}$ is the integral operator with kernel $K(x, t)$:

$$\boldsymbol{K}u(x) = \int_{\Omega} K(x, t)\, u(t)\, dt.$$

The converse of Theorem 2.1 is also valid:

THEOREM 2.2. *If the equation* (4.11) *has a positive solution for at least one positive function $f_0(x)$, then $\mu > \Lambda$, and thus, the equation* (4.11) *has a non-negative solution for any non-negative functions $f(x)$.*

Theorem 2.1 can be supplemented in the following way:

THEOREM 2.3. *If the kernel $K(x, t)$ is u_0-bounded, then for $\mu \leqslant \Lambda$ the equation* (4.11) *has no non-negative solutions for any non-negative function $f(x)$ which is not identically zero.*

In particular, (4.11) with a non-factorable kernel has no non-negative solutions for $\mu \leqslant \Lambda$ and $f(x) \geqslant 0$ $(f(x) \not\equiv 0)$.

2.2 *Convergence of successive approximations*

THEOREM 2.4. *Within the framework of the conditions of Theorem* 2.1, *the rate of convergence of the successive approximations* (4.12) *to the solution* $u^*(x)$ *of* (4.11) *is characterized by the inequality*

$$\|u^* - u_n\| \leqslant C(\mu)\left(\frac{\Lambda}{\mu}\right)^n \qquad (n = 1, 2, \ldots),$$

where $C(\mu)$ *is some constant.*

In this theorem,

$$\|y\| = \max_{x \in \Omega} |y(x)|,$$

if the kernel $K(x, t)$ and the function $f(x)$ are continuous. If the kernel satisfies the condition (4.1) and $f(x)$ is square integrable, then

$$\|y\| = \left(\int_\Omega y^2(x)\, dx\right)^{1/2}.$$

2.3 *Note*

The extension of the theorems of this section to a wider class of kernels demands the use of the general theory of linear operators, having an invariant cone (see [4], [5]).

§3 Estimates for the spectral radius

3.1 *Formulation of the problem*

The maximum modulus eigenvalue of a kernel $K(x, t)$ is called *the spectral radius* $\rho(\mathbf{K})$ *of this kernel or of the integral operator*

$$\mathbf{K}(u)(x) = \int_\Omega K(x, t)\, u(t)\, dt. \tag{4.13}$$

The role of the spectral radius is illustrated, for instance, by the fact that the integral equation (4.11) with a continuous kernel and arbitrary continuous function $f(x)$ has a continuous solution $u(x)$ which can be obtained using the method of successive approximations (4.12) when and only when $|\mu| > \rho(\boldsymbol{K})$. An analogous statement also holds for square summable kernels. In connection with this fact, estimates of the spectral radius become important.

In this section, we shall present some estimates for the spectral radius of non-negative kernels and describe general methods for determining such estimates (see [3]–[6]).

From Theorem 1.3, it follows that the spectral radius $\rho(\boldsymbol{K})$ of the non-negative kernel $\boldsymbol{K}(x, t)$ is either equal to zero or coincides with the largest positive eigenvalue. So the estimation of the spectral radius in this case coincides with the estimation of the largest positive eigenvalue.

Estimates for the spectral radius of non-negative kernels can be used in the study of oscillating kernels, since the spectral radius $\rho(\boldsymbol{K})$ of a non-negative kernel $K(x, t)$ is an upper estimate for the spectral radius of every kernel $R(x, t)$ which satisfies

$$|R(x, t)| \leqslant K(x, t) \qquad (x, t \in \Omega). \tag{4.14}$$

For simplicity of presentation, we assume that the kernel $K(x, t)$ and related functions are continuous. Nevertheless, all statements are valid if the kernel $K(x, t)$ is square summable.

3.2 *Upper estimates*

The following inequalities give the simplest upper estimates

$$\rho(\boldsymbol{K}) \leqslant \max_{x \in \Omega} \int_{\Omega} K(x, t)\, dt, \tag{4.15}$$

$$\rho(\boldsymbol{K}) \leqslant \sqrt{\int_{\Omega}\int_{\Omega} K^2(x, t)\, dx\, dt}. \tag{4.16}$$

More precise estimates are given by

$$\rho(\boldsymbol{K}) \leqslant \sqrt[m]{\max_{x \in \Omega} \int_{\Omega} K^{(m)}(x, t)\, dt}, \tag{4.17}$$

$$\rho(\boldsymbol{K}) \leqslant \sqrt[2m]{\int_{\Omega}\int_{\Omega} K^{(m)^2}(x, t)\, dx\, dt}. \tag{4.18}$$

We present two additional estimates. For some $\lambda_1 > 0$, assume that

$$\int_\Omega K(x, t)\left[\lambda_1 - \int_\Omega K^{(m)}(t, \tau)\, d\tau\right] dt \geqslant 0. \tag{4.19}$$

It follows that

$$\rho(\boldsymbol{K}) \leqslant \sqrt[m]{\lambda_1}.$$

If, for some $\lambda_2 > 0$,

$$\int_\Omega K^{(m)}(x, t)\left[\lambda_2 - \int_\Omega K(t, \tau)\, d\tau\right] dt \geqslant 0, \tag{4.20}$$

then $\rho(\boldsymbol{K}) \leqslant \lambda_2$.

3.3 *The general method*

THEOREM 3.1. *For some $\lambda_0 > 0$ and some non-negative function $u_0(x)$, which is positive on a set of positive measure, assume that*

$$\int_\Omega K(x, t)\, u_0(t)\, dt \geqslant \lambda_0 u_0(x) \qquad (x \in \Omega). \tag{4.21}$$

It follows that the largest eigenvalue Λ of the kernel $K(x, t)$ satisfies the inequality

$$\Lambda \geqslant \lambda_0. \tag{4.22}$$

The following theorems enable one to obtain upper estimates for $\rho(\boldsymbol{K})$.

THEOREM 3.2. *Assume that the non-negative kernel $K(x, t)$ is u_1-upper-bounded and, for some integer p and non-negative λ_0, that*

$$\int_\Omega K^{(p)}(x, t)\, u_1(t)\, dt \leqslant \lambda_0 u_1(x) \qquad (x \in \Omega), \tag{4.23}$$

where $K^{(p)}(x, t)$ is the iterated kernel.

Then

$$\Lambda \leqslant \sqrt[p]{\lambda_0}. \tag{4.24}$$

THEOREM 3.3. *Let the non-negative function $u_1(x)$, which is equal to zero only on a set of measure zero, satisfy the inequality*

$$\int_\Omega K^{(m)}(x, t)\, u_1(t)\, dt \leqslant \lambda_0 u_1(x) \qquad (x \in \Omega). \tag{4.25}$$

Then

$$\rho(\mathbf{K}) \leqslant \sqrt[m]{\lambda_0}. \tag{4.26}$$

All the theorems of this section can be regarded as defining a procedure for the estimation of Λ. The functions $u_0(x)$ and $u_1(x)$ must be chosen on the basis of the form of $K(x, t)$. The accuracy of the estimates obtained depends on the choice of these functions.

3.4 *The block method*

In this section, we assume that the continuous kernel $K(x, t)$ is defined on the square $a \leqslant x, t \leqslant b$. Let

$$a = x_0 < x_1 < \ldots < x_{n-1} < x_n = b \tag{4.27}$$

be an arbitrary partition of the interval $[a, b]$.

We write

$$m_{ik} = \max_{x_{i-1} \leqslant x \leqslant x_i} \int_{x_{k-1}}^{x_k} |K(x, t)|\, dt \qquad (i, k = 1, \ldots, n) \tag{4.28}$$

and form the matrix

$$S = \begin{pmatrix} m_{11} & m_{12} & \ldots & m_{1n} \\ m_{21} & m_{22} & \ldots & m_{2n} \\ \ldots & \ldots & \ldots & \ldots \\ m_{n1} & m_{n2} & \ldots & m_{nn} \end{pmatrix}. \tag{4.29}$$

THEOREM 3.4. *The spectral radius $\rho(\mathbf{K})$ is bounded above by the largest eigenvalue of the matrix S.*

An analogous statement is valid, if instead of S we consider the matrix

$$T = \begin{pmatrix} \alpha_{11} & \alpha_{12} & \dots & \alpha_{1n} \\ \alpha_{21} & \alpha_{22} & \dots & \alpha_{2n} \\ \dots & \dots & \dots & \dots \\ \alpha_{n1} & \alpha_{n2} & \dots & \alpha_{nn} \end{pmatrix}, \tag{4.30}$$

where

$$\alpha_{ik} = \left(\int_{x_{i-1}}^{x_i} \int_{x_{k-1}}^{x_k} K^2(x, t) \, dx \, dt \right)^{1/2}. \tag{4.31}$$

As an example, we shall consider the Green's function $G(x, t)$ of a vibrating string with ends clamped:

$$G(x, t) = \begin{cases} x(1 - t) & \text{for} \quad 0 \leqslant x \leqslant t \leqslant 1, \\ t(1 - x) & \text{for} \quad 0 \leqslant t \leqslant x \leqslant 1. \end{cases}$$

We construct the matrix (4.30) setting $n = 5$ and $x_i = \frac{1}{5} i$ $(i = 0, 1, \dots, 5)$. The largest eigenvalue of the matrix T is equal to 0.10216. This gives the estimate $\rho(\boldsymbol{K}) \leqslant 0.10216$. The exact value of the largest eigenvalue is $1/\pi^2 \approx 0.10132$.

3.5 *Supplementary notes*

Let the kernel $K(x, t)$ be $u_0(x)$-upper-bounded. We denote by δ_p, the smallest of all the values λ_0 for which the inequality (4.23) holds, and write $\beta_p = \sqrt[p]{\delta_p}$. Then $\beta_p \geqslant \Lambda$ and $\beta_p \to \Lambda$.

Let the kernel $K(x, t)$ be $u_0(x)$-bounded. We denote by γ_p the largest value γ for which the following inequality holds

$$\gamma u_0(x) \leqslant \int_{\Omega} K^{(p)}(x, t) \, u_0(t) \, dt$$

and write $\sqrt[p]{\gamma_p} = \alpha_p$. Then $\alpha_p \to \Lambda$ and $\alpha_p \leqslant \Lambda$. Under these conditions, the estimate $0 \leqslant \beta_n - \alpha_n \leqslant O(1/n)$ holds.

§4 Oscillating kernels

4.1 *The formulation of the problem*

In this section, we describe a class of non-negative kernels, all the eigenvalues of which are positive and simple, and the eigenfunctions of which form a system which is in many respects analogous to the system $\sin nx$ ($n = 1, 2, \ldots; 0 \leqslant x \leqslant \pi$) (see [1], [2]).

4.2 *Oscillating matrices* (see [1], [2])

A matrix A is called *strictly non-negative* (*strictly positive*), if all its minors of any order are non-negative (positive).

A strictly non-negative matrix A is called *oscillating*, if there exists an integer $k > 0$ such that A^k is strictly positive.

Before formulating the fundamental theorem concerning the properties of oscillating matrices, we introduce some notation.

Let

$$u = \{u_1, u_2, \ldots, u_n\} \tag{4.32}$$

be a sequence of real numbers. If some of these numbers are equal to zero then we shall attach arbitrary signs to them. Then, we can calculate the number of times the sequence (4.32) alternates in sign. This number will depend on the signs of the zero elements in (4.32). The largest and smallest values of this number will be called, respectively, the maximum and minimum number of sign changes in the sequence (4.32) and will be denoted by S_u^+ and S_u^-. If $S_u^+ = S_u^-$, then we have the exact number of sign changes in the sequence (4.32) and denote it simply by S_u. It is evident that $S_u^+ = S_u^-$ if and only if u_1 and u_n are different from zero, and $u_{i-1} \cdot u_{i+1} < 0$ when $u_i = 0$ $(1 < i < n)$.

THEOREM 4.1

1) *All characteristic values λ_i of an oscillating matrix $A = (a_{ij})(i,j = 1, \ldots, n)$ are positive and simple. Therefore, we can assume that*

$$\lambda_1 > \lambda_2 > \ldots > \lambda_n > 0. \tag{4.33}$$

2) *If $u_k = (u_{1k}, u_{2k}, \ldots, u_{nk})$ is an eigenvector of the oscillating matrix A corresponding to the k-th (in order of magnitude) characteristic value λ_k $(k = 1, 2, \ldots, n)$, then for any constants $c_p, c_{p+1}, \ldots, c_q$* $\Big(1 \leqslant p \leqslant q \leqslant n,$

$\left.\sum_{i=p}^{q} c_i^2 > 0\right)$ *the number of sign changes among the components of the vector*

$$u = c_p u_p + c_{p+1} u_{p+1} + \ldots + c_q u_q \tag{4.34}$$

is bounded by $p - 1$ *and* $q - 1$*:*

$$p - 1 \leqslant S_u^- \leqslant S_u^+ \leqslant q - 1. \tag{4.35}$$

In particular, among the components of the vector u_k $(k = 1, 2, \ldots, n)$ *there exist exactly* $k - 1$ *sign changes:*

$$S_{u_k}^+ = S_{u_k}^- = S_{u_k} = k - 1 \qquad (k = 1, 2, \ldots, n). \tag{4.36}$$

3) *The zero components of two successive eigenvectors* u_k *and* u_{k+1} $(k = 2, 3, \ldots, n - 1)$ *of an oscillating matrix are interlaced.*

For the proof that given matrices, occurring in applications, are oscillating, the following criterion plays a fundamental role.

THEOREM 4.2. *In order that a strictly non-negative matrix* $A = (a_{ik})$ $(i, k = 1, 2, \ldots, n)$ *be oscillating, it is necessary and sufficient that the following conditions be satisfied:*
1) *A is a non-degenerate matrix,*
2) $a_{i,i+1} > 0,\ a_{i+1,i} > 0 \qquad (i = 1, 2, \ldots, n - 1).$

4.3 *Vibrations of an elastic continuum with a discrete distribution of mass*

The properties of oscillating matrices are applicable, for example, in the study of small vibrations of a weightless linear elastic continuum (a string or a rod) which extends along the x-axis from $x = a$ to $x = b$, is not clamped (supported) inside (a, b), and carries n fixed masses (beads).

Let the masses $m_1, m_2, \ldots, m_n$ be fixed to the elastic continuum at the points $x_1, \ldots, x_n$ $(a < x_1 < \ldots < x_n < b)$. Vibrations of this continuum are described by the deflection $y(x, t)$ at the point x at time t. Let us denote by $K(x, \sigma)$ $(a \leqslant x, \sigma \leqslant b)$ the deflection of the continuum at the point x under the action of a unit force at the point σ. The function $K(x, \sigma)$ is called the *influence function*. According to d'Alembert's Principle, the deflection $y(x, t)$ satisfies the equation

$$y(x, t) = -\sum_{k=1}^{n} K(x, x_k) m_k \frac{d^2 y(x_k, t)}{dt^2}. \tag{4.37}$$

We require the harmonic vibration of the elastic continuum, i.e. the solution of the equation (4.37) in the form

$$y(x, t) = \varphi(x) \sin (pt + \alpha). \tag{4.38}$$

Such solutions are called *characteristic vibrations* of the elastic continuum, p is called the *characteristic frequency* and $\varphi(x)$ the *amplitude function.* In order to find p and $\varphi(x)$, we substitute (4.38) in (4.37) and as a consequence find

$$\varphi(x) = p^2 \sum_{k=1}^{n} K(x, x_k)\, m_k \varphi(x_k). \tag{4.39}$$

Then, setting x equal to $x_1, x_2, \ldots, x_n$, we obtain the following system of linear algebraic equations ($i = 1, 2, \ldots, n$)

$$\varphi(x_i) = p^2 \sum_{k=1}^{n} K(x_i, x_k)\, m_k \varphi(x_k). \tag{4.40}$$

This system enables one to find the value of $\varphi(x)$ only at the points x_i. The general value of this function is defined by (4.39).

In order that a non-zero solution $\varphi(x)$ of (4.39) exists, it is necessary and sufficient that

$$\begin{vmatrix} 1 - p^2K(x_1, x_1)m_1 & -p^2K(x_1, x_2)m_2 & \ldots & -p^2K(x_1, x_n)m_n \\ -p^2K(x_2, x_1)m_1 & 1 - p^2K(x_2, x_2)m_2 & \ldots & -p^2K(x_2, x_n)m_n \\ \cdots & \cdots & \cdots & \cdots \\ -p^2K(x_n, x_1)m_1 & -p^2K(x_n, x_2)m_2 & \ldots & 1 - p^2K(x_n, x_n)m_n \end{vmatrix} = 0.$$

The characteristic frequencies can be obtained from this last equation. Clearly, the number of characteristic frequencies cannot exceed n. Substituting the characteristic frequencies in (4.40), we obtain the values of the amplitude function $\varphi(x)$ at the points where the masses are fixed.

It is clear that the properties of the function $y(x_i)$ are completely defined by the properties of the matrix

$$(K(x_i, x_k)m_k) \qquad (i, k = 1, \ldots, n), \tag{4.41}$$

which is called the *influence matrix* with values $K(x_i, x_k)$ *the influence coefficients.*

It turns out that the matrix (4.41) is oscillating. The following basic properties of an elastic continuum follow from the properties of oscillating matrices:

1) All the characteristic frequencies p_i are different and their number is equal to n: $p_1 < p_2 < \ldots < p_n$. Each of them is a simple frequency, i.e. the amplitude function $\varphi_i(x)$ of the characteristic vibration, at the given frequency p_i, is uniquely defined up to a constant factor.

2) The characteristic vibration with the smallest frequency (the fundamental period) has no knots.

NOTE. A point t_0 is called a knot of $\varphi(t)$, if $\varphi(t_0) = 0$ and $\varphi(t)$ changes sign when passing through the point t_0.

3) The characteristic vibration with frequency p_{j+1} (the j-th overtone) has exactly j knots.

4) The knots of two neighbouring characteristic vibrations alternate.

5) The number of knots in the vibration which is obtained as the superposition of characteristic vibrations with frequencies $p_k < p_l < \ldots < p_m$, is contained between $k - 1$ and $m - 1$.

4.4 *Oscillating kernels* {*see* [2]}

Let the non-negative and continuous kernel $K(x, t)$ be defined for $a \leqslant x, t \leqslant b$. Let us denote by I the interval which is obtained from (a, b) by the addition of the extreme point a if $K(a, a) \neq 0$ and the extreme point b if $K(b, b) \neq 0$.

The kernel $K(x, t)$ $(a \leqslant x, t \leqslant b)$ is called *oscillating*, if for any points $t_1, t_2, \ldots, t_n \in I$, among which at least one is contained in (a, b), the matrix

$$(K(t_i, t_k)) \qquad (i, k = 1, 2, \ldots, n) \tag{4.42}$$

is oscillating.

If the kernel $K(x, t)$ is oscillating, then the following inequality holds

$$\begin{vmatrix} K(t_1, x_1) & \ldots & K(t_1, x_n) \\ \cdot & \cdot\ \cdot\ \cdot & \cdot \\ K(t_n, x_1) & \ldots & K(t_n, x_n) \end{vmatrix} > 0$$

for any $x_1, x_2, \ldots, x_n, t_1, t_2, \ldots, t_n \in I$ such that

$$x_1 < x_2 < \ldots < x_n, \qquad t_1 < t_2 < \ldots < t_n.$$

THEOREM 4.3. *Let the kernel $K(x, t)$ be oscillating. Then*

1) *All its eigenvalues are positive and simple.*

2) *The eigenfunction $\varphi_1(x)$ corresponding to the largest eigenvalue λ_1 has no zeros on the interval I.*

3) *The eigenfunction $\varphi_j(x)$ corresponding to the j-th (in order of decrease) eigenvalue λ_j has exactly $j - 1$ knots and has no other zeros on the interval I.*

4) *The knots of two neighbouring eigenfunctions $\varphi_j(x)$ and $\varphi_{j+1}(x)$ $(j = 2, \ldots)$ alternate.*

5) *The linear combination $\varphi(x) = \sum_{i=k}^{m} c_i \varphi_i(x)$ of eigenfunctions has on the interval I not more than $(m - 1)$ zeros and not less than $(k - 1)$ knots for any k and m $(0 \leqslant k \leqslant m)$ and arbitrary constants $c_k, \ldots, c_m \left(\sum_{i=k}^{m} c_i^2 > 0 \right)$.*

4.5 *Small vibrations of systems with infinite degrees of freedom*

The properties of small vibrations of a weightless linear elastic continuum with a discrete distribution of mass described in §4.3 extends naturally to the case of the vibration of a linear elastic continuum with continuous distribution of masses. In this case, the equation (4.37) is replaced by an integro-differential equation and the system of algebraic equations (4.40) by a linear equation for the amplitude function.

Let S be a linear elastic continuum, $K(s, \sigma)$ be its influence function, which is an oscillating kernel, if the ends of the continuum are fixed. In order to derive the equation for its vibration, we use d'Alembert Principle. As a result, we obtain the following integro-differential equation

$$y(x, t) = - \int_S K(x, \sigma) \frac{\partial^2 y(\sigma, t)}{\partial t^2} \rho(\sigma) \, d\sigma. \tag{4.43}$$

Here, $\rho(\sigma)$ is the linear density of the rod. We assume that $\rho(\sigma)$ is a piecewise continuous and positive function.

We require the characteristic harmonic vibrations of the continuum S, i.e. the vibrations of the form

$$y(x, t) = \varphi(x) \sin (pt + \alpha), \tag{4.44}$$

where $\varphi(x)$ is the amplitude function and p is the frequency of the vibration.

Substituting (4.44) in (4.43), we find that

$$\varphi(x) = p^2 \int_S K(x, \sigma)\, \varphi(\sigma)\, \rho(\sigma)\, d\sigma. \tag{4.45}$$

The kernel $K(x, \sigma)\, \rho(\sigma)$ is oscillating. Therefore, from Theorem 4.3, the following properties of small vibrations of a linear elastic continuum with a continuous distribution of mass are obtained:

1) The frequencies of characteristic harmonic vibrations are different.

2) The fundamental tone ($p = p_1$) has no knots.

3) The j-th-overtone ($p = p_j$) has $j - 1$ knots ($j = 1, 2, \ldots$)

4) In the vibration, which is the result of the superposition of the characteristic vibrations with frequencies $p_k < p_l < \ldots < p_m$, the deflection at any time t has not less than $k - 1$ and not more than $m - 1$ zeros.

5) The knots of two neighbouring overtones alternate.

Chapter V

CONTINUOUS AND COMPACT LINEAR OPERATORS

The general theory of operator equations contains the above theory of linear integral equations as a special case. This follows from the fact that the linear integral equation

$$u(x) = \int_{\Omega} K(x, t)\, u(t)\, dt + f(x)$$

can be regarded, under sufficiently general conditions, as a linear equation

$$u = \boldsymbol{K}u + f$$

with a continuous or compact operator $\boldsymbol{K}$ defined on some functional space E. Using such a representation for integral equations, we can apply the general results of functional analysis to their study.

Systems of integral equations can be represented as operator equations in vector-valued spaces. It is often useful to transform a system into a single equation.

Consider, for instance, the system

$$u_1(x) = \sum_{l=1}^{n} \int_{\Omega} K_{1l}(x, t)\, u_l(t)\, dt + f_1(x)$$

$$\cdots\cdots\cdots\cdots\cdots\cdots\cdots\cdots$$

$$u_n(x) = \sum_{l=1}^{n} \int_{\Omega} K_{nl}(x, t)\, u_l(t)\, dt + f_n(x),$$

where Ω is a bounded closed region of a finite-dimensional space. Let h be a vector such that $\Omega + ih\,(i = 1, \ldots, n - 1)$ and Ω are non-overlaping (i.e., the measure of their intersection is equal to zero). We introduce the set

$$\hat{\Omega} = \Omega \cup (\Omega + h) \cup \ldots \cup (\Omega + (n - 1)h),$$

and define the following kernel on it

$$\hat{K}(x, t) = K_{ml}(x - (m - 1)h, t - (l - 1)h),$$

if

$$x \in \Omega + (m - 1)h, \qquad t \in \Omega + (l - 1)h,$$

and the function

$$\hat{f}(x) = f_l(x - (l - 1)h),$$

if $x \in \Omega + (l - 1)h$, and consider the equation

$$\hat{u}(x) = \int_{\hat{\Omega}} \hat{K}(x, t)\, \hat{u}(t)\, dt + \hat{f}(x).$$

It is clear that this equation is equivalent to the given system.

§1 Continuity and compactness for linear integral operators

1.1 *The formulation of the problem*

Let Ω be a compact set of non-zero Lebesgue measure on a finite-dimensional space and let $K(x, t)$ be a function which is measurable with respect to the totality of variables on $\Omega \times \Omega$. The expression

$$\boldsymbol{K}u(x) = \int_{\Omega} K(x, t)\, u(t)\, dt \tag{5.1}$$

is called a linear integral operator, where the integral is interpreted in the sense of Lebesgue. The linear integral operator $\boldsymbol{K}$ is defined for any measurable function $u(x)$ such that $K(x, t) \cdot u(t)$ is a summable function of t on Ω for almost all $x \in \Omega$. Under these conditions, the function $\boldsymbol{K}u(x)$ is measureable. The fact that an operator $\boldsymbol{K}$ maps from some Banach space E_1 into another Banach space E_2 and has 'good' properties {continuity, compactness and others} plays an important and often crucial role in the study of integral, integro-differential and other equations. Consequently, we have the following problem: to find pairs (E_1, E_2) of Banach spaces such that the operator $\boldsymbol{K}$ maps from E_1 into E_2 and is continuous, compact or has other suitable properties.

Usually, pairs of spaces, in which a given operator has 'good' properties, are sought by the 'trial-and-error-method'. Many sufficient (and in some cases necessary) conditions on the kernel $K(x, t)$, which ensure that the corresponding linear operator $\boldsymbol{K}$ maps from some Banach space into another and has suitable properties, are known. Below, we shall give some of these results.

The most common 'trial' spaces are the space C of functions continuous on Ω, and the space L_p of functions which are summable with some order p on Ω. We recall that the space $C = C(\Omega)$ consists of functions u which are continuous on Ω and have the norm

$$\|u\|_C = \max_{x\in\Omega} |u(x)|. \tag{5.2}$$

The space $L_p = L_p(\Omega)$ $(p \in (1, \infty))$ is the set of functions which have finite measure on Ω with respect to the norm

$$\|u\|_{L_p} = \left\{\int_\Omega |u(x)|^p \, dx\right\}^{1/p} \tag{5.3}$$

Together with the space L_p one often considers the space $L_\infty = L_\infty(\Omega)$ of essentially bounded functions on Ω. The norm in this case is defined by:

$$\|u\|_{L_\infty} = \operatorname{vrai}\max_{x\in\Omega} |u(x)|. \tag{5.4}$$

The space C can be regarded as a subspace of L_∞. In this case, functions, which differs from continuous ones on a set of measure zero, can be regarded as elements of C.

When studying operators in the spaces C and L_p, it is convenient to consider the inner product

$$(u, v) = \int_\Omega u(x)\,\overline{v(x)}\,dx. \tag{5.5}$$

This inner product has a finite value if, for instance, $u \in L_p$ and $v \in L_{p'}$, where the numbers p and p' satisfy

$$\frac{1}{p} + \frac{1}{p'} = 1. \tag{5.6}$$

The expression

$$F(u) = (u, v_0) \tag{5.7}$$

is a linear continuous functional on L_p for a fixed $v_0 \in L_{p'}$ with $\|F\| = \|v_0\|_{L_{p'}}$. If $p < \infty$, then the converse is valid: every linear continuous functional F, defined on L_p, can be represented in the form (5.7) with the function v_0 in this representation uniquely defined. This is not a valid statement if $p = \infty$.

It follows that, if $p < \infty$, then the space $L_{p'}$ is the space conjugate[1]) to L_p. If $p = \infty$, then L_1 is a subspace of the space $(L_\infty)^*$ conjugate to L_∞.

For a fixed $v_0 \in L_1$, the expression (5.7) is a linear continuous functional in C with $\|F\|_C = \|v_0\|_{L_1}$. Hence, the space L_1 can also be regarded as a subspace of the space C^* which is conjugate to C.

Next, we give some fundamental definitions for linear operators:

A linear operator $\boldsymbol{K}$ mapping from E_1 into E_2 is called *continuous* if

$$\lim_{\|u - u_0\|_{E_1} \to 0} \|\boldsymbol{K}u - \boldsymbol{K}u_0\|_{E_2} = 0$$

for any $u_0 \in E_1$. The continuity of a linear operator $\boldsymbol{K}$ is equivalent to its *boundedness*

$$\|\boldsymbol{K}u\|_{E_2} \leqslant M \, \|u\|_{E_1} \qquad (u \in E_1). \tag{5.8}$$

The smallest M, for which the inequality (5.8) holds, is called the *norm* of the operator $\boldsymbol{K}$ and is denoted by $\|K\|_{E_1 \to E_2}$.

[1]) We recall that the space E^* is *conjugate* to the Banach space E, if it consists of linear continuous functionals f defined on E with the properties

$$(f_1 + f_2)(u) = f_1(u) + f_2(u), \qquad (\lambda f)(u) = \bar{\lambda} f(u),$$

and the norm

$$\|f\| = \sup_{\|u\| \leqslant 1} |f(u)|.$$

Let the linear operator $\boldsymbol{K}$ map from E_1 into E_2 and be continuous. We call an operator $\boldsymbol{K}^*$ conjugate to $\boldsymbol{K}$, if $\boldsymbol{K}^*$ maps from the space E_2^*, conjugate to E_2, into the space E_1^*, conjugate to E_1, and is defined by

$$\boldsymbol{K}^*f(u) = f(\boldsymbol{K}u) \qquad (u \in E_1). \tag{5.9}$$

If E_1 is the space L_∞ or C, then it is often inconvenient to consider $\boldsymbol{K}^*$ as an operator from E_2^* into E_1^*. In such cases, it often happens that $\boldsymbol{K}^*$ maps from E_2^* (or, if $E_2 = L_\infty$, at least from L_1) into L_1. Therefore it is possible to regard $\boldsymbol{K}^*$ as an operator mapping from $E_2^*(L_1)$ into L_1.

An operator $\boldsymbol{K}$, mapping from E_1 into E_2, is called *positive*, if it transforms non-negative functions of the space E_1 into non-negative functions of the space E_2. An operator $\boldsymbol{K}$, mapping from E_1 into E_2, is called *regular*, if for some linear positive operator $\boldsymbol{K}_0$, mapping from E_1 into E_2, the following inequality holds: $|\boldsymbol{K}u| \leqslant \boldsymbol{K}_0(|u|)$, $u \in E_1$. Regular operators, mapping from L_p into $L_q(C)$, are continuous.

Finally, a linear operator $\boldsymbol{K}$, mapping from E_1 into E_2, is called *completely continuous*, or *compact*, if it maps any bounded set of E_1 into a compact set of E_2. Every compact operator is bounded.

1.2 *Linear integral operators with their range in C*

In this section, we shall consider linear integral operators mapping from one of the spaces $L_p(1 \leqslant p \leqslant \infty)$ into the space C. The fundamental results concerning the theory of such operators are due to Radon [16].

THEOREM 1.1. *A linear integral operator $\boldsymbol{K}$ with kernel $K(x, t)$ maps from the space $L_p(1 \leqslant p \leqslant \infty)$ into the space C, if and only if*

a) *for all $x \in \Omega$, the function $K(x, t)$ belongs to the space $L_{p'}$, with $p' = p/(p-1)$;*
b) *the function*

$$\varphi(x) = \|K(x, t)\|_{L_{p'}} \qquad (x \in \Omega) \tag{5.10}$$

is bounded;
c) *for any measurable subset D of the set Ω and for any $x_0 \in \Omega$ the following equality holds*

$$\lim_{x \to x_0} \int_D K(x, t)\, dt = \int_D K(x_0, t)\, dt.$$

We note that conditions a) and b), for the case $p > 1$, are equivalent to

$$\int_{\Omega} |K(x, t)|^{p'} \, dt \leqslant k < \infty \qquad (x \in \Omega).$$

For the case $p = 1$, these conditions are equivalent to

$$\operatorname*{vrai\,max}_{t \in \Omega} |K(x, t)| \leqslant k < \infty \qquad (x \in \Omega).$$

THEOREM 1.2. *Let the linear integral operator* $\boldsymbol{K}$ *with kernel* $K(x, t)$ *map from the space* $L_p(1 \leqslant p \leqslant \infty)$ *into the space* C.

Then $\boldsymbol{K}$ *is continuous and*

$$\|\boldsymbol{K}\|_{L_p \to C} = \|\varphi(x)\|_{L_\infty}, \tag{5.11}$$

where $\varphi(x)$ *is the function* (5.10).

THEOREM 1.3. *Let the linear integral operator* $\boldsymbol{K}$ *with kernel* $K(x, t)$ *map from the space* $L_p(1 \leqslant p \leqslant \infty)$ *into the space* C.

Then the conjugate operator $\boldsymbol{K}^*$ *maps from the space* L_1 *into the space* $L_{p'}$ *and is defined for functions of the space* L_1 *by*

$$\boldsymbol{K}^* u(x) = \int_{\Omega} K^*(x, t)\, u(t)\, dt, \tag{5.12}$$

where $K^*(x, t) = \overline{K(t, x)}$.

THEOREM 1.4. *Let the linear operator* $\boldsymbol{K}$ *with kernel* $K(x, t)$ *map from the space* $L_p(1 \leqslant p \leqslant \infty)$ *into the space* C.

Then $\boldsymbol{K}$ *is compact if and only if for any* $x_0 \in \Omega$ *the following equality holds*

$$\lim_{x \to x_0} \|K(x, t) - K(x_0, t)\|_{L_{p'}} = 0. \tag{5.13}$$

The equality (5.13) can be rewritten for the case $p > 1$ in the form

$$\lim_{x \to x_0} \int_{\Omega} |K(x, t) - K(x_0, t)|^{p'} \, dt = 0,$$

and, for the case $p = 1$, in the form

$$\lim_{x \to x_0} \operatorname{vrai}\max_{t \in \Omega} |K(x, t) - K(x_0, t)| = 0. \tag{5.14}$$

Theorems 1.1–1.4 represent a full description of the properties of kernels of linear integral operators which map from $L_p(1 \leqslant p \leqslant \infty)$ into C.

The verification of the conditions of Theorem 1.1 is not difficult. For example, an integral operator $\boldsymbol{K}$ maps from any space $L_p(1 \leqslant p \leqslant \infty)$ into the space C and is compact, if the kernel $K(x, t)$ is continuous. We do not consider separately linear integral operators mapping from C into C, since such operators can be regarded as mappings from L_∞ into C (or from L_∞ into L_∞).

1.3 *General properties of integral operators in the L_p-spaces*

There exists no simply formulated necessary and sufficient conditions on the kernel $K(x, t)$ which ensure that the operator $\boldsymbol{K}$, with kernel $K(x, t)$, maps from L_p into L_q. However, different general properties of integral operators mapping from L_p into L_q as well as different sufficient conditions which ensure that $\boldsymbol{K}$ maps from L_p into L_q, are known.

THEOREM 1.5. *Let $K(x, t)$ $(x, t \in \Omega)$ be a measurable function with respect to both variables. Let the linear integral operator $\boldsymbol{K}$, with kernel $K(x, t)$, map from L_p into L_q.*

Then $\boldsymbol{K}$ is continuous.

THEOREM 1.6. *Let $K(x, t)$ $(x, t \in \Omega)$ be a measurable function with respect to both variables and $\boldsymbol{K}$ be the linear integral operator with kernel $K(x, t)$.*

Then $\boldsymbol{K}$ maps from L_p into L_q and is regular, if and only if the integral operator $|\boldsymbol{K}|$, with kernel $|K(x, t)|$, maps from L_p into L_q.

We observe that Theorems 1.5 and 1.6 do not contain estimates for the norm of the integral operator $\boldsymbol{K}$.

THEOREM 1.7. *Let a function $K(x, t)$ $(x, t \in \Omega)$ be measurable with respect to both variables and the integral operator $\boldsymbol{K}$, with kernel $K(x, t)$, map from L_p into L_q be regular.*

Then the conjugate of $\boldsymbol{K}$, the operator $\boldsymbol{K}^*$, maps from $L_{q'}$ into $L_{p'}$ and is *defined by*

$$\boldsymbol{K}^* u(x) = \int_{\Omega} K^*(x, t)\, u(t)\, dt,$$

where $K^*(x, t) = \overline{K(t, x)}$.

We shall now state a theorem concerning the compactness of integral operators. We denote by $\boldsymbol{P}_D$ the operator

$$\boldsymbol{P}_D u(x) = \kappa_D(x)\, u(x), \tag{5.15}$$

where κ_D is the characteristic function operating on the set $D \subset \Omega$.

THEOREM 1.8. *Let the function* $K(x, t)$ $(x, t \in \Omega)$ *be measurable with respect to both variables and let the integral operator with kernel* $K(x, t)$ *map from* L_p *into* L_q. *Let one of the following conditions hold:*
a) $1 \leqslant p \leqslant \infty, 1 \leqslant q < \infty, p > q$, *the operator* $\boldsymbol{K}$ *is regular;*
b) $1 < p \leqslant \infty, 1 \leqslant q < \infty, p \leqslant q$, *the operator* $\boldsymbol{K}$ *is regular and*

$$\lim_{\operatorname{mes} G + \operatorname{mes} F \to 0} \|\boldsymbol{P}_G \boldsymbol{K} \boldsymbol{P}_F\|_{L_p \to L_q} = 0;$$

c) $p > 1, q < \infty$, *the operator* $\boldsymbol{K}$ *satisfies the conditions*

$$\lim_{\operatorname{mes} G \to 0} \|\boldsymbol{P}_G \boldsymbol{K}\|_{L_p \to L_q} = \lim_{\operatorname{mes} F \to 0} \|\boldsymbol{K} \boldsymbol{P}_F\|_{L_p \to L_q} = 0;$$

d) $1 \leqslant p, q \leqslant \infty$, *for arbitrary* $\varepsilon > 0$, *there exist functions* $a_i(x) \in L_q$, $b_i(t) \in L_p$ $(i = 1, 2, \ldots, n)$ *such that*

$$\|\boldsymbol{K} - \boldsymbol{K}_{a_1, \ldots, a_n; b_1, \ldots, b_n}\|_{L_p \to L_q} < \varepsilon,$$

where

$$\boldsymbol{K}_{a_1, \ldots, a_n; b_1, \ldots, b_n} u = \sum_{i=1}^{n} (u, b_i) a_i.$$

Then the operator $\boldsymbol{K}$ *is compact.*

Usually, the verification of the conditions of Theorems 1.8 is not difficult.

We shall present some useful results about integral operators regarding estimates for their kernels.

Let $K(x, t)$ and $K_0(x, t)$ $(x, t \in \Omega)$ be measurable with respect to both variables and satisfy the inequality

$$|K(x, t)| \leqslant K_0(x, t) \qquad (x, t \in \Omega), \tag{5.16}$$

and let $\boldsymbol{K}$ and $\boldsymbol{K}_0$ be integral operators with the kernels $K(x, t)$ and $K_0(x, t)$, respectively. Then from Theorems 1.5–1.8, it follows that:

a) the operator $\boldsymbol{K}$ maps from L_p into L_q and is regular, if $\boldsymbol{K}_0$ is a continuous mapping from L_p into L_q, and

$$\|\boldsymbol{K}\|_{L_p \to L_q} \leqslant \|\boldsymbol{K}_0\|_{L_p \to L_q};$$

b) the operator $\boldsymbol{K}$ maps from L_p into L_q and is compact, if $\boldsymbol{K}_0$ is a compact mapping from L_p into L_q and if $p > 1, q < \infty$.

Let $K(x, t)$, $K_0(x, t)$, $Q(x, t)(x, t \in \Omega)$ be measurable with respect to both variables, $K_0(x, t)$, $Q(x, t) \geqslant 0$, and let $\boldsymbol{K}$, $\boldsymbol{K}_0$, $\boldsymbol{Q}$ be integral operators with the kernels $K(x, t)$, $K_0(x, t)$, $Q(x, t)$, respectively. The operator $\boldsymbol{K}$ maps from L_p into L_q with $p > 1$, $q > \infty$, and is compact, if the operator $\boldsymbol{K}_0$ maps from L_p into $L_{q'}$, the operator $\boldsymbol{Q}$ maps from L_p into L_q and is compact, and if for arbitrary $\varepsilon > 0$ there exists $h = h(\varepsilon)$ such that if

$$|K(x, t)| \geqslant hQ(x, t),$$

then

$$|K(x, t)| \leqslant \varepsilon K_0(x, t) \tag{5.17}$$

From this result it follows, in particular, that the integral operator with kernel $K(x, t)$ mapping from L_p into $L_q(p > 1, q > \infty)$ is compact, if the integral operator with kernel $M[K(x, t)]$ maps from L_p into L_q, where $M(u)$ denotes a non-negative function such that

$$\lim_{v \to \infty} \frac{M(x)}{u} = \infty. \tag{5.18}$$

1.4 *L-characteristics of linear integral operators*

The notion of the *L*-characteristics of an operator $\boldsymbol{K}$ (see [12]) is very useful in the study of linear operators in L_p.

§1 *Continuity for linear integral operators*

Let $\boldsymbol{K}$ be a given linear (for instance, integral) operator. The set of points $\{1/p, 1/q\}$ of the square $0 \leqslant \alpha, \beta \leqslant 1$ such that $\boldsymbol{K}$ maps from L_p into L_q and satisfies some property $\mathfrak{R}$ is called the *L-characteristic* $L(\boldsymbol{K}; \mathfrak{R})$ of the operator $\boldsymbol{K}$. As examples of $\mathfrak{R}$, we consider continuity, regularity and compactness with the corresponding L-characteristics defined by $L(\boldsymbol{K}; \textit{cont.})$, $L(\boldsymbol{K}; \textit{reg.})$ and $L(\boldsymbol{K}; \textit{comp.})$.

L-characteristics of linear operators have some important properties which follow from different interpolation theorems. In particular, every L-characteristic—$L(\boldsymbol{K}; \textit{cont.})$, $L(\boldsymbol{K}; \textit{reg.})$, $L(\boldsymbol{K}; \textit{comp.})$—is a convex set containing with every point $\{\alpha, \beta\}$ all points $\{\alpha', \beta'\}$ such that $\alpha' \leqslant \alpha$, $\beta' \geqslant \beta$. In addition, the L-characteristic $L(\boldsymbol{K}; \text{comp.})$ contains all interior points of the L-characteristic $L(\boldsymbol{K}; \text{cont.})$ or is the empty set.

THEOREM 1.9. *Let $\boldsymbol{K}$ be a linear integral operator with kernel $K(x, t)$ and $|\boldsymbol{K}|$ be the linear integral operator with kernel $|K(x, t)|$. Then*
a) $L(\boldsymbol{K}; \textit{cont.}) = L(\boldsymbol{K}; \textit{map})$.
b) $L(\boldsymbol{K}; \textit{reg.}) = L(|\boldsymbol{K}|; \textit{map})$.
c) *The L-characteristic $L(\boldsymbol{K}; \textit{comp.})$ contains the set of interior points of the L-characteristic $L(\boldsymbol{K}; \textit{cont.})$.*
d) $\{\alpha', \beta\} \in L(\boldsymbol{K}; \textit{comp.})$, if $\alpha' < \alpha$, $\beta > 0$ *and* $\{\alpha, \beta\} \in L(\boldsymbol{K}; \textit{reg.})$.
e) $\{\alpha, \beta'\} \in L(\boldsymbol{K}; \textit{comp.})$, if $\beta' > \beta$, $\alpha < 1$ *and* $\{\alpha, \beta\} \in L(\boldsymbol{K}; \textit{reg.})$.

The integral operator $\boldsymbol{K}^{\#}$ with kernel $K^*(x, t) = \overline{K(t, x)}$ will be called *the conjugate* of $\boldsymbol{K}$ with kernel $K(x, t)$. The integral operator $\boldsymbol{K}$ with a kernel $K(t, x)$ will be called *symmetric* if $K^{\#} = K$.

THEOREM 1.10. *The L-characteristics $L(\boldsymbol{K}; \textit{reg.})$ and $L(\boldsymbol{K}^{\#}; \textit{reg.})$ as well as $L(\boldsymbol{K}; \textit{reg. and comp.})$ and $L(\boldsymbol{K}^{\#}; \textit{reg. and comp.})$ are mutually symmetric with respect to the straight line $\alpha + \beta = 1$. In particular, every L-characteristic $L(\boldsymbol{K}; \textit{reg.})$ and $L(\boldsymbol{K}; \textit{reg. and comp.})$ of a symmetric operator $\boldsymbol{K}$ is symmetric with respect to the straight line $\alpha + \beta = 1$.*

1.5 *Linear U-bounded operators*

As already noted, simple conditions on $K(x, t)$ which ensure that $\boldsymbol{K}$ maps from L_p to L_q are not known. The sufficient conditions of continuity and compactness, examined in this section, define an important class of integral operators mapping in L_p—the class of U-bounded operators.

The linear operator $\boldsymbol{K}$ mapping from L_p into L_q is said to be *U-bounded* (see [6], [10], [12]), if for some $u_0 \in L_q$ the following inequality holds

$$|\boldsymbol{K}u(x)| \leqslant u_0(x) \, \|u\|_{L_p} \qquad (u \in L_p). \tag{5.19}$$

It is obvious that every U-bounded operator $\boldsymbol{K}$ is continuous and regular (as an operator from L_p into L_q); the converse is valid only if $q = \infty$.

For any integral operator $\boldsymbol{K}$, we can consider the L-characteristic $L(\boldsymbol{K}; U\text{-bound})$. This characteristic is a convex set.

THEOREM 1.11. *Let $K(x, t)$ $(x, t \in \Omega)$ be measurable with respect to both variables and satisfy the condition*

$$\varphi(x) = \|K(x, t)\|_{L_r} \in L_k,$$

where $1 \leqslant r, k \leqslant \infty$.

Then the linear integral operator $\boldsymbol{K}$, with kernel $K(x, t)$, maps from L_p, where $p = r/(r - 1)$, into the space L_q, where $q = k$, is U-bounded and

$$\|\boldsymbol{K}\|_{L_p \to L_q} \leqslant \|\varphi\|_{L_k}. \tag{5.20}$$

The inequality (5.20) reduces to an equality if $q = \infty$.

THEOREM 1.12. *Let $K(x, t)$ $(x, t \in \Omega)$ be measurable with respect to both variables and satisfy*

$$\varphi(x) = \|K(x, t)\|_{L_r} \in L_k,$$

where $1 \leqslant r, q \leqslant \infty$. Let either one of the following conditions be satisfied:

a) $r, k < \infty$;

b) $r = \infty$, $k < \infty$ *and, for arbitrary* $\varepsilon > 0$, *it is possible to construct a decomposition* $D_0 \cup D_1 \cup \ldots \cup D_l = \Omega$ *of the region* Ω *such that* mes $(D_0) = 0$ *and*

$$\int_\Omega |K(x, t') - K(x, t'')|\, dx < \varepsilon \qquad (t', t'' \in D_i;\ i = 1, \ldots, l);$$

c) $k = \infty$ *and, for arbitrary* $\varepsilon > 0$, *it is possible to construct a decomposition* $D_0 \cup D_1 \cup \ldots \cup D_l = \Omega$ *of the set* Ω *such that* mes $(D_0) = 0$ *and (if* $r < \infty$*)*

$$\int_\Omega |K(x', t) - K(x'', t)|^r\, dt < \varepsilon \qquad (x', x'' \in D_i;\ i = 1, \ldots, l)$$

or (if $r = \infty$*)*

$$\operatorname*{vrai\,max}_{x \in \Omega} |K(x', t) - K(x'', t)| < \varepsilon \qquad (x', x'' \in D_i,\ i = 1, \ldots, l).$$

Then the linear integral operator $\boldsymbol{K}$, *with kernel* $K(x, t)$, *maps from* L_p, *where* $p = r/(r-1)$, *into* L_q, *where* $q = k$, *and is compact.*

1.6 *Linear U-cobounded operators*

A linear operator $\boldsymbol{K}$ mapping from L_p into L_q is said to be *U-cobounded* (see [7], [12]), if for some function $v_0 \in L_{p'}$ the following inequality holds

$$\|\boldsymbol{K}u\|_{L_q} \leqslant (|u|, v_0) \qquad (u \in L_p). \tag{5.21}$$

Every U-cobounded operator is continuous. We can consider the L-characteristic $L(\boldsymbol{K}; U\text{-}cobound)$. This L-characteristic is a convex set.

The notion of U-boundedness and U-coboundedness of a linear operator are closely connected. Namely, a linear operator $\boldsymbol{K}$ mapping from L_p into L_q is U-bounded, if and only if the conjugate operator $\boldsymbol{K}^*$ mapping from $L_{q'}$ into $L_{p'}$ is U-cobounded. (In the case $p = \infty$, it is assumed that $\boldsymbol{K}^*$ maps from $L_{q'}$ into L_1. This condition always holds for regular integral operators.)

THEOREM 1.13. *Let* $K(x, t)$ $(x, t \in \Omega)$ *be measurable with respect to both variables and satisfy*

$$\psi(t) = \|K(x, t)\|_{L_r} \in L_k,$$

where $1 \leqslant r, k \leqslant \infty$.

Then the linear integral operator $\boldsymbol{K}$, *with kernel* $K(x, t)$, *maps from* L_p, *where* $p = k/(k-1)$, *into* L_q, *where* $q = r$, *is U-cobounded and*

$$\|\boldsymbol{K}\|_{L_p \to L_q} \leqslant \|\psi\|_{L_k}. \tag{5.22}$$

The inequality (5.22) reduces to an equality if $k = \infty$.

THEOREM 1.14. *Let* $K(x, t)$ $(x, t \in \Omega)$ *be measurable with respect to both variables and satisfy*

$$\psi(t) = \|K(x, t)\|_{L_r} \in L_k,$$

where $1 \leqslant r, k \leqslant \infty$. *Let either one of the following conditions be satisfied:*

a) $r, k < \infty$;

b) $r = \infty$, $k < \infty$ *and, for arbitrary* $\varepsilon > 0$, *it is possible to construct a decomposition* $D_0 \cup D_1 \cup \ldots \cup D_l = \Omega$ *of the region* Ω *such that* $\text{mes}\,(D_0) = 0$ *and*

$$\int_\Omega |K(x', t) - K(x'', t)|\, dt < \varepsilon \qquad (x', x'' \in D_i;\ i = 1, \ldots, l);$$

c) $k = \infty$ *and, for arbitrary* $\varepsilon > 0$, *it is possible to construct a decomposition* $D_0 \cup D_1 \cup \ldots \cup D_l = \Omega$ *of the region* Ω *such that* $\text{mes}\,(D_0) = 0$ *and (if* $r < \infty$)

$$\int_\Omega |K(x, t') - K(x, t'')|^r\, dx < \varepsilon \qquad (t', t'' \in D_i;\ i = 1, \ldots, l)$$

or (if $r = \infty$)

$$\operatorname*{vrai\,max}_{x \in \Omega} |K(x, t') - K(x, t'')| < \varepsilon \qquad (t', t'' \in D_i;\ i = 1, \ldots, l).$$

Then the linear integral operator $\boldsymbol{K}$, *with kernel* $K(x, t)$, *maps from* L_p *where* $p = k/(k - 1)$, *into* L_q, *where* $q = r$, *and is compact.*

1.7 *Theorems with two conditions*

The sufficient conditions for continuity and compactness, examined in the preceding section, do not include many common integral operators. Below, we give more complex continuity and compactness conditions.

THEOREM 1.15. *Let* $K(x, t)\ (x, t \in \Omega)$ *be a measureable function with respect to both variables satisfying*

$$\varphi(x) = \|K(x, t)\|_{L_{r_1}} \in L_{k_1},$$

$$\psi(t) = \|K(x, t)\|_{L_{r_2}} \in L_{k_2},$$

where $0 < r_1, r_2, k_1, k_2 \leqslant \infty$.[2])

[2]) By L_r, where $r < 1$ (analogously for the case $r \geqslant 1$), we denote the set of functions $Z(x)$ for which the 'norm' $\|Z\|_{L_r} = \left\{\int_\Omega |Z(x)|^r\, dx\right\}^{1/r}$ has sense and is finite.

If λ satisfies the inequalities

$$0 \leqslant \lambda, \qquad \frac{1-\lambda}{r_1} + \frac{\lambda}{r_2}, \qquad \frac{1-\lambda}{r_1} + \frac{\lambda}{k_2}, \qquad \frac{1-\lambda}{k_1} + \frac{\lambda}{r_2} \leqslant 1,$$

then the linear integral operator $\boldsymbol{K}$, *with kernel* $K(x, t)$, *maps from* L_p *into* L_q, *where*

$$\frac{1}{p} = 1 - \frac{1-\lambda}{r_1} - \frac{\lambda}{k_2}, \qquad \frac{1}{q} = \frac{1-\lambda}{\lambda_1} + \frac{\lambda}{r_2},$$

is regular and

$$\|\boldsymbol{K}\|_{L_p \to L_q} \leqslant \|\varphi\|_{L_{k_1}}^{1-\lambda} \|\psi\|_{L_{k_2}}^{\lambda}. \tag{5.23}$$

THEOREM 1.16. *Let* $K(x, t)$ $(x, t \in \Omega)$ *be a measurable function with respect to both variables satisfying*

$$\varphi(x) = \|K(x, t)\|_{L_{r_1}} \in L_{k_1},$$

$$\psi(t) = \|K(x, t)\|_{L_{r_2}} \in L_{k_2},$$

where $0 < r_1, r_2, k_1, k_2 \leqslant \infty$. *If* λ *satisfies the inequalities*

$$0 \leqslant \lambda, \qquad \frac{1-\lambda}{r_1} + \frac{\lambda}{r_2}, \qquad \frac{1-\lambda}{r_1} + \frac{\lambda}{k_2}, \qquad \frac{1-\lambda}{k_1} + \frac{\lambda}{r_2} \leqslant 1,$$

and at least one of the following conditions is satisfied:

a) $0 < \lambda < 1$, $\min\{k_1, k_2\} < \infty$;

b) $0 < \lambda < 1$, $k_1 = k_2 = \infty$, $r_1, r_2 < \infty$ *and*

$$\lim_{\operatorname{mes} D \to 0} \left\| \int_D |K(x, t)|^{r_1}\, dt \right\|_{L_\infty} = 0;$$

c) $0 < \lambda < 1$, $k_1 = k_2 = \infty$, $r_1, r_2 < \infty$ *and*

$$\lim_{\operatorname{mes} D \to 0} \left\| \int_D |K(x, t)|^{r_2}\, dx \right\|_{L_\infty} = 0;$$

then the linear integral operator **K**, *with kernel* $K(x, t)$, *maps from* L_p *into* L_q, *where*

$$\frac{1}{p} = 1 - \frac{1-\lambda}{r_1} - \frac{\lambda}{k_2}, \qquad \frac{1}{q} = \frac{1-\lambda}{k_1} + \frac{\lambda}{r_2},$$

and is compact.

We conclude by stating a theorem for symmetric kernels $K(x, t)$.

THEOREM 1.17. *Let* $K(x, t)$ $(x, t \in \Omega)$ *be a function which is measurable with respect to both variables and let it satisfy the condition*

$$\varphi(x) = \|K(x, t)\|_{L_r} \in L_k,$$

where $1 \leqslant r \leqslant \infty, 0 < k < \infty$. *If*

$$|K(x, t)| = |K(t, x)|$$

and λ *satisfies the inequalities*

$$0 \leqslant \lambda, \qquad \frac{1-\lambda}{r} + \frac{\lambda}{k} \leqslant 1,$$

then the linear integral operator **K**, *with kernel* $K(x, t)$, *maps from* L_p *into* L_q *with*

$$\frac{1}{p} = 1 - \frac{1-\lambda}{r} - \frac{\lambda}{k}, \qquad \frac{1}{q} = \frac{1-\lambda}{k} + \frac{\lambda}{r},$$

is regular and

$$\|\boldsymbol{K}\|_{L_q \to L_q} \leqslant \|\varphi\|_{L_k}. \tag{5.24}$$

In addition, if $0 < \lambda < 1$ *and one of the following conditions*

a) $k < \infty$,

b) $k = \infty, r < \infty$ and

$$\lim_{\operatorname{mes} D \to 0} \left\| \int_D |K(x, t)|^r \, dx \right\|_{L_\infty} = 0$$

is satisfied then **K** *is compact.*

1.8 *A subtle continuity and compactness condition*

In this section, we present some sufficient continuity and compactness conditions from which the well-known Sobolev limit theorems (see [19]) for potential operators follow:

Let $K(x, t)$ $(x, t \in \Omega)$ be measurable with respect to both variables. We say that $K(x, t)$ satisfies the condition $\mathfrak{N}(\alpha, \beta)$ (or $\mathfrak{N}^0(\alpha, \beta)$) if it is summable and

$$\int_G \int_F K(x, t)\, dx\, dt \leqslant k(\text{mes } F)^\alpha (\text{mes } G)^\beta \tag{5.25}$$

(or

$$\int_G \int_F K(x, t)\, dx\, dt \leqslant k(\text{mes } F, \text{mes } G)(\text{mes } F)^\alpha (\text{mes } G)^\beta, \tag{5.26}$$

where $k(u, v)$ is a function satisfying the condition

$$\lim_{u,v \to 0} k(u, v) = 0 \Big).$$

Here, α and β are non-negative numbers.

THEOREM 1.18. *Let $K(x, t)$ $(x, t \in \Omega)$ be measurable with respect to both variables and satisfy the conditions $\mathfrak{N}(\alpha_1, \beta_1)$ {or $\mathfrak{N}^0(\alpha_1, \beta_1)$} and $\mathfrak{N}(\alpha_2, \beta_2)$, with*

$$\alpha_1 \neq \alpha_2, \qquad \beta_1 \neq \beta_2.$$

If λ satisfies

$$0 < \lambda < 1, \qquad 0 \leqslant (1 - \lambda)\alpha_1 + \lambda\alpha_2, \qquad (1 - \lambda)\beta_1 + \lambda\beta_2 \leqslant 1,$$

and, either

$$(1 - \lambda)(\alpha_1 + \beta_1) + \lambda(\alpha_2 + \beta_2) < 1,$$

*or $\alpha_1 + \beta_1 = \alpha_2 + \beta_2 = 1$, then the linear integral operator **K**, with kernel*

$K(x, t)$, maps from L_p into L_q with

$$1/p = 1 - (1 - \lambda)\alpha_1 - \lambda\alpha_2, \qquad 1/q = (1 - \lambda)\beta_1 + \lambda\beta_2,$$

and is continuous {compact}.

1.9 *Some particular classes of integral operators*

In this subsection, we examine the *L*-characteristics of some important classes of linear integral operators (see [8], [9], [19], [22]).

a) Linear integral operators with kernels of the form

$$K(x, t) = \frac{Q(x, t)}{|x - t|^{\lambda}}, \tag{5.27}$$

where λ is some non-negative number, $Q(x, t)$ is a continuous function for $t \neq x$ and satisfies for arbitrary $\varepsilon > 0$ the conditions

$$\begin{aligned} &\lim_{|x-t|\to 0} |x - t|^{\varepsilon} Q(x, t) = 0, \\ &\lim_{|x-t|\to 0} |x - t|^{-\varepsilon} Q(x, t) = \infty, \end{aligned} \tag{5.28}$$

are called *potential operators*.

We say that an operator $\boldsymbol{K}$ has *exponent* λ, if

$$0 < m_1 \leqslant |Q(x, t)| \leqslant m_2 < \infty; \tag{5.29}$$

exponent $\lambda - 0$, if

$$\lim_{|x-t|\to 0} |Q(x, t)| = 0; \tag{5.30}$$

and *exponent* $\lambda + 0$, if

$$\lim_{|x-t|\to 0} |Q(x, t)| = \infty. \tag{5.31}$$

In particular, operators with the following kernels are of potential type:

$$K(x, t) = \frac{1}{|x - t|^{\lambda} \ln^{\nu}|x - t|}. \tag{5.32}$$

The exponents of such operator are equal to $\lambda - \operatorname{sign} \nu \cdot 0$.

The L-characteristics $L(\mathbf{K}; reg.)$ and $L(\mathbf{K}; comp.)$ of potential operators can be studied using the theorems of subsections 1.2–1.8. In the theorems below, we examine such L-characteristics.

THEOREM 1.19. *Let $\mathbf{K}$ be a potential operator with exponent λ. Then the L-characteristic $L(\mathbf{K}; reg.)$ coincides with the point set $\{\alpha, \beta\}$:*

$$0 \leqslant \alpha, \beta \leqslant 1, \qquad \beta \leqslant \alpha - 1 + \lambda/n.$$

from which points $\{1 - \lambda/n, 0\}$ and $\{1, \lambda/n\}$ are excluded. The L-characteristic $L(\mathbf{K}; comp.)$ coincides with the point set $\{\alpha, \beta\}$:

$$0 \leqslant \alpha, \beta \leqslant 1, \qquad \beta < \alpha - 1 + \lambda/n.$$

THEOREM 1.20. *Let $\mathbf{K}$ be a potential operator with exponent $\lambda + 0$, then the L-characteristics $L(\mathbf{K}; reg.)$, $L(\mathbf{K}; comp.)$ coincide with the point set $\{\alpha, \beta\}$ for which*

$$0 \leqslant \alpha, \beta \leqslant 1, \qquad \beta < \alpha - 1 + \lambda/n.$$

THEOREM 1.21. *Let $\mathbf{K}$ be a potential operator with exponent $\lambda - 0$, then each of the L-characteristics $L(\mathbf{K}; reg.)$ and $L(\mathbf{K}; comp.)$ coincide with the point set*

$$0 \leqslant \alpha, \beta \leqslant 1, \qquad \beta \leqslant \alpha - 1 + \lambda/n,$$

from which the points $\{1 - \lambda/n, 0\}$, $\{1, \lambda/n\}$ are, possibly, excluded.

b) An integral operator with the kernel

$$K_r(x, t) = \begin{cases} \dfrac{1}{x^r}(x - t)^{r-1}, & \text{if} \quad 0 \leqslant t \leqslant x \leqslant 1, \\ 0, & \text{if} \quad x < t \leqslant 1, \end{cases} \tag{5.33}$$

where $r > 0$, is called a *Riemann-Liouville operator*. The number r is called the order of this operator.

THEOREM 1.22. *The L-characteristic $L(\mathbf{K}; reg.)$ of a Riemann-Liouville operator coincides with the point set $\{\alpha, \beta\}$ for which*

$$0 \leqslant \alpha \leqslant \beta \leqslant 1,$$

with the points $\{0, 0\}$ and $\{1, 1\}$ *excluded; the L-characteristic* $L(\boldsymbol{K}; comp.)$ *of a Riemann-Liouville operator coincides with the point set* $\{\alpha, \beta\}$ *for which*

$$0 \leqslant \alpha < \beta \leqslant 1.$$

c) An integral operator with the kernel

$$K_r(x, t) = \begin{cases} x^{r-1}t^{-r}, & \text{if } 0 \leqslant t \leqslant x \leqslant 1, \\ 0, & \text{if } x < t \leqslant 1, \end{cases} \tag{5.34}$$

where $0 \leqslant r < 1$, is called a *Hardy operator*.

THEOREM 1.23. *The L-characteristic* $L(\boldsymbol{K}; reg.)$ *of a Hardy operator coincides with the point set* $\{\alpha, \beta\}$ *for which*

$$0 \leqslant \alpha \leqslant \beta \leqslant 1, \qquad 0 \leqslant \alpha \leqslant 1 - r.$$

The L-characteristic $L(\boldsymbol{K}; comp.)$ *of a Hardy operator coincides with the point set for which*

$$0 \leqslant \alpha < \beta \leqslant 1, \qquad 0 \leqslant \alpha \leqslant 1 - r,$$

with the points $\{0, 0\}$ *and* $\{1 - r, 1 - r\}$ *excluded.*

We pause to note that the kernel of a Hardy operator does not satisfy the conditions of Theorem 1.1, and therefore, it does not map from L_∞ into C. Also, it does not map from C into C. But if C is a subspace of the space L_∞ (i.e. C is the space of functions which coincide almost everywhere with continuous functions) then the Hardy operator maps from C (but not from L_∞) into C and is continuous.

1.10 *Additional notes*

The theorems for linear integral operators given in the preceding sections can be extended (see [6]) to general Banach spaces of measurable functions in which the cone of non-negative functions is reproducing, normal and strongly miniedral (see [11], [14]). Examples are Orlicz, Lorentz, Marcinkievicz (see, for instance, [6], [12], [13]) and other spaces.

For the sufficient conditions, presented above, the specific character of the L_p spaces played an important role. Full generalizations of the above sufficient conditions has only been obtained for different classes of Orlicz spaces (see [13]).

We also note that in some problems we have to consider integral operators mapping from a space of functions defined on a set Ω into a space of functions defined on another set Ω_0. The above results extend naturally to cover this case.

§2 Equations of the second kind. The resolvent of an integral operator

2.1 *The formulation of the problem*

The equation

$$u(x) = \int_{\Omega} K(x, t)\, u(t)\, dt + f(x) \tag{5.35}$$

is called a linear Fredholm equation of the second kind. Here, Ω is a bounded closed set of a finite-dimensional space of non-zero Lebesque measure, $K(x, t)$ $(x, t \in \Omega)$ is a given function which is measureable with respect to both variables, $f(x)$ is a known measurable function, and $u(x)$ is unknown.[3]) If $f(x) \equiv 0$, then (5.35) is said to be *homogeneous*, and if $f(x) \not\equiv 0$, then (5.35) is said to be *non-homogeneous*.

The study of important classes of Fredholm equations of the second kind is based on the Riesz-Schauder-Banach theory (see, for instance, [1], [5], [9], [23]) for linear operator equations. This is bound up with the fact that the integral operator

$$\boldsymbol{K}u(x) = \int_{\Omega} K(x, t)\, u(t)\, dt, \tag{5.36}$$

defined by the r.h.s. of (5.35) usually turns out to be continuous or compact in some Banach space E of measurable functions on Ω. In this case, it is natural to consider the linear operator equation

$$u = \boldsymbol{K}u + f \tag{5.37}$$

in the space E, instead of the equation (5.35) (under the additional supposition that $f \in E$).

[3]) Editors' Note: This is not the usual definition of an integral equation of the second kind.

We note that the study of (5.37) is not completely equivalent to that of (5.35). For instance, if (5.37) has a unique solution, then equation (5.35) is solvable, but it can have several solutions (among which, one will certainly belong to the space E). On the other hand, if (5.37) is solvable, then there exists the solution of (5.35) from the space E, i.e. a solution with some special properties.

It often turns out that the operator (5.36) is continuous (or compact) in several spaces at the same time. Consequently the problem arises as to which space should be chosen for the study of the integral equation (5.35). It often happens that for the study of different kinds of problems it is necessary to choose different kinds of spaces. For instance, when studying the existence of a solution, it is desirable to choose the 'narrowest' space on which the operator (5.36) acts; when studying the uniqueness the 'widest' is required.

Suppose that the integral operator $\boldsymbol{K}$, with kernel $K(x, t)$, is continuous in the space E. In this section, we shall mainly be interested in the problem: under what additional conditions on the operator $\boldsymbol{K}$ {i.e. on the kernel $K(x, t)$} will (5.37) have a unique solution $u \in E$ for arbitrary $f \in E$. This problem is closely connected with the properties of the resolvent $\boldsymbol{R}(\lambda, \boldsymbol{K})$ of a linear integral operator $\boldsymbol{K}$.

2.2 *The resolvent of a linear operator and the spectrum*

We recall the definition of the resolvent of an arbitrary linear operator.

Let $\boldsymbol{K}$ be a linear continuous operator acting in a complex Banach space E. The complex value λ is called *regular* for the operator $\boldsymbol{K}$, if

$$\lambda u - \boldsymbol{K}u = f \tag{5.38}$$

has the unique solution $u \in E$ for any $f \in E$. This solution, written in the form

$$\boldsymbol{R}(\lambda, \boldsymbol{K})f = u, \tag{5.39}$$

defines the linear continuous operator $\boldsymbol{R}(\lambda, \boldsymbol{K})$ in the space E. The totality of all regular values of the operator $\boldsymbol{K}$ forms its *resolvent set* $\Lambda(\boldsymbol{K})$; the complementary set $\Sigma(\boldsymbol{K})$ is called the *spectrum* of the operator $\boldsymbol{K}$.

The operator-function $\boldsymbol{R}(\lambda, \boldsymbol{K})$ defined on $\Lambda(\boldsymbol{K})$ is called the *resolvent* of the operator $\boldsymbol{K}$. According to the definition, the resolvent $\boldsymbol{R}(\lambda, \boldsymbol{K})$ is the operator satisfying the identities

$$\boldsymbol{R}(\lambda, \boldsymbol{K})\boldsymbol{K} = \boldsymbol{K}\boldsymbol{R}(\lambda, \boldsymbol{K}) = \lambda \boldsymbol{R}(\lambda, \boldsymbol{K}) - \boldsymbol{I}. \tag{5.40}$$

For any $\lambda, \mu \in \Lambda(\boldsymbol{K})$, the resolvent satisfies the important Hilbert identity

$$\boldsymbol{R}(\lambda, \boldsymbol{K}) - \boldsymbol{R}(\mu, \boldsymbol{K}) = (\mu - \lambda)\boldsymbol{R}(\lambda, \boldsymbol{K})\boldsymbol{R}(\mu, \boldsymbol{K}). \tag{5.41}$$

The resolvent set $\Lambda(\boldsymbol{K})$ of any linear continuous operator is non-empty: It contains the exterior of the circle with center at the origin and radius

$$\rho(\boldsymbol{K}) = \lim_{n\to\infty} \sqrt[n]{\|\boldsymbol{K}^n\|} \tag{5.42}$$

(where the limit always exists). The number $\rho(\boldsymbol{K})$ is called the *spectral radius* of the operator $\boldsymbol{K}$. Further, for every λ, $|\lambda| > \rho(\boldsymbol{K})$, the resolvent $\boldsymbol{R}(\lambda, \boldsymbol{K})$ can be represented in the form

$$\boldsymbol{R}(\lambda, \boldsymbol{K}) = \sum_{n=0}^{\infty} \frac{1}{\lambda^{n+1}} \boldsymbol{K}^n; \tag{5.43}$$

where the series converges with respect to the norm of the space of operators acting in E.

The resolvent set of an operator $\boldsymbol{K}$ is open and the resolvent $\boldsymbol{R}(\lambda, \boldsymbol{K})$ is an analytic function. What is more, for arbitrary $\lambda_0 \in \Lambda(\boldsymbol{K})$, the resolvent $\boldsymbol{R}(\lambda, \boldsymbol{K})$ has the representation

$$\boldsymbol{R}(\lambda, \boldsymbol{K}) = \sum_{n=0}^{\infty} (-1)^n(\lambda - \lambda_0)^n[\boldsymbol{R}(\lambda_0, \boldsymbol{K})]^n, \tag{5.44}$$

$$(|\lambda - \lambda_0| < \{\rho[\boldsymbol{R}(\lambda_0, \boldsymbol{K})]\}^{-1}).$$

where the series converges with respect to the norm of the operators acting in E.

Below, it will be convenient to consider along with the resolvent $\boldsymbol{R}(\lambda, \boldsymbol{K})$ the so-called *solution operator*

$$L(\lambda, \boldsymbol{K}) = \lambda \boldsymbol{K}\boldsymbol{R}(\lambda, \boldsymbol{K}) \qquad (\lambda \in \Lambda(\boldsymbol{K})), \tag{5.45}$$

This operator usually has better properties than $\boldsymbol{K}$. Obviously,

$$\boldsymbol{R}(\lambda, \boldsymbol{K}) = \frac{1}{\lambda}\boldsymbol{I} + \frac{1}{\lambda^2} L(\lambda, \boldsymbol{K}) \qquad (\lambda \in \Lambda(\boldsymbol{K}), \lambda \neq 0). \tag{5.46}$$

The solution operator $\boldsymbol{L}(\lambda, \boldsymbol{K})$ satisfies an identity analogous to (5.41):

$$\boldsymbol{L}(\mu, \boldsymbol{K}) - \boldsymbol{L}(\lambda, \boldsymbol{K}) = (1/\lambda - 1/\mu)\, \boldsymbol{L}(\lambda, \boldsymbol{K})\, \boldsymbol{L}(\mu, \boldsymbol{K}) \tag{5.47}$$

$$(\lambda, \mu \in \Lambda(\boldsymbol{K});\ \lambda, \mu \neq 0).$$

Like the resolvent, the solution operator is an analytic function of λ on $\Lambda(\boldsymbol{K})$. From (5.43) and (5.44), it follows that the solution operator has the representations:

$$\boldsymbol{L}(\lambda, \boldsymbol{K}) = \sum_{n=0}^{\infty} \frac{1}{\lambda^n} \boldsymbol{K}^{n+1} \left(\frac{1}{|\lambda|} < [\rho(\boldsymbol{K})] \right), \tag{5.48}$$

$$\boldsymbol{L}(\lambda, \boldsymbol{K}) = \sum_{n=0}^{\infty} (-1)^n (1/\lambda - 1/\lambda_0)^n [\boldsymbol{L}(\lambda_0, \boldsymbol{K})]^n, \tag{5.49}$$

$$(|1/\lambda - 1/\lambda_0| < \{\rho[\boldsymbol{L}(\lambda_0, \boldsymbol{K})]\}^{-1}).$$

2.3 *The space of kernels*

Before we pass onto the study of the solution operator of a linear integral operator, we examine some special Banach spaces of kernels. For simplicity, we limit considerations to kernels of operators mapping from some space L_p into another space L_q.

Let $p, q \in [1, \infty]$. We shall denote by $B_{p,q}$ the set of functions $K(x, t)$ which are measurable with respect to both variables and, for any $u \in L_p$ and $v \in L_{q'}$, satisfy the inequality

$$\int_\Omega \int_\Omega K(x, t)\, u(x)\, v(t)\, dx\, dt < \infty. \tag{5.50}$$

The set $B_{p,q}$ along with the norm

$$\|\boldsymbol{K}\|_{B_{p,q}} = \sup_{\|u\|_{L_p}, \|v\|_{L_q} \leq 1} \int_\Omega \int_\Omega |K(x, t)\, u(x)\, v(t)|\, dx\, dt \tag{5.51}$$

is a Banach space. The functions $K(x, t) \in B_{p,q}$ are the kernels of regular operators (see, §1.1, 1.3) mapping from L_p into L_q.

Let $K_1(x, t) \in B_{p,q}$, $K_2(x, t) \in B_{q,r}$, where $1 \leq p, q, r \leq \infty$ and let $\boldsymbol{K}_1$ be the integral operator with the kernel $K_1(x, t)$ and $\boldsymbol{K}_2$ be the integral operator with the kernel $K_2(x, t)$. We define the kernel $K = K_2 * K_1$ by

$$K(x, t) = \int_{\Omega} K_2(x, \sigma)\, K_1(\sigma, t)\, d\sigma. \tag{5.52}$$

The function $K(x, t)$ belongs to the space $B_{p,r}$ and is the kernel of the regular integral operator $\boldsymbol{K} = \boldsymbol{K}_2 \cdot \boldsymbol{K}_1$ mapping from L_p into L_r.

Below, we mainly consider spaces of kernels of operators which map a given L_p space onto itself. In such spaces, the operation (5.52) corresponds to 'associative multiplication' in the space $B_{p,p}$. This is certainly a non-commutative operation.

Let $K(x, t)$ belong to $B_{p,p}$ and $\boldsymbol{K}$ be an integral operator with kernel $K(x, t)$. We introduce the kernels

$$K_{(1)} = K, \qquad K_{(2)} = K * K, \ldots \tag{5.53}$$

where $K_{(n)}$ is the iterated kernel constructed from the kernel $K(x, t)$. Clearly, the n-th iterated kernel is the kernel of the integral operator $\boldsymbol{K}^n$ acting in L_p.

2.4 *The resolvent of a linear integral operator*

In general, the point $\lambda = 0$ does not belong to the resolvent set $\Lambda(\boldsymbol{K})$ of a linear integral operator $\boldsymbol{K}$, so the study of their resolvents is completely equivalent to the study of the solution operators $\boldsymbol{L}(\lambda, \boldsymbol{K})$. It turns out that the solution operator $\boldsymbol{L}(\lambda, \boldsymbol{K})$ of an integral operator $\boldsymbol{K}$ is, generally speaking, also an integral operator. As above, we limit consideration to operators acting in the space $L_p(1 \leqslant p \leqslant \infty)$.

Below, we denote by $\Lambda^0(\boldsymbol{K})$ the set of complex values $\lambda \in \Lambda(\boldsymbol{K})$ for which the solution operator $\boldsymbol{L}(\lambda, \boldsymbol{K})$ can be expressed in the form

$$\boldsymbol{L}(\lambda, \boldsymbol{K})\, u(x) = \int_{\Omega} L(\lambda; x, t)\, u(t)\, dt, \tag{5.54}$$

and by $\Lambda^{00}(K)$ the subset $\lambda \in \Lambda^0(K)$ for which $\boldsymbol{L}(\lambda, \boldsymbol{K})$ is a regular integral operator. Obviously, $\Lambda^{00}(\boldsymbol{K}) \subseteq \Lambda^0(\boldsymbol{K}) \subseteq \Lambda(\boldsymbol{K})$.

THEOREM 2.1. *Let $\boldsymbol{K}$ be an integral operator, acting in L_p $(1 \leqslant p \leqslant \infty)$, with kernel $K(x, t)$. Then:*

a) *The set $\Lambda^{00}(\boldsymbol{K})$ is open; if $\lambda_0 \in \Lambda^{00}(\boldsymbol{K})$, then {we recall that $|\boldsymbol{K}|$ denotes an* *integral operator with kernel* $|K(x, t)|\}$

$$L(\lambda; x, t) = \sum_{n=0}^{\infty} (-1)^n (1/\lambda - 1/\lambda_0)^n L_{(n)}(\lambda_0; x, t), \tag{5.55}$$

$$(|1/\lambda - 1/\lambda_0| < \{\rho[|L(\lambda_0, \boldsymbol{K})|]\}^{-1}),$$

where the series on the r.h.s. converges in the norm of $B_{p,p}$ *and, in particular, in measure.*

b) *For any* $\lambda, \mu \in \Lambda^{00}(\boldsymbol{K})$ *the following identity holds*

$$\mu\lambda[L(\lambda; x, t) - L(\mu; x, t)] = (\mu - \lambda)\int_\Omega L(\lambda; x, \sigma)\, L(\mu; \sigma, t)\, d\sigma. \tag{5.56}$$

THEOREM 2.2. *Let* $\boldsymbol{K}$ *be an integral operator, acting in* $L_p (1 \leqslant p \leqslant \infty)$, *with kernel* $K(x, t)$. *Then:*

a) *The set* $\Lambda^{00}(\boldsymbol{K})$ *contains the set*

$$|\lambda| > \{\rho(|\boldsymbol{K}|)\}^{-1}, \tag{5.57}$$

with

$$L(\lambda; x, t) = \sum_{n=0}^{\infty} \frac{1}{\lambda^n} K_{(n+1)}(x, t), \tag{5.58}$$

where on series on the r.h.s. converges in the norm of $B_{p,p}$ *and, in particular, in measure.*

b) *For any* $\lambda \in \Lambda^{00}(\boldsymbol{K})$ *the following identities hold*

$$L(\lambda; x, t) = \lambda\int_\Omega K(x, \sigma)\, L(\lambda; \sigma, t)\, d\sigma + K(x, t)$$

$$= \lambda\int_\Omega L(\lambda; x, \sigma)\, K(\sigma, t)\, d\sigma + K(x, t). \tag{5.59}$$

In conclusion, we present a theorem concerning the solution of $\boldsymbol{U}$-bounded and $\boldsymbol{U}$-cobounded operators (see §§1.5–1.6).

THEOREM 2.3. *Let* $\boldsymbol{K}$ *be a* $\boldsymbol{U}$*-bounded* ($\boldsymbol{U}$*-cobounded*) *integral operator with kernel* $K(x, t)$ *acting in* L_p $(1 \leqslant p \leqslant \infty)$. *Then* $\Lambda(\boldsymbol{K}) = \Lambda^{00}(\boldsymbol{K})$; *and for any* $\lambda \in \Lambda(\boldsymbol{K})$, *the operator* $\boldsymbol{L}(\lambda, \boldsymbol{K})$ *is* $\boldsymbol{U}$*-bounded* ($\boldsymbol{U}$*-cobounded*).

2.5 *The resolvent of 'improving' operators*

Let E, E_1, E_2 be Banach spaces with $E_2 \subset E \subset E_1$ and

$$\|u\|_{E_1} \leqslant a\,\|u\|_E \quad (u \in E), \qquad \|u\|_E \leqslant a\,\|u\|_{E_2} \quad (u \in E_2). \tag{5.60}$$

An operator $\boldsymbol{K}$, acting in E, will be called *improving* if simultaneously it maps from E_1 into E_2 and is continuous.

To the theorems formulated above we can add some for improving operators.

THEOREM 2.4. *Let E, E_1, E_2 be Banach spaces satisfying* (5.60) *and let $\boldsymbol{K}$ be an improving operator. Then, for any $\lambda \in \Lambda(\boldsymbol{K})$, the operator $\mathbf{L}(\lambda, \boldsymbol{K})$ maps from E_1 into E_2 and is continuous. Moreover, the decompositions* (5.48) *and* (5.49) *converge in the norm of the space of operators mapping from E_1 into E_2.*

For integral operators, this theorem can be extended in the following way.

THEOREM 2.5. *Let $\boldsymbol{K}$ map from L_p into L_q* ($1 \leqslant p \leqslant q \leqslant \infty$) *and be regular. Then, for $\lambda \in \Lambda^{00}(K)$, the operator*

$$\mathbf{L}(\lambda, \boldsymbol{K})\, u(x) = \int_\Omega L(\lambda; x, t)\, u(t)\, dt \tag{5.61}$$

also maps from L_p into L_q and is regular. Further, the decompositions (5.55) *and* (5.58) *converge in the norm of the space $B_{p,q}$.*

2.6 *Conditions for unique solvability*

We return to the study of Fredholm equations of the second kind. The second kind equation

$$\lambda u = \boldsymbol{K}u + f, \tag{5.62}$$

where $\boldsymbol{K}$ is a linear continuous operator, has a unique solution $u \in E$ for arbitrary $f \in E$, if and only if λ belongs to the resolvent set $\Lambda(\boldsymbol{K})$ of $\boldsymbol{K}$. This solution can be written as

$$u = \frac{1}{\lambda} f + \frac{1}{\lambda^2}\, \mathbf{L}(\lambda, \boldsymbol{K}) f, \tag{5.63}$$

where $\boldsymbol{L}(\lambda, \boldsymbol{K})$ is the solution operator corresponding to $\boldsymbol{K}$. If $\boldsymbol{K}$ is an improving operator and $f \in E_2$, then $u \in E_2$. This solution is also unique in the space E_1.

The main difficulty consists in showing that $\lambda \in \Lambda(\boldsymbol{K})$. This takes place, if $\rho(\boldsymbol{K}) < |\lambda|$. In this connection, upper estimates for the spectral radius are of interest. Such estimates have been discussed in Chapter 4, §3 {see also §5 of the present chapter}.

2.7 *Equations with iterated kernels*

Consider the equation

$$\lambda u = \boldsymbol{K}u + f. \tag{5.64}$$

It is clear that every solution $u \in E$ of this equation will simultaneously be a solution of the equation

$$\lambda^n u = \boldsymbol{K}^n u + \boldsymbol{H}(\lambda) f, \tag{5.65}$$

where the operator $\boldsymbol{H}(\lambda)$ is defined by

$$\boldsymbol{H}(\lambda) f = \lambda^{n-1} f + \lambda^{n-2} \boldsymbol{K} f + \ldots + \boldsymbol{K}^{n-1} f. \tag{5.66}$$

If $\lambda^n \in \Lambda(\boldsymbol{K}^n)$, then the converse statement is also valid: every solution of (5.65) is also a solution of (5.64).

The resolvent $\boldsymbol{R}(\lambda, \boldsymbol{K})$ can be defined in terms of the resolvent $\boldsymbol{R}(\lambda^n, \boldsymbol{K}^n)$ in the following way

$$\boldsymbol{R}(\lambda, \boldsymbol{K}) = \boldsymbol{R}(\lambda^n, \boldsymbol{K}^n)\, \boldsymbol{H}(\lambda). \tag{5.67}$$

The following formula which defines the resolvent $\boldsymbol{R}(\lambda, \boldsymbol{K}^n)$ of $\boldsymbol{K}^n$ in terms of $\boldsymbol{R}(\lambda, \boldsymbol{K})$ is of interest

$$\boldsymbol{R}(\lambda, \boldsymbol{K}^n) = \frac{1}{n\lambda^{1-(1/n)}} [\boldsymbol{R}(\lambda, \boldsymbol{K}) + \omega \boldsymbol{R}(\omega\lambda, \boldsymbol{K}) + \ldots + \omega^{n-1} \boldsymbol{R}(\omega^{n-1}\lambda, \boldsymbol{K})], \tag{5.68}$$

where ω is an arbitrary primitive root of unity of order n.

Analogous formulas exist for solution operators. Namely, it follows from (5.67) that

$$\boldsymbol{L}(\lambda, \boldsymbol{K}) = \frac{1}{\lambda^{n-1}} \boldsymbol{H}(\lambda)\, \boldsymbol{K} - \frac{1}{\lambda^{2n-1}} \boldsymbol{L}(\lambda^n, \boldsymbol{K}^n)\, \boldsymbol{H}(\lambda)\, \boldsymbol{K} \tag{5.69}$$

and analogously from (5.68) that

$$\mathbf{L}(\lambda^n, \mathbf{K}^n) = \frac{\lambda^n}{n} \sum_{i=0}^{n-1} \frac{1}{\omega^i \lambda} \mathbf{L}(\omega^i \lambda, \mathbf{K}). \tag{5.70}$$

For the case when $\mathbf{K}$ is an integral operator (acting in the space L_p) and $\lambda^n \in \Lambda^{00}(\mathbf{K}^n)$, the following formulas for kernels correspond to (5.69) and (5.70) are valid:

$$L(\lambda; x, t) = \frac{1}{\lambda^{n-1}} \Delta(\lambda; x, t) - \frac{1}{\lambda^{2n-1}} \int_\Omega L^{[n]}(\lambda; x, \sigma)\, \Delta(\lambda, \sigma, t)\, d\sigma, \tag{5.71}$$

where

$$\Delta(\lambda, x, t) = \sum_{k=1}^{n} \lambda^{n-k} K_{(k)}(x, t)$$

and

$$L^{[n]}(\lambda; x, t) = \frac{\lambda^n}{n} \sum_{i=0}^{n-1} \frac{1}{\omega^i \lambda} L(\omega^i \lambda; x, t), \tag{5.72}$$

where $L^{[n]}(\lambda; x, t)$ denotes the kernel of the solution operator $\mathbf{L}(\lambda^n, \mathbf{K}^n)$.

The shift of emphasis from (5.64) to (5.65) is often justified by the fact that the operator $\mathbf{K}^n$ has better properties than $\mathbf{K}$. Let, for example, (see §1.9) $K(x, t)$ be the kernel of a potential operator $\mathbf{K}$ with exponent λ, $\lambda - 0$, or $\lambda + 0$. Then, the n-th iterated kernel $K_n(x, t)$ (n is the dimension of the space in which the region Ω lies) is also a potential kernel with exponent $r\lambda - (r-1)n$, $[r\lambda - (r-1)n] - 0$ or $[r\lambda - (r-1)n] + 0$, if $r < n/(n-\lambda)$, for $r = n/(n-\lambda)$, it is a potential kernel with exponent 0, -0 or $+0$. It is a bounded kernel, if $r > n/(n-\lambda)$.

2.8 *The conjugate equation*

The equations

$$\lambda u(x) = \int_\Omega K(x, t)\, u(t)\, dt + f(x), \tag{5.73}$$

$$\bar{\lambda} v(x) = \int_\Omega K^*(x, t)\, v(t)\, dt + g(x), \tag{5.74}$$

where $K^*(x, t) = \overline{K(t, x)}$ are usually called *conjugate*. It is convenient to study them simultaneously. We consider (5.73) in a Banach space E and (5.74) in another space F. Naturally, the spaces E and F must be connected in some definite way.

The Banach spaces E and F will be called *dual*, if the following conditions hold

$$\|u\|_E = \sup_{\|v\|_F \leqslant 1} |(u, v)|, \qquad \|v\|_F = \sup_{\|u\|_E \leqslant 1} |(u, v)| \tag{5.75}$$

$$(u \in E, v \in F),$$

where

$$(u, v) = \int_\Omega u(x)\, \overline{v(x)}\, dx. \tag{5.76}$$

The spaces $\{L_p, L_{p'}\}$, where $1/p + 1/p' = 1$, $\{C, L_1\}$ are examples of dual spaces. If E and F are dual, then F can be regarded as a subspace of the space E^* conjugate to E.

Let E and F be dual. Let $\boldsymbol{K}$ be an operator acting in E. We assume that $\boldsymbol{K}^*F \subset F$, and denote by $\boldsymbol{K}'$ the restriction of the operator $\boldsymbol{K}^*$ to the space F. In this case, the operators $\boldsymbol{K}$ and $\boldsymbol{K}'$ are called *dual*.

THEOREM 2.6. *Let E and F be dual spaces and $\boldsymbol{K}$ and $\boldsymbol{K}'$ be dual operators acting in them. Then the components of the resolvent sets $\Lambda_\infty(\boldsymbol{K})$ and $\Lambda_\infty(\boldsymbol{K}')$ of the operators $\boldsymbol{K}$ and $\boldsymbol{K}'$ which contain the point ∞ are symmetric with respect to the abscissa axis. If $\lambda \in \Lambda_\infty(\boldsymbol{K})$, then the operators $\boldsymbol{R}(\lambda, \boldsymbol{K})$ and $\boldsymbol{R}(\bar{\lambda}, \boldsymbol{K}')$ are also dual.*

The following theorem is valid for conjugate integral equations.

THEOREM 2.7. *Let $\boldsymbol{K}$ be a regular operator acting in $L_p (1 \leqslant p \leqslant \infty)$ (or in C) with kernel $K(x, t)$, and $\boldsymbol{K}^{\#}$ be the operator with the adjoint kernel $K^*(x, t) = \overline{K(t, x)}$. Then the resolvent sets $\Lambda^{00}(\boldsymbol{K})$ and $\Lambda^{00}(\boldsymbol{K}^{\#})$ of the operators $\boldsymbol{K}$ and $\boldsymbol{K}^{\#}$ are symmetric with respect to the abscissa axis and, what is more, the kernels $L(\lambda; x, t)$ and $\tilde{L}(\lambda; x, t)$ of the solution operators $\boldsymbol{L}(\lambda, \boldsymbol{K})$ and $L(\lambda, \boldsymbol{K}^{\#})$ are connected by the equality $L(\lambda; x, t) = \tilde{L}(\bar{\lambda}; t, x)$.*

2.9 *Supplementary notes*

The general theory, presented above, concerned operators acting in the spaces L_p. After certain incidental changes, it applies to equations with operators acting in arbitrary spaces E in which the cone of non-negative functions is reproducing, normal and strongly miniedral (see, for instance, [23], where equations with operators acting in Orlicz spaces are studied).

§3 Equations of the second kind with compact operators in a Banach space

3.1 *The formulation of the problem*

In this section, we continue the study of the linear equation

$$\lambda u = \boldsymbol{K}u + f \tag{5.77}$$

with a compact (for instance, integral) operator $\boldsymbol{K}$ acting in some space E.

We recall that a linear operator $\boldsymbol{K}$ mapping from E_1 into E_2 is *compact*, if it maps bounded sets of E_1 into compact sets of E_2. Different tests for compactness of integral operators have been presented in §1.

Compactness of the operator $\boldsymbol{K}$ enables a detailed study of (5.77) to be made. We examine the cases when λ is a regular value of the operator $\boldsymbol{K}$ and when λ is a point of its spectrum.

3.2 *The spectrum of a compact operator*

Let $\boldsymbol{T}$ be a linear continuous operator acting in E. By $N(\boldsymbol{T})$ and $R(\boldsymbol{T})$ we denote the null space and range of $\boldsymbol{T}$, respectively. Let U be some set in E and denote by $U^{\perp}$ the set of all linear continuous functionals which annihilate U, and by ${}^{\perp}V$ the set of elements of E which functionals from V annihilate, where V is some set in E^*.

The value λ is called an *eigenvalue* of the operator $\boldsymbol{K}$ (acting in a Banach space), if the equation $\lambda u = \boldsymbol{K}u$ has non-zero solutions. These solutions are called *eigenvectors*. The set $E(\lambda) = N(\lambda \boldsymbol{I} - \boldsymbol{K})$ is called the *eigen-subspace* of the operator $\boldsymbol{K}$ corresponding to the eigenvalue λ. The dimension of the subspace $E(\lambda)$ is called the *order* of the eigenvalue λ. The set of all eigenvalues of the operator $\boldsymbol{K}$ is called its *eigenspectrum*.

Let λ be an eigenvalue of the operator $\boldsymbol{K}$. The subspace

$$\hat{E}(\lambda) = \overline{\bigcup_{n=1}^{\infty} N[(\lambda \boldsymbol{I} - \boldsymbol{K})^n]} \tag{5.78}$$

is called the *root subspace* of the operator $\boldsymbol{K}$ corresponding to the eigenvalue λ. The dimension of the subspace $\hat{E}(\lambda)$ is called the *multiplicity* of this eigenvalue.

Vectors x from $\hat{E}(\lambda)$ are called *root vectors*. The smallest n for which $(\lambda \boldsymbol{I} - \boldsymbol{K})^n x = 0$ is called the *rank* of the root vector $x \in \hat{E}(\lambda)$. If n does not exist, then the rank of the root vector is equal to infinity. The maximum rank of the root vectors corresponding to the eigenvalue λ is called the *rank of the eigenvalue*. The rank of the eigenvalue λ (if it is finite) coincides with the smallest integer n_0 for which

$$N[(\lambda \boldsymbol{I} - \boldsymbol{K})^n] = N[(\lambda \boldsymbol{I} - \boldsymbol{K})^{n_0}] \qquad (n > n_0). \tag{5.79}$$

The value λ is called the *defect value* of the operator $\boldsymbol{K}$ acting in a Banach space E, if there exist non-zero continuous functionals f, satisfying the identity

$$f[(\lambda u - \boldsymbol{K})] = 0 \qquad (u \in E). \tag{5.80}$$

Such functionals are called *defective*. The set $F(\lambda) = [R(\lambda \boldsymbol{I} - \boldsymbol{K})]^{\perp}$ of all defect functionals corresponding to a defect value λ is called the *defect subspace*. Its dimension is called the *order of the defect value*.

Let λ be a defect value. The subspace

$$\hat{F}(\lambda) = \overline{\bigcup_{n=1}^{\infty} \{R[(\lambda \boldsymbol{I} - \boldsymbol{K})^n]\}^{\perp}} \tag{5.81}$$

is called the *adjoint subspace* of an operator $\boldsymbol{K}$ corresponding to the defect value λ. The dimension of $\hat{F}(\lambda)$ is called the *multiplicity* of this defect value.

Functionals f from $\hat{F}(\lambda)$ are called *adjoint functionals*. The smallest integer n for which

$$f[(\lambda \boldsymbol{I} - \boldsymbol{K})^n u] = 0 \qquad (u \in E) \tag{5.82}$$

is called the *rank* of the adjoint functional f. If such an integer n does not exist then the rank is infinite. The maximum rank of adjoint functionals, corresponding to a defect value λ, is called the *rank* of this defect value. The rank of a defect value λ (if it is finite) coincides with the smallest integer n_0 for which

$$R[\lambda \boldsymbol{I} - \boldsymbol{K})^n]^{\perp} = R[(\lambda \boldsymbol{I} - \boldsymbol{K})^{n_0}]^{\perp} \qquad (n > n_0). \tag{5.83}$$

It follows from the above definitions that the value $\bar{\lambda}$ is an eigenvalue of the operator $\boldsymbol{K}^*$, if λ is a defect value of $\boldsymbol{K}$. The order, the rank and the multiplicity of $\bar{\lambda}$ and λ coincide.

THEOREM 3.1. *Let $\boldsymbol{K}$ be a compact operator acting in an infinite-dimensional space E.*

Then:

a) *The spectrum of $\boldsymbol{K}$ consists of zero and a denumerable set $\lambda_1, \ldots, \lambda_n, \ldots$ with zero as its only limit point.*

b) *Every non-zero point λ of the spectrum is an eigenvalue of finite multiplicity. At the same time, it is a defect value of the same order, multiplicity and rank.*

c) *The spectrums of the operator $\boldsymbol{K}$ and of the conjugate operator $\boldsymbol{K}^*$ are symmetric with respect to the abscissa axes.*

3.3 *The splitting of compact operators*

A linear, continuous operator $\boldsymbol{P}$ acting in a Banach space E is called a *projection operator*, if $\boldsymbol{P}^2 = \boldsymbol{P}$. If $\boldsymbol{P}$ is a projection operator then the operator $\boldsymbol{Q} = \boldsymbol{I} - \boldsymbol{P}$ is also a projector operator. $\boldsymbol{P}$ and $\boldsymbol{Q}$ are called *complementary* projection operators.

Let $\boldsymbol{P}$, $\boldsymbol{Q}$ be complementary projection operators, $E_1 = R(\boldsymbol{P})$, $E_2 = R(\boldsymbol{Q})$. The subspaces E_1 and E_2 are called *mutually complementary*, i.e., every $u \in E$ can be uniquely represented in the form $u = u_1 + u_2$, where $u_1 \in E_1$, $u_2 \in E_2$. Conversely, if two complementary subspaces are given in E, then there exist two complementary projection operators $\boldsymbol{P}$ and $\boldsymbol{Q}$ which project E onto E_1 and E_2. These projection operators are defined by

$$\boldsymbol{P}u = u_1, \qquad \boldsymbol{Q}x = u_2, \tag{5.84}$$

with $u = u_1 + u_2$ ($u_1 \in E_1, u_2 \in E_2$). In fact, we say that $\boldsymbol{P}$ projects E onto E_1 parallel to E_2 and $\boldsymbol{Q}$ projects E onto E_2 parallel to E_1.

THEOREM 3.2. *Let $\boldsymbol{K}$ be a compact operator acting in E and λ_0 be a non-zero point of its spectrum.*

Then:

a) *The subspaces $\hat{E}(\lambda_0)$ and $\check{E}(\lambda_0) = {}^{\perp}\hat{F}(\lambda_0)$ are mutually complementary.*

b) *The subspaces $\hat{E}(\lambda_0)$ and $\check{E}(\lambda_0)$ are invariant for the operator $\boldsymbol{K}$. The complementary projection operators $\boldsymbol{P}(\lambda_0)$ and $\boldsymbol{Q}(\lambda_0)$ corresponding to $\hat{E}(\lambda_0)$ and $\check{E}(\lambda_0)$ commute with the operator $\boldsymbol{K}$ and its resolvent $\boldsymbol{R}(\lambda, \boldsymbol{K})$.*

c) *The operator $\lambda_0\boldsymbol{I} - \boldsymbol{K}$ is nilpotent as an operator in $\hat{E}(\lambda_0)$, i.e.*

$$(\lambda_0\boldsymbol{I} - \boldsymbol{K})^{r(\lambda_0)} = 0,$$

where $r(\lambda_0)$ is the rank of the eigenvalue (defect value) λ_0.

d) *The operator $\lambda_0\boldsymbol{I} - \boldsymbol{K}$ has a continuous inverse as an operator in $\check{E}(\lambda_0)$.*

e) *The operator* $\boldsymbol{K}$ *has the representation*

$$\boldsymbol{K} = \boldsymbol{K}(\lambda_0) + \boldsymbol{K}_0(\lambda_0), \tag{5.85}$$

where $\boldsymbol{K}(\lambda_0) = \boldsymbol{K}\boldsymbol{P}(\lambda_0)$, $\boldsymbol{K}_0(\lambda_0) = \boldsymbol{K}\boldsymbol{Q}(\lambda_0)$. *The spectrum of the operator* $\boldsymbol{K}(\lambda_0)$ *consists of the unique point* λ_0. *The spectrum of the operator* $\boldsymbol{K}_0(\lambda_0)$ *coincides with that of the operator* $\boldsymbol{K}$ *from which the point* λ_0 *has been excluded.*

Let $\boldsymbol{K}$ be a compact operator acting in E and λ_0 be a non-zero point of its spectrum. From Theorem 3.1 it follows that the spaces $\hat{E}(\lambda_0)$ and $\hat{F}(\lambda_0)$ have the same finite dimensionality which coincides with the multiplicity $\kappa(\lambda_0)$ of the eigenvalue (and simultaneously, the defect value) λ_0. The set of elements

$$e_1^{(0)}, e_1^{(1)}, \ldots, e_1^{(\kappa_1 - 1)}, \ldots, e_n^{(0)}, e_n^{(1)}, \ldots, e_n^{(\kappa_n - 1)} \tag{5.86}$$

and the set of functionals

$$f_1^{(0)}, f_1^{(1)}, \ldots, f_1^{(\kappa_1 - 1)}, \ldots, f_n^{(0)}, f_n^{(1)}, \ldots, f_n^{(\kappa_n - 1)} \tag{5.87}$$

($\kappa_1 + \ldots + \kappa_n = \kappa(\lambda_0)$) are called *the canonical bases* of the operator $\boldsymbol{K}$, corresponding to the eigenvalue λ_0, if

$$\begin{aligned} (\lambda_0 \boldsymbol{I} - \boldsymbol{K})e_i^{(j+1)} &= e_i^{(j)} \qquad (j = 0, \ldots, \kappa_i - 2;\ i = 1, \ldots, n), \\ (\lambda_0 \boldsymbol{I} - \boldsymbol{K})e_i^{(0)} &= 0 \qquad (i = 1, \ldots, n), \end{aligned} \tag{5.88}$$

$$\begin{aligned} (\bar{\lambda}_0 \boldsymbol{I} - \boldsymbol{K}^*)f_i^{(j+1)} &= f_i^{(j)} \qquad (j = 0, \ldots, \kappa_i - 2;\ i = 1, \ldots, n), \\ (\bar{\lambda}_0 \boldsymbol{I} - \boldsymbol{K}^*)f_i^{(0)} &= 0 \qquad (i = 1, \ldots, n), \end{aligned} \tag{5.89}$$

and if

$$f_{i'}^{(j')}(e_{i''}^{(j'')}) = \begin{cases} 1, & \text{if } i' = i'' \text{ and } j' + j'' = \kappa_i - 1, \\ 0, & \text{if } i' \neq i'' \text{ or} \\ & \quad i' = i'' \text{ and } j' + j'' \neq \kappa_i - 1. \end{cases} \tag{5.90}$$

THEOREM 3.3 *Let* $\boldsymbol{K}$ *be a compact operator acting in* E *and* λ_0 *be a non-zero point of its spectrum.*

Then:

a) *There exist canonical bases for the operator* $\boldsymbol{K}$, *corresponding to* λ_0.

b) *The operator* $\boldsymbol{P}(\lambda_0)$, *projecting onto* $\hat{E}(\lambda_0)$ *parallel to* $E(\lambda_0)$, *has the representation*

$$P(\lambda_0)u = \sum_{i=1}^{n} \sum_{j=0}^{\kappa_i - 1} f_i^{(\kappa_i - j - 1)}(u)e_i^{(j)}. \tag{5.91}$$

c) *For the operator* $\boldsymbol{K}(\lambda_0)$ *the following decomposition holds:*

$$\boldsymbol{K}(\lambda_0)u = \lambda_0 u - \sum_{i=1}^{n} \sum_{j=1}^{\kappa_i - 1} f_i^{(\kappa_i - 1)}(u)e_i^{(j)}. \tag{5.92}$$

Let us denote by $\lambda_1, \ldots, \lambda_n, \ldots$ the set of non-zero points of the spectrum of a compact operator $\boldsymbol{K}$ acting in E. Then, for any n, we can write down the following decomposition for $\boldsymbol{K}$:

$$\boldsymbol{K} = \sum_{i=1}^{n} \boldsymbol{K}(\lambda_i) + \boldsymbol{R}_n, \tag{5.93}$$

where $\boldsymbol{R}_n$ is a compact operator, acting in E, the non-zero spectrum of which consists of the points $\lambda_{n+1}, \lambda_{n+2}, \ldots$. Consequently, the operator $\boldsymbol{K}$ has the following (perhaps, finite) series representation

$$\boldsymbol{K} \sim \sum_{i=1}^{\infty} \boldsymbol{K}(\lambda_i). \tag{5.94}$$

This series 'converges to the operator $\boldsymbol{K}$ with respect to its spectrum'

$$\lim_{n\to\infty} \rho\left(\boldsymbol{K} - \sum_{i=1}^{n} \boldsymbol{K}(\lambda_i)\right) = 0. \tag{5.95}$$

We stress that this series does not in general converge with respect to the norm of the operators acting in E.

We note in passing that the spectrum of a compact operator $\boldsymbol{K}$ (even when the range of its values is dense in E) may consist of only the zero point. As examples, consider different Volterra operators (see §6).

3.4 *The spectrum of a compact integral operator*

It should be noted that the integral operator

$$\boldsymbol{R}u(x) = \int_{\Omega} R(x, t)\, u(t)\, dt \tag{5.96}$$

acting in a Banach space E is a projection operator, if

$$R(x, t) = \sum_{i=1}^{n} a_i(x)\, b_i(t), \tag{5.97}$$

where a_i and b_i are functions satisfying the conditions

$$\int_{\Omega} a_i(t)\, b_j(t)\, dt = \delta_{ij} \qquad (i, j = 1, \ldots, n). \tag{5.98}$$

Now, let us consider the integral operator

$$\boldsymbol{K}u(x) = \int_{\Omega} K(x, t)\, u(t)\, dt. \tag{5.99}$$

Suppose that this operator is compact and acts in some space of measurable functions on Ω. In this case, Theorems 3.1 and 3.2 are applicable. In fact, these theorems can be made much sharper for integral operators. In particular, the operators $\boldsymbol{K}(\lambda_n)$ define integral operators, if $\boldsymbol{K}$ is an integral operator.

THEOREM 3.4. *Let $\boldsymbol{K}$ be a compact, regular, integral operator, with kernel $K(x, t)$, acting in the space L_p $(1 \leqslant p \leqslant \infty)$. Let $\boldsymbol{K}^{\#}$ be the transposed operator, and λ_0 be a non-zero point of the spectrum of the operator $\boldsymbol{K}$.*

Then, in the space L_p (or, in C), there exist functions

$$e_1^{(0)}(x), e_1^{(1)}(x), \ldots, e_1^{(\kappa_1 - 1)}(x), \ldots, e_n^{(0)}(x), \ldots, e_n^{(\kappa_n - 1)}(x), \tag{5.100}$$

and, in the space $L_{p'}$ (or in L_1), functions

$$f_1^{(0)}(x), f_1^{(1)}(x), \ldots, f_1^{(\kappa_1 - 1)}(x), \ldots, f_n^{(0)}(x), \ldots, f_n^{(\kappa_n - 1)}(x), \tag{5.101}$$

for which the following equalities hold

$$\lambda_0 e_i^{(j+1)}(x) = \int_{\Omega} K(x, t) e_i^{(j+1)}(t)\, dt + e_i^{(j)}(x)$$

$(i = 1, \ldots, n;\ j = 1, \ldots, \kappa_i - 2)$,

$$\lambda_0 e_i^{(0)}(x) = \int_{\Omega} K(x, t) e_i^{(0)}(t)\, dt \qquad (i = 1, \ldots, n), \tag{5.102}$$

$$\bar{\lambda}_0 f_i^{(j+1)}(x) = \int_\Omega K^*(x, t)\, f_i^{(j+1)}(t)\, dt + f_i^{(j)}(x)$$

$$(i = 1, \ldots, n;\ j = 1, \ldots, \kappa_i - 2),$$

$$\bar{\lambda}_0 f_i^{(0)}(x) = \int_\Omega K^*(x, t)\, f_i^{(0)}(t)\, dt \qquad (i = 1, \ldots, n), \tag{5.103}$$

and

$$\int_\Omega e_{i'}^{(j')}(t)\, \overline{f_{i''}^{(j'')}(t)}\, dt = \begin{cases} 1, & \text{if } i' = i'' \text{ and } j' + j'' = \kappa_{i'} - 1, \\ 0, & \text{if } i' \neq i'' \text{ or } i' = i'' \\ & \text{and } j' + j'' \neq \kappa_{i'} - 1. \end{cases} \tag{5.104}$$

Here, the operator $\mathbf{P}(\lambda_0)$ *is defined by*

$$\mathbf{P}(\lambda_0)\, u(x) = \int_\Omega \Delta_{\lambda_0}(x, t)\, u(t)\, dt,$$

where

$$\Delta_{\lambda_0}(x, t) = \sum_{i=1}^{n} \sum_{j=0}^{\kappa_i - 1} \overline{f_i^{(\kappa_i - j - 1)}(t)}\, e_i^{(j)}(x),$$

and the operator $\mathbf{K}(\lambda_0)$ *is defined by*

$$\mathbf{K}(\lambda_0)\, u(x) = \int_\Omega K(\lambda_0; x, t)\, u(t)\, dt,$$

where

$$K(\lambda_0; x, t) = \lambda_0\, \Delta_{\lambda_0}(x, t) - \sum_{i=1}^{n} \sum_{j=1}^{\kappa_i - 1} \overline{f_i^{(\kappa_i - j - 1)}(t)}\, e_i^{(j-1)}(x).$$

Analogous statements are valid for any pair of dual Banach spaces (see §2.9).

3.5 *Fredholm theorems*

In this section, we present general conditions for the solvability of

$$\lambda u - \mathbf{K} u = f \tag{5.105}$$

for non-zero λ and compact $\boldsymbol{K}$ acting in E. In particular, these conditions contain tests for the solvability of the integral equation (5.35) with a compact operator.

The conjugate operator $\boldsymbol{K}^*$, acting in the conjugate space E^*, plays a basic role in the study of equation (5.105).

THEOREM 3.5. *Let $\boldsymbol{K}$ be a compact operator acting in E, $\boldsymbol{K}^*$ be conjugate to $\boldsymbol{K}$, acting in E^*, and $\lambda \neq 0$.*

Then the equation $\lambda u = Ku + f$ is solvable, if and only if $v(f) = 0$ for any $v \in E^*$ satisfying the condition $\bar{\lambda} v = K^* v$. The number of linearly independent solutions of the equation $\lambda u = Ku + f$ coincides with the number of linearly independent solutions of the equation $\bar{\lambda} v = K^* v$.

Now we pass to integral equations.

THEOREM 3.6. *Let $\boldsymbol{K}$ be a compact, regular, integral operator, with kernel $K(x, t)$ acting in the space L_p (or in C). Let $\lambda \neq 0$.*

Then

$$\lambda u(x) = \int_\Omega K(x, t)\, u(t)\, dt + f(x), \tag{5.106}$$

where $f \in L_p$ (or C), has a solution in L_p (or C), if and only if

$$\int_\Omega f(t)\, \overline{y(t)}\, dt = 0 \tag{5.107}$$

for any function y from $L_{p'}$ (or from L_1) satisfying the equation

$$v(x) = \int_\Omega K^*(x, t)\, v(t)\, dt, \tag{5.108}$$

where $K^*(x, t) = \overline{K(t, x)}$. The number of linearly independent solutions of (5.107) and (5.108) is finite and the same.

3.6 *The resolvent of a compact operator*

The points of the spectrum of a linear operator $\boldsymbol{K}$ are singular points of the analytic function $\boldsymbol{R}(\lambda; \boldsymbol{K})$. For compact operators, the character of these points is defined by the properties of the corresponding eigenvalues.

THEOREM 3.7. *Let* $\boldsymbol{K}$ *be a compact operator acting in a space E. Let* $\lambda_0 \neq 0$ *be a point of its spectrum.*

Then λ_0 *is a pole of the resolvent* $\boldsymbol{R}(\lambda; \boldsymbol{K})$. *The order of this pole coincides with the rank* $r(\lambda_0)$ *of the eigenvalue* λ_0. *The Laurent expansion for* $\boldsymbol{R}(\lambda; \boldsymbol{K})$ *in the neighbourhood of the point* λ_0 *is*

$$\boldsymbol{R}(\lambda, \boldsymbol{K}) = \sum_{k=1}^{r(\lambda_0)} \frac{\boldsymbol{B}_{k-1}}{(\lambda - \lambda_0)^k} + \boldsymbol{R}(\lambda, \boldsymbol{K})\, \boldsymbol{Q}(\lambda_0), \tag{5.109}$$

where

$$\boldsymbol{B}_{k-1} = (\lambda_0 \boldsymbol{I} - \boldsymbol{K})^k\, \boldsymbol{P}(\lambda_0) \qquad (k = 1, \ldots, r(\lambda_0)), \tag{5.110}$$

and $\boldsymbol{P}(\lambda_0)$ *and* $\boldsymbol{Q}(\lambda_0)$ *are complementary projection operators for* $\boldsymbol{K}$, *corresponding to the eigenvalue* λ_0.

If

$$e_1^{(0)}, e_1^{(1)}, \ldots, e_1^{(\kappa_1 - 1)}, \ldots, e_n^{(0)}, e_n^{(1)}, \ldots, e_n^{(\kappa_n - 1)},$$
$$f_1^{(0)}, f_1^{(1)}, \ldots, f_1^{(\kappa_1 - 1)}, \ldots, f_n^{(0)}, f_n^{(1)}, \ldots, f_n^{(\kappa_n - 1)}$$

are canonical bases corresponding to the eigenvalue λ_0, then the operators $\boldsymbol{B}_{k-1}$ are defined by

$$\boldsymbol{B}_{k-1} u = \sum_{i=1}^{n} \sum_{j=k}^{\kappa_i - 1} f_i^{(\kappa_i - j - 1)}(u)\, e_i^{(j-k)}.$$

If $\boldsymbol{K}$ is an integral operator, then the operators $\boldsymbol{B}_{k-1}$ are also integral operators

$$\boldsymbol{B}_{k-1} u(x) = \int_{\Omega} B_k(x, t)\, u(t)\, dt,$$

with

$$B_{k-1}(x, t) = \sum_{i=1}^{n} \sum_{j=k}^{\kappa_i - 1} \overline{f^{(\kappa_i - j - 1)}(t)}\, e_i^{(j-k)}(x).$$

3.7 *Equations with improving operators*

Let E, E_1, E_2 be Banach spaces with $E_2 \subset E \subset E_1$ and $\|x\|_{E_1} \leqslant a\, \|x\|_E$ $(x \in E)$, $\|x\|_E \leqslant a\, \|x\|_{E_2}$ $(x \in E_2)$.

Suppose that some iteration $\boldsymbol{K}^n$ of a compact operator $\boldsymbol{K}$, acting in E, is a continuous (compact) operator, acting from E_1 into E_2. Then the eigenvectors and root vectors of $\boldsymbol{K}$, corresponding to non-zero points of the spectrum, lie in E_2 and, analogously, the defect and adjoint functionals of $\boldsymbol{K}$, corresponding to non-zero points of the spectrum, can be extended to linear continuous functionals on E_1.

This simple fact enables one to determine in certain basic cases the properties of the eigenvectors, root vectors and adjoint functionals of linear integral operators. To do this, it is convenient to use the L-characteristics of the corresponding operators and their iterations (see §5.4).

3.8 *Supplementary notes*

The basic results of this section can be transferred without change to operators $\boldsymbol{K}$, some power $\boldsymbol{K}^n$ of which is a compact operator. In particular, this fact is useful in the study of weakly compact (i.e. transforming bounded sets into weakly compact sets) operators, acting in spaces L_1 and L_∞, since the square of a weakly compact operator, acting in the space L_1 or L_∞, is weakly compact. We note that a regular integral operator $\boldsymbol{K}$ with kernel $K(x, t)$, acting in L_1 or in L_∞, is weakly compact if and only if

$$\lim_{\operatorname{mes} D \to 0} \operatorname{vrai\,max}_{t \in \Omega} \int_D |K(x, t)| \, dx = 0$$

or, correspondingly,

$$\lim_{\operatorname{mes} D \to 0} \operatorname{vrai\,max}_{x \in \Omega} \int_D |K(x, t)| \, dt = 0.$$

Theorems, relating to integral operators, can be transferred to operator equations, acting in Banach spaces of measurable functions, in which the cone of non-negative functions is reproducing, normal and miniedral.

§4 Equations of the second kind with compact operators in a Hilbert space

4.1 *Preliminary notes*

In this section, linear Fredholm equations of the second kind with operators

$$\boldsymbol{K}u(x) = \int_\Omega K(x, t)\, u(t)\, dt, \tag{5.111}$$

acting in the space L_2 will be studied. Such operators can be regarded as operator equations in Hilbert spaces.

Such an approach enables one to formulate some important statements for integral equations with operators acting in L_2 which are not valid for arbitrary integral equations. This particularly concerns equations with self-adjoint (or normal) operators. It is these operators which will be mainly considered in the present section.

4.2 *Equations with self-adjoint operators*

Here, we consider the equation

$$\lambda \boldsymbol{u} = \boldsymbol{K}u + f \tag{5.112}$$

where the integral operator

$$\boldsymbol{K}u(x) = \int_{\Omega} K(x, t)\, u(t)\, dt, \tag{5.113}$$

acting in L_2, has a symmetric kernel

$$K(x, t) = \overline{K(t, x)} \qquad (x, t \in \Omega). \tag{5.114}$$

Suppose, further, that the kernel $K(x, t)$ is summable with respect to both variables. Then the operator $\boldsymbol{K}$ will be symmetric in L_2, i.e.

$$(\boldsymbol{K}u, v) = (u, \boldsymbol{K}v) \qquad (u, v \in L_2). \tag{5.115}$$

It should be noted that the kernel $K(x, t)$ of the operator $\boldsymbol{K}$ is summable, if $\boldsymbol{K}$ is a regular operator (see §1.1) in L_2.

Solvability conditions can be simplified for equations with compact self-adjoint operators.

THEOREM 4.1. *Let* $\boldsymbol{K}$ *be a compact self-adjoint operator acting in* L_2. *Then*

a) *for a real* λ $(\lambda \neq 0)$, *the equation* $\lambda u = \boldsymbol{K}u + f$ *has a solution in* L_2 *if and only if*

$$(f, v) = 0 \tag{5.116}$$

for any solution v *of the homogeneous equation* $\lambda v = \boldsymbol{K}v$;

b) *for a complex* λ *the equation* $\lambda u = Ku + f$ *has a unique solution in* L_2 *for arbitrary* $f \in L_2$.

4.3 *The resolvent and the spectrum of a self-adjoint integral operator*

For compact self-adjoint operators in a Hilbert space the following fundamental theorem holds.

THEOREM 4.2. *Let* $\boldsymbol{K}$ *be a compact self-adjoint operator in* L_2. *Then the following statements are valid:*

a) *The spectrum of* $\boldsymbol{K}$ *consists of zero and a countable set of real values* λ_n, $(n = 1, 2, \ldots, -1, -2, \ldots)$:

$$\lambda_{-1} < \lambda_{-2} < \ldots < \lambda_{-n} < \ldots < 0 < \ldots < \lambda_n < \ldots < \lambda_2 < \lambda_1,$$

the only limit point of which is zero.

b) *Every non-zero point* $\lambda = \lambda_n$ *of the spectrum has a corresponding subspace* $E(\lambda_n)$ *of eigenvectors (and simultaneously, defect vectors). The subspaces* $E(\lambda)$, *corresponding to different* λ, *are mutually orthogonal.*

c) *The operator* $\boldsymbol{K}$ *has the spectral decomposition*[4])

$$\boldsymbol{K}u = \sum_n \lambda_n \boldsymbol{P}(\lambda_n)\, u, \tag{5.117}$$

where the operator $\boldsymbol{P}(\lambda_n)$ *denotes orthogonal projection onto* $E(\lambda_n)$, *viz.*

$$\boldsymbol{P}(\lambda_n)\, u(x) = \int_\Omega \varphi_n(x, t)\, u(t)\, dt, \tag{5.118}$$

where the series

$$\varphi_n(x, t) = \sum_{i=1}^{\kappa_n} \overline{e_i^{(n)}(t)}\, e_i^{(n)}(x), \tag{5.119}$$

---

[4]) The notation $\sum_n$ is interpreted as $\lim\limits_{-n_1, n_2 \to \infty} \sum\limits_{n=n_1}^{n_2}$, if both sequences $\lambda_1, \ldots, \lambda_n$ and $\lambda_{-1}, \ldots, \lambda_{-n}$ are infinite; as $\lim\limits_{n_2 \to \infty} \sum\limits_{n=-n_0}^{n_2}$, if among the values of n the number of negative values, n_0, is finite and the number of positive values is infinite; as $\lim\limits_{-n_1 \to \infty} \sum\limits_{n=n_1}^{n_0}$, if the number of positive values, n_0, is finite and the number of negative values is infinite; and, simply, as a finite sum, if the number of eigenvalues is finite.

converges with respect to the norm of the operators acting in L_2 *with* κ_n *the dimension of* $E(\lambda_n)$ *(the multiplicity of the eigenvalue* λ_n*), where also*

$$e_1^{(n)}(x), \ldots, e_{\kappa_n}^{(n)}(x) \tag{5.120}$$

is an arbitrary orthogonal basis in $E(\lambda_n)$.

d) *The resolvent* $\boldsymbol{R}(\lambda; \boldsymbol{K})$ $(\lambda \neq 0, \lambda_{\pm 1}, \lambda_{\pm 2}, \ldots)$ *of the operator* $\boldsymbol{K}$ *has the spectral decomposition*

$$\boldsymbol{R}(\lambda, \boldsymbol{K})\, u = \sum_n \frac{1}{\lambda - \lambda_n} \boldsymbol{P}(\lambda_n)\, u, \tag{5.121}$$

where the series converges with respect to the norm of the operators acting in L_2.

Theorem 4.2 gives a full description of the resolvent and the spectrum of a compact self-adjoint operator. In particular, it enables one to write down formulas for the solutions of (5.111), if the operators $\boldsymbol{P}(\lambda_n)$ are known.

THEOREM 4.3. *Let* $\boldsymbol{K}$ *be a compact self-adjoint operator, acting in* L_2. *Let* λ_n *be an eigenvalue of the operator* $\boldsymbol{K}$ *and* $\varphi_n(x, t)$ *be the corresponding kernel of the projection operator* $\boldsymbol{P}(\lambda_n)$.

Then:

a) *If* $\lambda \neq \lambda_n$ *for all n then the unique solution u of the equation* $\lambda u = Ku + f$ *in the space* L_2 *has the form*

$$\boldsymbol{u}(x) = \sum_n \frac{1}{\lambda - \lambda_n} \int_\Omega \varphi_n(x, t)\, f(t)\, dt, \tag{5.122}$$

where the series converges with respect to the norm of the space L_2.

b) *If* $\lambda = \lambda_{n_0}$, *then the solutions u of the equation* $\lambda u = Ku + f$ *exist if and only if*

$$\int_\Omega \varphi_{n_0}(x, t)\, f(t)\, dt = 0. \tag{5.123}$$

They are defined by

$$u(x) = h(x) + \sum_{n \neq n_0} \frac{1}{\lambda - \lambda_n} \int_\Omega \varphi_n(x, t)\, f(t)\, dt, \tag{5.124}$$

where h denotes an arbitrary function from $E(\lambda_n)$. *The series converges with respect to the norm of the space* L_2.

Suppose, now, that $\boldsymbol{K}$ is a regular integral operator. We put

$$\boldsymbol{K}(u, v) = \int_{\Omega}\int_{\Omega} K(x, t)\, u(t)\, \overline{v(x)}\, dx\, dt. \tag{5.125}$$

If, in addition, $\boldsymbol{K}$ is compact and self-adjoint, then

$$\boldsymbol{K}(u, v) = \sum_n \lambda_n \int_{\Omega}\int_{\Omega} \varphi_n(x, t)\, u(t)\, \overline{v(t)}\, dx\, dt. \tag{5.126}$$

From this formula, we obtain the important theorem of Courant.

THEOREM 4.4. *Let* $\boldsymbol{K}$ *be a compact self-adjoint regular integral operator, with kernel* $K(x, t)$, *acting in* L_2.

Then, for an eigenvalue λ_n, *the following expressions hold:*

a) *If* $n > 0$, *then*

$$\lambda_n = \min_{v_1,\ldots,v_{\mu(n)}} \max_{\|u\|_{L_2} \leqslant 1, (u,v_1) = \ldots = (u, x_{\mu(n)}) = 0} \boldsymbol{K}(u, v), \tag{5.127}$$

where $\mu(n)$ *is the sum of multiplicities of the eigenvalues* λ_i, *which exceed* λ_n *and* $u, v_1, \ldots, v_{\mu(n)} \in L_2$; *the minimax on the r.h.s. of* (5.128) *is attained only if* $v_1, \ldots, v_{\mu(n)} \in \bigcup_{\lambda_i > \lambda_n} E(\lambda_i)$ and $u \in E(\lambda_n)$.

b) *If* $n < 0$, *then*

$$\lambda_n = \max_{v_1,\ldots,v_{\nu(n)}} \min_{\|u\|_{L_2} \leqslant 1, (u,v_1) = \ldots = (u, v_{\nu(n)}) = 0} \boldsymbol{K}(u, v), \tag{5.128}$$

where $\nu(n)$ *is the sum of multiplicities of eigenvalues* λ_i, *which are less than* λ_n *and* $u, v_1, \ldots, v_{\nu(n)} \in L_2$; *the minimax on the r.h.s. is attained only if* $v_1, \ldots, v_{\nu(n)} \in \bigcup_{\lambda_i < \lambda_n} E(\lambda_i)$ *and* $u \in E(\lambda_n)$.

4.4 *Hilbert-Schmidt operators*

We continue the study of self-adjoint operators acting in L_2.

THEOREM 4.5. *Let* $\boldsymbol{K}$ *be a compact self-adjoint integral operator with kernel* $K(x, t)$.
Then

$$\int_{\Omega} |K(x, t)|^2 \, dt = \sum_n \lambda_n^2 \int_{\Omega} |\varphi_n(x, t)|^2 \, dt \qquad (x \in \Omega) \tag{5.129}$$

and

$$\int_{\Omega}\int_{\Omega} |K(x, t)|^2 \, dx \, dt = \sum_n \kappa_n \lambda_n^2. \tag{5.130}$$

From this theorem it immediately follows that a self-adjoint integral operator $\boldsymbol{K}$, acting in L_2, is $\boldsymbol{U}$-bounded (see §1.5) when and only when

$$\sum_n \kappa_n \lambda_n^2 < \infty \tag{5.131}$$

or, what is just the same, when the kernel $K(x, t)$ is square summable. Operators with square summable kernels are *Hilbert-Schmidt operators* (see Chapter III, §1).

An important bilinear expression holds for the kernels of Hilbert-Schmidt operators.

THEOREM 4.6. *Let* $\boldsymbol{K}$ *be a Hilbert-Schmidt operator with kernel* $K(x, t)$.

Then

$$K(x, t) = \sum_n \lambda_n \varphi_n(x, t). \tag{5.132}$$

This series converges in the norm of the space $L_2(\Omega \times \Omega)$.

It is evident that (5.132) is only another form of the bilinear series (3.9).

It should be noted that the solution operator

$$\boldsymbol{L}(\lambda, \boldsymbol{K}) = \lambda \boldsymbol{K} \boldsymbol{R}(\lambda, \boldsymbol{K})$$

of a Hilbert-Schmidt operator is also a Hilbert-Schmidt operator.

4.5 Mercer operators

A symmetric kernel $K(x, t)$ will be called a *Mercer kernel*, if it can be written in the form

$$K(x, t) = \int_\Omega a(x, \sigma)\, b(\sigma, t)\, d\sigma,$$

where $a(x, t)$ and $b(x, t)$ are measurable with respect to both variables and are square summable functions. Operators with Mercer kernels will be called *Mercer operators*. Mercer kernels satisfy the inequality

$$|K(x, t)| \leqslant a(x)\, b(t) \qquad (x, t \in \Omega), \tag{5.133}$$

where

$$a(x) = \sqrt{\left(\int_\Omega |a(x, \sigma)|^2\, d\sigma\right)} \in L_2,$$

$$b(t) = \sqrt{\left(\int_\Omega |b(\sigma, t)|^2\, d\sigma\right)} \in L_2.$$

For any operator $\boldsymbol{K}$ with a Mercer kernel $K(x, t)$, we put

$$\delta(\boldsymbol{K}) = \int_\Omega \int_\Omega a(x, t)\, b(t, x)\, dx\, dt. \tag{5.134}$$

If the functions $a(x, t)$ and $b(x, t)$ are continuous then, obviously,

$$\delta(\boldsymbol{K}) = \int_\Omega K(\sigma, \sigma)\, d\sigma. \tag{5.135}$$

THEOREM 4.7. *Let $\boldsymbol{K}$ be a compact self-adjoint operator, with kernel $K(x, t)$, acting in L_2.*

Then:

a) *The kernel $K(x, t)$ is a Mercer kernel, if and only if $\boldsymbol{K}$ is a kernel operator, i.e.*

$$\sum_n \kappa_n |\lambda_n| < \infty.$$

b) *If $K(x, t)$ is a Mercer kernel, then*

$$\delta(\boldsymbol{K}) = \sum_n \kappa_n |\lambda_n|.$$

THEOREM 4.8. *Let* $\boldsymbol{K}$ *be an integral operator with Mercer kernel $K(x, t)$. Then there exist $a_0(x)$ and $b_0(t) \in L_2$ such that the bilinear decomposition*

$$K(x, t) = \sum_n \lambda_n \varphi_n(x, t)$$

converges uniformly with weight $a_0(x)\, b_0(t)$ almost everywhere:

$$\lim_{-n_1, n_2 \to \infty} \operatorname{vrai\,max}_{x \in \Omega} \operatorname{vrai\,max}_{t \in \Omega} \frac{\left| K(x, t) - \sum_{n=n_1}^{n_2} \lambda_n \varphi_n(x, t) \right|}{a_0(x)\, b_0(t)} = 0. \tag{5.136}$$

4.6 *Self-adjoint operators with their range in a Banach space*

In this section, we study self-adjoint integral operators in L_2 which are at the same time continuous or compact and map from L_2 into some Banach space E (for example, the space L_q, where $q > 2$, or the space C). For such operators the statements of the previous subsections can be supplemented.

THEOREM 4.9. *Let* $\boldsymbol{K}$ *be a compact self-adjoint operator acting in L_2. Let* $\boldsymbol{K}$ *map from L_2 into a Banach space E and be continuous.*

Then the following statements are valid:

a) *The spectral decomposition*

$$\boldsymbol{K}u = \sum_n \lambda_n \boldsymbol{P}(\lambda_n)\, u \tag{5.137}$$

of the operator $\boldsymbol{K}$ *converges for any fixed u in the norm of the space E.*

b) *The spectral decomposition*

$$\boldsymbol{L}(\lambda, \boldsymbol{K})\, u = \sum_n \frac{\lambda \lambda_n}{\lambda - \lambda_n} \boldsymbol{P}(\lambda_n)\, u \tag{5.138}$$

of the solution operator $\boldsymbol{L}(\lambda, \boldsymbol{K}) = \lambda \boldsymbol{K} \boldsymbol{R}(\lambda, \boldsymbol{K})$ also converges, for any fixed u, in the norm of the space E.

In the case when E is the space C of continuous functions, the above theorem contains the classical Hilbert-Schmidt theorem regarding the decomposition of sourcewise representable functions into uniformly convergent series.

THEOREM 4.10. *Let $\boldsymbol{K}$ be a compact self-adjoint operator acting in L_2. Let $\boldsymbol{K}$ map from L_2 into some Banach space E and be compact {map from some Banach space E into L_2, and be compact}.*

Then the following statements are valid:

a) *The spectral decomposition*

$$\boldsymbol{K}u = \sum_n \lambda_n \boldsymbol{P}(\lambda_n)\, u \tag{5.139}$$

of the operator $\boldsymbol{K}$ converges with respect to the norm of the space of operators mapping from L_2 into E (with respect to the norm of the space of operators mapping from E into L_2).

b) *The spectral decomposition*

$$\boldsymbol{L}(\lambda, \boldsymbol{K})\, u = \sum_n \frac{\lambda\lambda_n}{\lambda - \lambda_n} \boldsymbol{P}(\lambda_n)\, u \tag{5.140}$$

of the solution operator $\boldsymbol{L}(\lambda; \boldsymbol{K}) = \lambda \boldsymbol{K}\boldsymbol{R}(\lambda; \boldsymbol{K})$ converges in the norm of operators mapping from L_2 into E (in the norm of operators mapping from E into L_2).

4.7 *Positive definite self-adjoint operators*

We recall that an operator $\boldsymbol{K}$ acting in the Hilbert space L_2 is called positive definite if

$$(\boldsymbol{K}u, u) \geqslant 0 \qquad (u \in L_2). \tag{5.141}$$

The positive definiteness of a regular integral operator $\boldsymbol{K}$, with kernel $K(x, t)$, acting in L_2 is equivalent to the positive definiteness of $K(x, t)$

$$\int_\Omega \int_\Omega K(x, t)\, u(t)\, \overline{u(x)}\, dx\, dt \geqslant 0 \qquad (u \in L_2). \tag{5.142}$$

In particular, for any positive kernel $K(x, t)$, we have the inequality

$$\int_D \int_D K(x, t)\, dx\, dt \geqslant 0 \qquad (D \subset \Omega). \tag{5.143}$$

For any continuous positive definite kernel $K(x, t)$, we have the inequality

$$K(x, x) \geqslant 0 \qquad (x \in \Omega). \tag{5.144}$$

Suppose that $\boldsymbol{K}$ is a compact self-adjoint operator acting in L_2. In this case, the positive definiteness of $\boldsymbol{K}$ is obviously equivalent to the condition that $\lambda_n \geqslant 0$ for all n.

THEOREM 4.11. *Let $\boldsymbol{K}$ be a compact self-adjoint operator acting in L_2. Let E and F be a pair of dual spaces and let the operator $\boldsymbol{K}$ map from E into F and be compact.*

Then the following statements are valid:

a) *The spectral decomposition*

$$\boldsymbol{K}u = \sum_n \lambda_n \boldsymbol{P}(\lambda_n)u$$

of the operator $\boldsymbol{K}$ converges in the norm of the space of operators mapping from E into F.

b) *The spectral decomposition*

$$\boldsymbol{L}(\lambda, \boldsymbol{K})u = \sum_n \frac{\lambda\lambda_n}{\lambda - \lambda_n} \boldsymbol{P}(\lambda_n)u$$

of the solution operator $\boldsymbol{L}(\lambda; \boldsymbol{K}) = \lambda\boldsymbol{K}\boldsymbol{R}(\lambda; \boldsymbol{K})$ converges in the norm of operators mapping from E into F.

In particular, if E is the space L_1 and F is the space C of continuous functions, then the statement of the theorem reduces to the statement of the Mercer theorems (see §4.5), since the norm of an integral operator $\boldsymbol{K}$ with a continuous kernel $K(x, t)$ (mapping from L_1 into C) is defined by

$$\|\boldsymbol{K}\|_{L_1 \to C} = \max_{x,t \in \Omega} |K(x, t)|.$$

THEOREM 4.12. *Let the kernel $K(x, t)$ $(x, t \in \Omega)$ of a compact self-adjoint operator $\boldsymbol{K}$ acting in L_2 satisfy the inequality*

$$|K(x, t)| \leqslant \varphi(x)\, \varphi(t) \qquad (x, t \in \Omega).$$

Then the kernel $K(x, t)$ is a Mercer kernel and can be represented in the *form*

$$K(x, t) = \int_{\Omega} H(x, \sigma)\, H(\sigma, t)\, d\sigma,$$

where $H(x, t)$ *is the square summable kernel*

$$H(x, t) = \sum_n \sqrt{(\lambda_n)}\, \varphi_n(x, t). \tag{5.145}$$

The series (5.145) *converges in the norm of the space* $L_2(\Omega \times \Omega)$.

4.8 *Hilbert-Schmidt decomposition for compact operators*

In this section, we study arbitrary compact integral operators acting in the space L_2. Let $\boldsymbol{K}$ be an arbitrary compact operator acting in the space L_2. We introduce positive definite self-adjoint operators

$$\boldsymbol{B}_1 = \boldsymbol{K}^*\boldsymbol{K}, \qquad \boldsymbol{B}_2 = \boldsymbol{K}\boldsymbol{K}^* \tag{5.146}$$

having the same spectrum. Let

$$s_1^2, s_2^2, \ldots \qquad (s_n > 0) \tag{5.147}$$

be the eigenvalues of the operators $\boldsymbol{B}_1$ and $\boldsymbol{B}_2$. The numbers $s_1, \ldots, s_n, \ldots$ are called the *Schmidt numbers* of the operator $\boldsymbol{K}$ (and of the conjugate operator $\boldsymbol{K}^*$). The multiplicity of the eigenvalue s_n^2 is the same for both $\boldsymbol{B}_1$ and $\boldsymbol{B}_2$. It is called the multiplicity of the Schmidt number s_n. Let $e_i^{(n)}$ $(i = 1, \ldots, \kappa_n)$ be an orthonormal basis in the eigenspace $E_1(s_n)$ of the operator $\boldsymbol{B}_1$, and $g_i^{(n)}$ $(i = 1, 2, \ldots, \kappa_n)$ be an orthonormal basis in the eigenspace $E_2(s_n)$ of the operator $\boldsymbol{B}_2$. We assume that these bases are chosen so that

$$s_n^{-1}\boldsymbol{K}e_i^{(n)} = g_i^{(n)}, \qquad s_n^{-1}\boldsymbol{K}^*g_i^{(n)} = e_i^{(n)} \qquad (i = 1, \ldots, \kappa_n).$$

We put

$$\psi_n(x, t) = \sum_{i=1}^{\kappa_n} e_i^{(n)}(x)\, g_i^{(n)}(t) \tag{5.148}$$

and introduce the projection operator

$$Q(\lambda_n)u = \int_{\Omega} \psi_n(x, t)\, u(t)\, dt \qquad (n = 1, 2, \ldots). \tag{5.149}$$

THEOREM 4.13. *Let* $\boldsymbol{K}$ *be a compact operator. Let E be a Banach space and let* $\boldsymbol{K}$ *map from* L_2 *into E and be continuous.*

Then $\boldsymbol{K}$ *can be represented in the form*

$$\boldsymbol{K}x = \sum_n s_n Q(s_n)u. \tag{5.150}$$

The series on the r.h.s. converges for fixed u in the norm of the space E.

Theorem 4.12 is analogous to Theorem 4.9. The analogue of Theorem 4.10 is the following:

THEOREM 4.14. *Let* $\boldsymbol{K}$ *be a compact operator acting in* L_2. *Let E be a Banach space and* $\boldsymbol{K}$ *map from* L_2 *into E (from E into* L_2*) and be compact.*

Then $\boldsymbol{K}$ *has the decomposition*

$$\boldsymbol{K}x = \sum_n s_n Q(s_n)u.$$

The series on the r.h.s. converges in the norm of operators acting from L_2 *into E (from E into* L_2*).*

Below we give two more statements concerning the decomposition of kernels of compact integral operators. These theorems are analogous to the corresponding ones concerning self-adjoint operators.

THEOREM 4.15. *Let* $K(x, t)$ *be the kernel of a compact integral operator acting in the space* L_2.

Then the following relationships hold

$$\int_{\Omega} |K(x, t)|^2\, dt = \sum_n s_n^2 \int_{\Omega} |\psi_n(x, t)|^2\, dt, \tag{5.151}$$

$$\int_{\Omega} |K(x, t)|^2\, dx = \sum_n s_n^2 \int_{\Omega} |\psi_n(x, t)|^2\, dx, \tag{5.152}$$

$$\int_{\Omega}\int_{\Omega} |K(x, t)|^2\, dx\, dt = \sum_n \kappa_n s_n^2. \tag{5.153}$$

THEOREM 4.16. *Let the kernel $K(x, t)$ be square summable. Then*

$$K(x, t) = \sum_n \lambda_n \psi_n(x, t). \tag{5.154}$$

The series on the r.h.s. converges in the norm of the space $L_2(\Omega \times \Omega)$.

§5 Positive operators

The theory of linear positive operators contains the theory for linear integral equations with positive kernels, presented in Chapter IV. It simultaneously contains analogous statements for systems of integral equations.

5.1 *Semi-ordered spaces*

Let E be a real Banach space. A closed convex set $K \subset E$ is called *a cone* if for $u \in K$ it follows that $\alpha u \in K$ for any $\alpha \geqslant 0$; and for $u, -u \in K$, it follows that $u = \theta$.

The cone K enables one to introduce in E a *semi-order;* we write $u \geqslant v$ if $u - v \in K$. The sign $\geqslant$ has the usual properties of the sign $\geqslant$ (inequalities may be multiplied by positive constants, inequalities of the same sign may be summed up, one may pass to the limit in inequalities and so on). It must be remembered that there exist pairs of *incomparable* elements u, v (when $u - v \notin K$ and $v - u \notin K$).

In this section, we assume without loss of generality that the norm is montonic in the sense that for $\theta \leqslant u \leqslant v$ it follows that $\|u\| \leqslant \|v\|$ and that the span of the cone coincides with the whole space.

A basic example is the cone of non-negative functions in the space C and in the spaces L_p. In the spaces of vector-functions (with continuous or L_p components), we can consider the cone of functions with non-negative components.

The cone K is called *solid*, if it contains internal elements. The cone of non-negative functions in the space C is solid, while the cone of non-negative functions in the space L_p is not solid.

5.2 *General theorems relating to positive operators*

An operator $\boldsymbol{K}$, acting in the space E with cone K, is called *positive* if $\boldsymbol{K}K \subset K$. A linear positive operator is *monotonic*, if the inequality $u \leqslant v$ implies $\boldsymbol{K}u \leqslant \boldsymbol{K}v$.

As an example, consider the space E of vector-functions

$$u(x) = \{u_1(x), \ldots, u_n(x)\}$$

and the operator $\boldsymbol{K}$ defined by

$$\boldsymbol{K}u = \{\boldsymbol{K}_1 u, \ldots, \boldsymbol{K}_n u\}, \tag{5.155}$$

where

$$\boldsymbol{K}_i u = \sum_{j=1}^{n} \int_{\Omega} \boldsymbol{K}_{ij}(x, t)\, u_j(t)\, dt, \qquad i = 1, 2, \ldots, n.$$

Suppose that $\boldsymbol{K}$ acts in E. Obviously, it is additive and linear

$$\boldsymbol{K}(\alpha u + \beta v) \equiv \alpha \boldsymbol{K}u + \beta \boldsymbol{K}v \qquad (u, v \in E).$$

If we introduce the cone of vector-functions with non-negative components in E, then $\boldsymbol{K}$ is positive, if and only if the kernels $K_{ij}(x, t)$ are non-negative.

THEOREM 5.1. *Every positive linear operator is continuous.*

THEOREM 5.2. *The spectral radius is a point of the spectrum of a linear positive operator.*

5.3 *Estimates for the spectral radius*

The main results of this section rest upon the possibility of estimating the spectral radius using inequalities connecting the value of $\boldsymbol{K}$ on certain elements of the space E.

THEOREM 5.3. *Let* $\boldsymbol{K}v \geqslant \alpha_1 v$, *where* $v \neq 0$ *is an element from* $\boldsymbol{K}$. *Then* $\rho(\boldsymbol{K}) \geqslant \alpha_1$.

Analogously, the relationship $\boldsymbol{K}^m v \geqslant \alpha_2 v$ *implies the inequality* $\rho(\boldsymbol{K}) \geqslant \sqrt[m]{\alpha_2}$.

THEOREM 5.4. *Let the cone K be solid and let the following inequality hold for some internal element u_1 of this cone*

$$\boldsymbol{K}^m u_1 \leqslant \lambda_0 u_1. \tag{5.156}$$

Then

$$\rho(\boldsymbol{K}) \leqslant \sqrt[m]{\lambda_0}. \tag{5.157}$$

The element $v \in K$ will be called quasi-internal, if every non-zero, linear and non-negative functional on K takes a positive value at v. In the case of a solid cone, every internal point is quasi-internal and conversely.

THEOREM 5.5. *Let the linear positive operator* $\boldsymbol{K}$ *be compact, and, at a quasi-internal element* u_1, *the inequality* (5.156) *be valid.*
Then the inequality (5.157) *holds.*

A linear positive operator $\boldsymbol{K}$ is called *u_0-upper-bounded*, if for every $u \in E$ there exist an $\alpha > 0$ such that

$$-\alpha u_0 \leqslant \boldsymbol{K}u \leqslant \alpha u_0. \tag{5.158}$$

Theorem 5.5 is valid if the assumption of the compactness of $\boldsymbol{K}$ is replaced by the assumption of u_0-upper-boundedness.

In conclusion, we describe an abstract scheme for estimating the spectral radius by generalizing Theorem 3.4. Consider two Banach spaces E_1 and E_2 with cones K_1 and K_2 respectively. Let φ be an operator mapping from E_1 into K_2 and satisfying the two conditions:

1° $\varphi(u + v) \leqslant \varphi(u) + \varphi(v) \qquad (u, v \in E_1)$,

2° $\|\varphi(u_n)\| \to 0$ implies that $\|u_n\| \to 0 \; (n \to \infty)$.

THEOREM 5.6. *Let* $\boldsymbol{K}$ *be a linear operator acting in* E_1 *and* $\boldsymbol{B}$ *be a linear positive operator acting in* E_2. *Further, let*

$$\varphi(\boldsymbol{K}u) \leqslant \boldsymbol{B}\varphi(u) \qquad (u \in K). \tag{5.159}$$

Then the spectral radius of $\boldsymbol{K}$ *and* $\boldsymbol{B}$ *are related by the inequality* $\rho(\boldsymbol{K}) \leqslant \rho(\boldsymbol{B})$.

5.4 *The non-homogeneous equation*

Consider the equation

$$\mu u = \boldsymbol{K}u + f \tag{5.160}$$

where $\boldsymbol{K}$ is a linear positive operator.

THEOREM 5.7. *A necessary and sufficient condition that* (5.160) *has a solution* $u^* \in K$ *for arbitrary* $f \in K$ *is that the following inequality holds:*

$$\mu > \rho(\boldsymbol{K}). \tag{5.161}$$

This theorem can be generalized in different ways. For instance, if $\boldsymbol{K}$ is compact and (5.160) has a solution in K for at least one quasi-internal element f, then the inequality (5.161) holds.

5.5 *Existence of an eigenvector*

A positive operator may have no eigenvectors. As an example, we consider the operator

$$\boldsymbol{K}u(x) = xu(x)$$

in the space C of continuous functions on $[0, 1]$. A second example is the Volterra operator (see §6).

Nevertheless, for many classes of positive operators, we can establish the existence of an eigenvector in K corresponding to the eigenvalue $\rho(\boldsymbol{K})$.

THEOREM 5.8. *Let* $\boldsymbol{K}$ *be a positive continuous operator and let there exist a non-zero element* $u \in K$ *such that*

$$\boldsymbol{K}u \geqslant cu, \tag{5.162}$$

where $c > 0$.

Then $\boldsymbol{K}$ *has an eigenvector corresponding to the eigenvalue* $\rho(\boldsymbol{K})$ *in the cone* K.

If $\boldsymbol{K}$ is a compact operator and $\rho(\boldsymbol{K}) > 0$, then the eigenvector corresponding to the eigenvalue $\rho(\boldsymbol{K})$ exists even in the case when the span of the cone K does not coincide with E (but only the closure of the span coincides with E).

From Theorem 5.8 it follows, for instance, that (5.155) has an eigenvector with positive components, if the kernels $K_{ij}(x, t)$ are positive and continuous. Instead of the positiveness of the kernels $K_{ij}(x, t)$, it is sufficient to assume that they are non-factorable (see §1, Chapter IV).

5.6 *Properties of the eigenvalue* $\rho(\boldsymbol{K})$

In this section, we suppose that the eigenvalue $\lambda_0 = \rho(\boldsymbol{K})$ of the positive operator $\boldsymbol{K}$ has a corresponding eigenvector in the cone K.

A positive operator $\boldsymbol{K}$ is called u_0-*positive*, if for any $u \in K$ $(u \neq 0)$, there exist $m = m(u)$, $\alpha = \alpha(u) > 0$ and $\beta = \beta(u) > 0$ such that $\alpha u_0 \leqslant K^m u \leqslant \beta u_0$.

THEOREM 5.9. *Let the linear operator* $\boldsymbol{K}$ *be* u_0*-positive. Then the following three statements are valid:*

1° *The eigenvalue* λ_0 *is simple.*
2° *The moduli of the eigenvalues of* $\boldsymbol{K}$ *which differ from* λ_0 *are less than* λ_0.
3° *There are no eigenvectors of* $\boldsymbol{K}$ *in* K *corresponding to eigenvalues which differ from* λ_0.

This theorem has different generalizations.

5.7 *Eigenvalues of the conjugate operator*

A linear functional l which has no non-negative values in K is called *positive*. Positive linear functionals form a cone K^* in the conjugate space E^*. The cone K^* is called *conjugate*.

THEOREM 5.10. *If a linear operator* $\boldsymbol{K}$ *is positive, then the conjugate operator* $\boldsymbol{K}^*$ *is positive in* E^*.

From the fact that positive operators are conjugate to positive operators, it follows that they have many properties analogous to those of compact operators.

THEOREM 5.11. *Let the linear positive operator* $\boldsymbol{K}$ *leave the solid cone* K *invariant, and let* $\rho(\boldsymbol{K}) > 0$. *Then the operator* $\boldsymbol{K}^*$ *has an eigenvector in the cone* K *corresponding to the eigenvalue* $\rho(\boldsymbol{K})$.

Along with Theorem 5.3, Theorem 5.11 is useful in applications.

If the cone K is not solid, then Theorem 5.11 is not valid. In such a case, the following fact is useful:

THEOREM 5.12. *Let* $\boldsymbol{K}$ *be a* U_0*-upper-bounded operator. Then the operator* $\boldsymbol{K}^*$ *has an eigenvector in the cone* K *corresponding to the eigenvalue* $\rho(\boldsymbol{K}) = \rho(\boldsymbol{K}^*)$.

5.8 *Operators which leave a miniedral cone invariant*

The cone K is called *miniedral* if, for any $u, v \in K$, there exists an element w such that $w \geqslant u$, $w \geqslant v$ and such that the inequalities $z \geqslant u$, $z \geqslant v$ imply

$w \leqslant z$. The element w is called the *maximum* of the elements u and v, and is written as following

$$w = \sup \{u, v\}.$$

The cones of non-negative functions and cones of vector-functions with non-negative components considered above are miniedral.

An operator $\boldsymbol{U}(\boldsymbol{U}E_1 \subset E_1)$ is called a *permutator*, if its range is finite-dimensional and there exists a basis $e_1, e_2, \ldots, e_s$ in this range (the basis of the permutator) such that

$$\boldsymbol{U}e_j = e_{k_j} \qquad (j = 1, 2, \ldots, s).$$

We say that a set of elements $g_1, g_2, \ldots, g_n$ is *properly contained in* K, if the linear combination $c_1 g_1 + c_2 g_2 + \ldots + \ldots c_n g_n$ is contained in K when and only when all the c_i are non-negative.

THEOREM 5.13. *Let K be a miniedral solid cone, and $\boldsymbol{K}$ be a linear positive compact operator having an eigenvector inside K:$\boldsymbol{K}e_0 = e_0$.*

Then the operator $\boldsymbol{K}$ has the decomposition

$$\boldsymbol{K} = \boldsymbol{U}_1 + \boldsymbol{K}_1, \tag{5.163}$$

where $\boldsymbol{K}_1\boldsymbol{U}_1 = \boldsymbol{U}_1\boldsymbol{K}_1 = \theta$,

$$\lim_{n\to\infty} \sqrt[n]{\|\boldsymbol{K}_1^n\|} < 1, \tag{5.164}$$

and $\boldsymbol{U}_1$ is a permutator with a basis which is properly contained in K.

The set of eigenvalues of the operator $\boldsymbol{K}$, the moduli of which are equal to unity, coincides with the set of roots of the algebraic equation

$$(z^\alpha - 1)(z^\beta - 1) \ldots (z^\tau - 1) = 0.$$

The set of stationary vectors of the operators $\boldsymbol{K}$ and $\boldsymbol{K}^*$ has the basis $v_1, v_2, \ldots, v_r$ and $\psi_1, \psi_2, \ldots, \psi_r$, respectively, with the following properties:

1° The systems $v_1, v_2, \ldots, v_r$ and $\psi_1, \psi_2, \ldots, \psi_r$ are properly contained in K and K^*, respectively.

2° The systems $v_1, v_2, \ldots, v_r$ and $\psi_1, \psi_2, \ldots, \psi_r$ are biorthogonal:

$$\psi_i(v_j) = \delta_{ij} = \begin{cases} 1 & (i = j) \\ 0 & (i \neq j) \end{cases} \qquad (i, j = 1, 2, \ldots, r).$$

3° For any pair $i \neq j$ $(i, j = 1, 2, \ldots, r)$

$\inf(\psi_i, \psi_j) = \theta$.

§6 Volterra equations of the second kind

6.1 *The formulation of the problem*

The equation of the form

$$\lambda u(x) = \int_a^x K(x, t)\, u(t)\, dt + f(x), \tag{5.165}$$

where the variables x and t vary over the interval $[a, b]$, is called a *linear Volterra equation of the second kind*.

Suppose that the Volterra operator

$$\boldsymbol{K}u(x) = \int_a^x K(x, t)\, u(t)\, dt \tag{5.166}$$

acts in a space E and is continuous. It turns out that, under additional restrictions, the spectral radius $\rho(\boldsymbol{K})$ of $\boldsymbol{K}$ is equal to zero and so (5.165) has a unique solution in E for arbitrary r.h.s. from E.

In the example of (5.165), where the operator

$$\boldsymbol{K}_0 u(x) = \frac{1}{x} \int_0^x u(t)\, dt \tag{5.167}$$

acts in L_p $(1 < p \leqslant \infty)$, we see that additional conditions on $K(x, t)$ are necessary: in this example, the spectral radius of $\boldsymbol{K}$ is equal to 1.

6.2 *Basic theorems*

We examine (5.166) in either L_p or C. Setting $a \leqslant t \leqslant x \leqslant b$, we denote by $\boldsymbol{P}_{[t,x]}$ the linear operator of multiplication by the characteristic function of the interval $[t, x]$. Below, we denote by $\delta(\Psi)$ the infimum of the values $\max |\Psi(t_{i-1}, t_i)|$ with respect to all partitions of $a = t_0 < t_1 < \ldots < t_n = b$, where $\Psi = \Psi(t, x)$ is a non-negative function of two variables, defined on the square $a \leqslant t \leqslant x \leqslant b$.

Let $\boldsymbol{K}$ be a Volterra operator. We introduce the functions

$$\Phi_n(t, x) = \sqrt[n]{\|\{P_{[t,x]}\boldsymbol{K}P_{[t,x]}\}^n\|} \qquad (n = 1, 2, \ldots),$$

$$\Phi_\infty(t, x) = \lim_{n\to\infty} \Phi_n(t, x)$$

(the limit in the last equality always exists).

THEOREM 6.1. *Let the Volterra integral operator* $\boldsymbol{K}$, *with kernel* $K(x, t)$, *act in* L_p $(1 \leqslant p \leqslant \infty)$ *or in* C.

Then $\rho(\boldsymbol{K}) \leqslant \delta(\Phi_n)$ $(n = 1, 2, \ldots)$ *and* $\rho(\boldsymbol{K}) = \delta(\Phi_\infty)$.

From this theorem it follows, in particular, that

THEOREM 6.2. *Let* $\boldsymbol{K}$ *be a compact Volterra integral operator, with kernel* $K(x, t)$, *acting in* L_p $(1 \leqslant p \leqslant \infty)$ *or in* C. *Then* $\rho(\boldsymbol{K}) = 0$.

6.3 *Supplementary notes*

Results analogous to those presented above are also valid for Volterra integral operators acting in arbitrary spaces of measurable functions (in particular, Orlicz spaces).

The decay of the sequence $\|\boldsymbol{K}^n\|^{1/n}$ is an important characteristic of Volterra operators with zero spectral radius. For U-bounded operators in L_p $(1 \leqslant p \leqslant \infty)$ spaces and U-cobounded operators in L_p $(1 \leqslant p \leqslant \infty)$ spaces and C, the following interesting inequality holds

$$\|\boldsymbol{K}^n\| \leqslant \frac{c}{n!}.$$

In the general case, convergence can be slow.

Finally, it is worth noting that there exist different definitions for Volterra operators in multi-dimensional spaces. For such operators, different theorems about the zero value of the spectral radius are known.

§7 Equations of the first kind

7.1 *The formulation of the problem*

The equation

$$\int_{\Omega} K(x, t)\, u(t)\, dt = f(x) \tag{5.168}$$

is called a *linear integral equation of the first kind*. Here $K(x, t)(x, t \in \Omega)$ is a measurable function of both variables, $f(x)$ is a known function, defined on Ω, and $u(t)$ is an unknown function.

The theory for first kind equations is more difficult than that for second kind. The reason is that the linear integral operator

$$\boldsymbol{K}u(x) = \int_{\Omega} K(x, t)\, u(t)\, dt, \tag{5.169}$$

defined by the l.h.s. of the equation (5.168) is in general not normally solvable, i.e. the range of its values is not a bounded region. Hence, the solvability conditions for first kind equations differ, in principle, from those for second kind.

7.2 *Equations in Hilbert space*

The most complete theory for linear first kind equations is that for equations with self-adjoint operators. For such equations, both solvability conditions and the general form of solutions {in terms of a convergent eigenfunction expansion} can be established.

THEOREM 7.1. *Let* $\boldsymbol{K}$ *be a compact self-adjoint operator, acting in the space* L_2, *and let*

$$\boldsymbol{K}u = \sum_n \lambda_n \boldsymbol{P}(\lambda_n) u$$

be its spectral decomposition. Let $f \in L_2$, *then the equation*

$$\boldsymbol{K}u = f \tag{5.170}$$

has a solution in L_2, *if and only if*

a) $(u, f) = 0$ *for all* $u \in L_2$, *satisfying the homogeneous equation* $\boldsymbol{K}u = 0$;

b) $$\sum_n \frac{1}{\lambda_n^2} \|\boldsymbol{P}(\lambda_n)f\|^2 < \infty. \qquad (5.171)$$

If the conditions (a) *and* (b) *are satisfied, then the solutions* $u \in L_2$ *of* (5.170) *are defined by*

$$u = h + \sum_n \frac{1}{\lambda_n} \boldsymbol{P}(\lambda_n)f, \qquad (5.172)$$

where h is an arbitrary solution of the homogeneous equation $\boldsymbol{K}u = 0$.

We now consider equations with non self-adjoint operators.

THEOREM 7.2. *Let* $\boldsymbol{K}$ *be a compact operator, acting in* L_2, *and let its Schmidt decomposition be*

$$\boldsymbol{K}u = \sum_n s_n \boldsymbol{Q}(s_n)u. \qquad (5.173)$$

The equation $\boldsymbol{K}u = f$, *where* $f \in L_2$, *has a solution in* L_2, *if and only if*

a) $(u, f) = 0$ *for all* $u \in L_2$, *satisfying the equation* $\boldsymbol{K}^*u = 0$;

b) $$\sum_n \frac{1}{s_n^2} \|\boldsymbol{Q}^*(s_n)f\|^2 < \infty. \qquad (5.174)$$

If the conditions (a) *and* (b) *are satisfied, then the solutions* $u \in L_2$ *of* $\boldsymbol{K}u = f$ *are defined by*

$$u = h + \sum_n \frac{1}{s_n} \boldsymbol{Q}^*(s_n)f, \qquad (5.175)$$

where h is an arbitrary solution of the homogeneous equation $\boldsymbol{K}u = 0$.

7.3 *The regularization method*

The basic method used in the study of first kind integral equations is the regularization method {see §5, Chapter VI}.

Consider the equations

$$\boldsymbol{K}x = f, \qquad (5.176)$$

where $\boldsymbol{K}$ is an operator acting in a Banach space E.

A linear operator $\boldsymbol{S}$, defined on the range of values $R(\boldsymbol{K})$, is called a *left equivalent regularizator*, if the operator $\boldsymbol{C} = \boldsymbol{I} - \boldsymbol{SK}$ is compact and if $\boldsymbol{S}x = \theta$ implies $x = \theta$. If $\boldsymbol{S}$ is the left equivalent regularizator of the operator $\boldsymbol{K}$ and f belongs to the domain of the operator $\boldsymbol{S}$, then (5.176) is equivalent to the second kind equation

$$x - \boldsymbol{C}x = \boldsymbol{S}f. \tag{5.177}$$

One can apply the general method of the theory for second kind equations with compact operators to this equation.

A linear operator $\boldsymbol{T}$ is called a *right equivalent regularizator*, if the operator $\boldsymbol{D} = \boldsymbol{I} - \boldsymbol{KT}$ admits an extention to a compact operator, acting in E, with range contained in the range of $\boldsymbol{T}$, and if $\boldsymbol{T}x = \theta$ implies $x = \theta$. If $\boldsymbol{T}$ is the right equivalent regularizator of the operator $\boldsymbol{K}$ and $f \in D(\boldsymbol{T})$, then (5.176) is equivalent to the second kind equation

$$y - \boldsymbol{D}y = f. \tag{5.178}$$

The solutions x of (5.176) are determined by $x = \boldsymbol{T}y$, where the y are solutions of (5.178).

We examine two examples.

Consider the following equation in the space C

$$\int_0^t K(t, s)\, x(s)\, ds = f(t) \tag{5.179}$$

with continuous kernel $K(t, s)$, such that $K(t, t)$ does not vanish. Let $r(t) = 1/K(t, t)$ and C_0 be the set of continuously differentiable functions which equal zero for zero values of the argument.

Suppose that $K(t, s)$ is differentiable with respect to t. Then we can take the operator

$$\boldsymbol{S}x(t) = r(t) \cdot x'(t).$$

defined on C_0, as the left regularizator. If $f(t) \in C_0$, then (5.179) is equivalent to

$$x(t) = \int_0^t L_1(t, s)\, x(s)\, ds + \varphi(t), \tag{5.180}$$

where $L_1(t, s) = -r(t)\, K_t'(t, s)$, and $\varphi(t) = r(t)\, f'(t)$.

Suppose now, that $K(t, s)$ is differentiable with respect to t and s. Then we can take the operator

$$\boldsymbol{T}x(t) = [r(t)\, x(t)]',$$

defined on C_0, as the right regularizator. If $f(t) \in C_0$, then (5.179) is equivalent to

$$y(t) = \int_0^t L_2(t, s)\, y(s)\, ds + f(t), \tag{5.181}$$

where $L_2(t, s) = r(s)\, K_s'(t, s)$. The solutions $x(t)$ of (5.179) are determined by $x(t) = [r(t)\, y(t)]'$, where the $y(t)$ are the solutions of (5.181).

As the second example, consider the equation

$$\int_a^b K(t, s)\, x(s)\, ds = f(t). \tag{5.182}$$

We assume that the function $K(t, s)$ is continuous and satisfies the conditions $K(a, s) = K(b, s) = K(t, a) = K(t, b) = 0$. We denote by C_0 the set of twice continuously differentiable functions which equal zero at the points $x = a$ and $x = b$.

We assume that $K(t, s)$ has continuous first and second derivatives with respect to t (or s) when $t \neq s$ and that the function

$$\frac{1}{r_1(t)} = K_t'(t + 0, t) - K_t'(t - 0, t)$$

$$\left(\frac{1}{r_2(t)} = K_s'(t + 0, t) - K_s'(t - 0, t)\right)$$

is continuous and does not vanish (both these conditions are satisfied, for example, for the Green's function of the Dirihlet boundary value problem for ordinary differential equations of the second order). Here we can take the operator

$$\boldsymbol{S}x(t) = r_1(t)\, x''(t) \qquad (\boldsymbol{T}x(t) = r_2(t)\, x''(t))$$

as the left (right) regularizator. If $f(t) \in C_0$, then (5.182) is equivalent to

$$x(t) = \int_a^b L_1(t, s)\, x(s)\, ds + \varphi(t),$$

$$y(t) = \int_a^b L_2(t, s)\, y(s)\, ds + f(t),$$

where

$$L_1(t, s) = -r_1(t)\, K''_{tt}(t, s), \qquad \varphi(t) = r_1(t)\, f''(t),$$
$$L_2(t, s) = -r_2(t)\, K''_{ss}(t, s), \qquad x(t) = [r_2(t)\, y(t)]''.$$

Chapter VI

ONE-DIMENSIONAL SINGULAR EQUATIONS[1])

§1 Basic notions

1.1 *The singular integral*

For functions of one independent variable, which we deal with in the present chapter, the singular integral is defined in the following way.

Let Γ be a closed or open curve in the complex plane. Let ζ denote the complex coordinates of points which vary along Γ. Usually, we do not distinguish between the points of the complex plane and their complex coordinates. Suppose that a function $f(\zeta)$, defined almost everywhere on Γ, has the following property: on Γ, there exists a point t such that, if this point is removed using a circle with centre at t and arbitrary radius $\varepsilon > 0$, then on the remainder of Γ, viz. Γ_ε, the function $f(\zeta)$ is summable. If the following limit exists

$$\lim_{\varepsilon \to 0} \int_{\Gamma_\varepsilon} f(\zeta)\, d\zeta,$$

it is called the *singular integral*[2]) of the function $f(\zeta)$ along the contour Γ and is denoted by the usual symbol

$$\lim_{\varepsilon \to 0} \int_{\Gamma_\varepsilon} f(\zeta)\, d\zeta = \int_{\Gamma} f(\zeta)\, d\zeta. \tag{6.1}$$

If the integral

$$\int_{\Gamma} f(\zeta)\, d\zeta$$

[1]) The topic of the present chapter is examined in the books [4], [5], [25] as well as in the review paper [22], Chapter I.
[2]) Cauchy referred to this limit as the *principal value* of the integral $\int_\Gamma f(\zeta)\, d\zeta$.

exists as the Lebesgue integral or at least as an improper one, then the singular integral (6.1) exists and coincides with the mentioned Lebesgue or improper integral. Nevertheless, singular integrals can exist in a more general framework.

Example 1.1. Let Γ be the interval $-1 \leqslant \zeta \leqslant 1$ and let $f(\zeta) = \zeta^{-n}$, where $n > 1$ is an integer. In this case, $t = 0$. We have

$$\int_{\Gamma_\varepsilon} f(\zeta)\, d\zeta = \int_{-1}^{-\varepsilon} \frac{d\zeta}{\zeta^n} + \int_{\varepsilon}^{1} \frac{d\zeta}{\zeta^n} = \frac{1-(-1)^{n-1}}{n-1}\left(1 - \frac{1}{\varepsilon^{n-1}}\right).$$

Hence it is clear that the singular integral

$$\int_{-1}^{1} \frac{d\zeta}{\zeta^n}, \qquad n > 1,$$

does not exist for even n; while, for odd n, it exists and takes the value

$$\int_{-1}^{1} \frac{d\zeta}{\zeta^n} = 0.$$

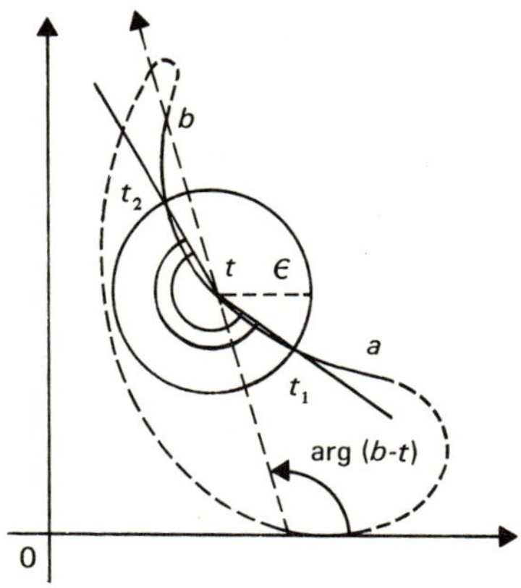

Figure 2

Example 1.2. Let Γ be a simple open Liapunov[3]) arc with end points a and b (see Figure 2) and t an internal point of this arc. We examine the existence of the singular integral

$$\int_{\Gamma} \frac{d\zeta}{\zeta - t}. \tag{6.2}$$

We draw a circle of radius ε with centre at t. Let t_1 and t_2 be the points of intersection with Γ. According to the definition of singular integral, we must evaluate the integral

$$\int_{\Gamma_\varepsilon} \frac{d\zeta}{\zeta - t} = \int_a^{t_1} \frac{d\zeta}{\zeta - t} + \int_{t_2}^{b} \frac{d\zeta}{\zeta - t} = \ln \frac{t_1 - t}{a - t} + \ln \frac{b - t}{t_2 - t}. \tag{6.3}$$

This last result is valid, if we choose the branch of the logarithm so that $\ln(\zeta - t)$ changes continuously along the arc (a, t_1) as well as along the arc (t_2, b). We make this choice as following: we complete Γ to form a smooth closed contour Γ_1 (in Figure 2, the additional section is shown as a dotted line) so that the direction of Γ from a to b corresponds to a counter-clockwise direction for Γ_1. We fix the branch of $\ln(\zeta - t)$ so that $\arg(b - t)$ has the smallest possible non-negative value. This defines a branch of $\ln(\zeta - t)$ which is single valued and continuous on the contour Γ_1. We assume that this is the branch of the logarithm which is chosen in (6.3) and rewrite this formula as

$$\int_{\Gamma_\varepsilon} \frac{d\zeta}{\zeta - t} = \ln \frac{b - t}{a - t} + \ln \frac{t_1 - t}{t_2 - t}.$$

Using the equality $|t_1 - t| = |t_2 - t| = \varepsilon$, we obtain

$$\ln \frac{t_1 - t}{t_2 - t} = i[\arg(t_1 - t) - \arg(t_2 - t)].$$

We see from Figure 2 that as ζ moves along the contour Γ_1 from $\zeta = t_2$ to $\zeta = t_1$, the value of $\arg(\zeta - t)$ increases by α (the angle shown in Figure 2

[3]) An arc is called *Liapunov*, if it has a well-defined tangent at every point and the angle θ between the tangents at the points t_1 and t_2 of this arc satisfies the inequality $\theta \leqslant A\,|t_1 - t_2|^\alpha$ where A and α are positive constants.

by the double arc). If $\varepsilon \to 0$ then, obviously, $\alpha \to \pi$. Now

$$\int_{\Gamma_\varepsilon} \frac{d\zeta}{\zeta - t} = \ln \frac{b - t}{a - t} + i\alpha,$$

and it is clear that the integral (6.2) exists and is equal to

$$\int_{\Gamma} \frac{d\zeta}{\zeta - t} = \ln \frac{b - t}{a - t} + i\pi. \tag{6.4}$$

If we complete Γ so that the direction on the closed contour Γ_1 is clockwise, then instead of (6.4) we obtain

$$\int_{\Gamma} \frac{d\zeta}{\zeta - t} = \ln \frac{b - t}{a - t} - i\pi. \tag{6.5}$$

The difference between the formulas (6.4) and (6.5) is, certainly, only visual: each corresponds to its own branch of the logarithm and these branches differ by the term $2\pi i$.

Example 1.3. If the contour Γ is closed and its direction is counter-clockwise, then we must take $a = b$ in (6.4). Because of the branch chosen for the logarithm, the function $\ln(\zeta - t)$ is single valued and continuous on Γ for $\zeta \neq t$. Hence, $\ln(b - t) = \ln(a - t)$, and the first term in (6.4) disappears and we obtain the simple and important result

$$\int_{\Gamma} \frac{d\zeta}{\zeta - t} = i\pi. \tag{6.6}$$

If the direction of the closed contour Γ is clockwise, then we have to insert a minus sign in (6.6).

Example 1.4. Let Γ be the interval $[a, b]$ of the real axis. In this case, (6.4) as well as (6.5) can be written in the form

$$\int_a^b \frac{d\zeta}{\zeta - t} = \ln \frac{b - t}{t - a}. \tag{6.7}$$

1.2 *The singular Cauchy integral*

Through the examples given above one can introduce an important class of singular integrals which we shall study in this chapter. These are integrals of the form

$$\frac{1}{\pi i}\int_{\Gamma}\frac{u(\zeta)}{\zeta - t}\,d\zeta, \qquad t \in \Gamma, \tag{6.8}$$

where Γ is a Liapunov contour.

It is evident that the integral (6.5) exists, if $u(\zeta)$ satisfies a Lipschitz condition on Γ with exponent α (this condition is often called a Hölder condition):

$$|u(\zeta_1) - u(\zeta_2)| \leqslant A\,|\zeta_1 - \zeta_1|^{\alpha},$$

where A and α are positive constants. Indeed,

$$\frac{1}{\pi i}\int_{\Gamma_\varepsilon}\frac{u(\zeta)}{\zeta - t}\,dt = \frac{1}{\pi i}\int_{\Gamma_\varepsilon}\frac{u(\zeta) - u(t)}{\zeta - t}\,d\zeta + \frac{u(t)}{\pi i}\int_{\Gamma_\varepsilon}\frac{d\zeta}{\zeta - t}.$$

Using, for instance, (6.4) and the corresponding choice of the branch for the logarithm, we obtain

$$\frac{1}{\pi i}\int_{\Gamma}\frac{u(\zeta)}{\zeta - t}\,d\zeta = \frac{1}{\pi i}\int_{\Gamma}\frac{u(\zeta) - u(t)}{\zeta - t}\,d\zeta + \frac{u(t)}{\pi i}\left[\ln\frac{b - t}{a - t} + \pi i\right]. \tag{6.9}$$

If Γ is a closed contour with a counter-clockwise direction, then

$$\frac{1}{\pi i}\int_{\Gamma}\frac{u(\zeta)}{\zeta - t}\,d\zeta = \frac{1}{\pi i}\int_{\Gamma}\frac{u(\zeta) - u(t)}{\zeta - t}\,d\zeta + u(t). \tag{6.10}$$

The integrals on the r.h.s. of (6.9) and (6.10) are absolutely convergent.

The integral (6.8) is called the *singular Cauchy integral* and the expression $1/(\zeta - t)$ the *Cauchy kernel*. The function $u(\zeta)$ is called the density of the integral (6.8).

Clearly, the integral (6.8) exists when Γ is a finite collection of closed or open curves, bounding some region (generally speaking, multi-connected) in the complex plane.

1.3 *The singular Hilbert integral*

The singular Hilbert integral

$$\frac{1}{2\pi}\int_{-\pi}^{\pi} u(\sigma)\operatorname{ctg}\frac{\sigma - s}{2}\,d\sigma, \qquad -\pi \leqslant s \leqslant \pi, \tag{6.11}$$

is closely connected with the singular Cauchy integral. If the function $u(\sigma)$ is 2π-periodic and satisfies a Lipschitz condition with positive exponent, then the integral (6.11) exists for any s and can be represented in the form of an absolutely convergent integral

$$\frac{1}{2\pi}\int_{-\pi}^{\pi} u(\sigma)\operatorname{ctg}\frac{\sigma - s}{2}\,ds = \frac{1}{2\pi}\int_{-\pi}^{\pi}[u(\sigma) - u(s)]\operatorname{ctg}\frac{\sigma - s}{2}\,d\sigma. \tag{6.12}$$

There exists a close connection between the Cauchy and Hilbert integrals. Let γ be the unit circle with centre at the origin and let ζ and t be points of its circumference. Then $\zeta = e^{i\sigma}$, $t = e^{is}$, where the real variables σ and s vary over the interval $[-\pi, \pi]$. If we put

$$\frac{d\zeta}{\zeta - t} = \left(\frac{1}{2}\operatorname{ctg}\frac{\sigma - s}{2} + \frac{i}{2}\right)d\sigma,$$

and $u(\zeta) = u(e^{i\sigma}) = \hat{u}(\sigma)$, we obtain

$$\frac{1}{\pi i}\int_{\gamma}\frac{u(\zeta)}{\zeta - t}\,d\zeta = \frac{1}{2\pi i}\int_{-\pi}^{\pi}\hat{u}(\sigma)\operatorname{ctg}\frac{\sigma - s}{2}\,d\sigma + \frac{1}{2\pi}\int_{-\pi}^{\pi}\hat{u}(\sigma)\,d\sigma. \tag{6.13}$$

Note. The Hilbert integral is closely connected with the theory of conjugate trigonometrical series. For details see [13], Chapter VII.

§2 Some properties of singular integrals

2.1 *Assumptions about the contour*

Henceforth, we shall assume that Γ is a contour which bounds some region D of the complex plane which is in general multi-connected. We take the direction of the contour to be counter-clockwise; i.e., the region D always remains on the left of Γ. The contour Γ consists of some finite or countable set of closed or open curves $\Gamma_0, \Gamma_1, \Gamma_2, \ldots$. Throughout this chapter we assume that this set is finite. For ease of presentation, we assume that each of the curves Γ_j lies in a finite region of the plane and has continuous curvature. However, many of the results presented below are valid for Liapunov curves.

2.2 *On the existence of singular Cauchy integrals*

In the preceding section, we have established that the singular integral (6.8) exists for any t, if the density $u(\sigma)$ of this integral satisfies a Lipschitz condition on Γ with some positive exponent α (this is written symbolically in the form $u \in \mathrm{Lip}_\alpha(\Gamma)$). If we only require the existence of $u(\sigma)$ almost everywhere on Γ, then we can formulate the following much more general result:

THEOREM 2.1. *If, on the contour Γ, the convexity and concavity changes only a finite number of times and the density $u(\zeta)$ is summable on Γ, then the singular integral* (6.8) *exists for almost all $t \in \Gamma$.*

Theorem 2.1 has been proved by Privalov [30]. The proof of a more general theorem is given in [31], pp. 192–194.

2.3 *Limit formulas*

Theorems about the limit values of Cauchy type integrals are fundamental in many applications of one-dimensional singular integrals. Such theorems have been proved under different conditions by many authors. The first result is due to Sokhotsky[4]) (1873). The most general proof is due to Privalov[5]). We formulate his result here.

Consider the Cauchy integral

$$\Phi(z) = \frac{1}{2\pi i} \int_\Gamma \frac{u(\zeta)}{\zeta - z}\, d\zeta, \tag{6.14}$$

where z is a point lying either in the region D, bounded by the contour Γ, or in one of the regions complementary to D. In the first case, we denote the integral (6.14) by $\Phi^+(z)$ and in the second by $\Phi^-(z)$. Suppose now that

---

[4]) See [20].

[5]) See [30] as well as [31], pp. 190–194.

$z \to t$, $t \in \Gamma$, and consider the problem of the existence of the limits

$$\Phi^+(t) = \lim_{z \to t} \Phi^+(z), \qquad \Phi^-(t) = \lim_{z \to t} \Phi^-(z). \tag{6.15}$$

The answer is given by the following theorem of Privalov.

THEOREM 2.2. *If the contour Γ and the density $u(\zeta)$ satisfy the conditions of Theorem* 2.1 *and $z \to t$ along a path which is not tangent to Γ, then for almost all $t \in \Gamma$ the limits* (6.15) *exist and, almost everywhere on Γ,*

$$\Phi^+(t) = \tfrac{1}{2}u(t) + \frac{1}{2\pi i}\int_\Gamma \frac{u(\zeta)}{\zeta - t}\,d\zeta, \tag{6.16$_1$}$$

$$\Phi^-(t) = -\tfrac{1}{2}u(t) + \frac{1}{2\pi i}\int_\Gamma \frac{u(\zeta)}{\zeta - t}\,dt. \tag{6.16$_2$}$$

If $u \in \mathrm{Lip}_\alpha(\Gamma)$, $\alpha > 0$, and Γ is a Liapunov contour, then the limits (6.15) *exist everywhere on Γ and the formulas* (6.16) *are valid* (*see, for example,* [25]).

In many important cases, the limit results (6.16) enable one to calculate the singular Cauchy integral rather simply. We develop some such formulas.

If the function $u(z)$ is holomorphic in the region D, bounded by the contour Γ and continuous on $D \cup \Gamma$, then

$$\frac{1}{\pi i}\int_\Gamma \frac{u(\zeta)}{\zeta - t}\,d\zeta = u(t). \tag{6.17$_1$}$$

If the function $u(z)$ is holomorphic in every region, which is complementary to D, continuous in the corresponding closed domain and vanishes at infinity, then

$$\frac{1}{\pi i}\int_\Gamma \frac{u(\zeta)}{\zeta - t}\,d\zeta = -u(t). \tag{6.17$_2$}$$

Example 2.1. If n is a non-negative integer, then

$$\frac{1}{\pi i}\int_\Gamma \frac{\zeta^n}{\zeta - t}\,d\zeta = t^n.$$

If n is a positive integer and the point $z = 0$ belongs to D, then

$$\frac{1}{\pi i}\int_{\Gamma}\frac{d\zeta}{\zeta^n(\zeta - t)} = -\frac{1}{t^n}.$$

For the calculation of the Hilbert integrals (6.11) the following formula is useful (n is an integer):

$$\frac{1}{2\pi}\int_{-\pi}^{\pi} e^{in\sigma}\operatorname{ctg}\frac{\sigma - s}{2}\,d\sigma = \begin{cases} ie^{ins}, & n > 0, \\ 0, & n = 0, \\ -ie^{ins}, & n < 0. \end{cases} \tag{6.18}$$

2.4 *Integration by parts*

For the Hilbert integral, integration by parts gives

$$\frac{1}{2\pi}\int_{-\pi}^{\pi} u(\sigma)\operatorname{ctg}\frac{\sigma - s}{2}\,d\sigma = -\frac{1}{\pi}\int_{-\pi}^{\pi} u'(\sigma)\ln\left|\sin\frac{\sigma - s}{2}\right|d\sigma. \tag{6.19}$$

It is valid for almost all $s \in [-\pi, \pi]$, if the function $u(\sigma)$ is absolutely continuous on $[-\pi, \pi]$ and $u(-\pi) = u(\pi)$.

If the function $u(\zeta)$ is absolutely continuous on a closed contour Γ which satisfies the assumptions of §2.1 and has a counter-clockwise direction, then for almost all $t \in \Gamma$

$$\frac{1}{\pi i}\int_{\Gamma} u'(\zeta)\ln(\zeta - t)\,d\zeta = u(t) - \frac{1}{\pi i}\int_{\Gamma}\frac{u(\zeta)}{\zeta - t}\,d\zeta. \tag{6.20}$$

If Γ is open and is a simple arc which begins at a and ends at b, then we choose the same branch of $\ln(\zeta - t)$ as in §1.2. Then the following formula is valid almost everywhere on Γ

$$\int_{\Gamma} u'(\zeta)\ln(\zeta - t)\,d\zeta$$
$$= \pi i u(t) + u(b)\ln(b - t) - u(a)\ln(a - t) - \int_{\Gamma}\frac{u(\zeta)}{\zeta - t}\,d\zeta. \tag{6.21}$$

The integration by parts formulas (6.19)–(6.21) are valid everywhere on Γ, if not only $u(\zeta)$ is absolutely continuous on Γ, but also $u' \in L_p(\Gamma)$ with the constant $p > 1$.

2.5 *The distortion of the removed arc*

Let $t \in \Gamma$ and, as above, t_1 and t_2 be points of intersection of Γ with the circumference of the circle of radius ε with centre at t. Further, let t_1' and t_2' be points on Γ such that the ratios

$$\frac{|t_1' - t_1|}{\varepsilon^{p'}}, \qquad \frac{|t_2' - t_2|}{\varepsilon^{p'}}, \tag{6.22}$$

where p' is a constant, $1 < p' < \infty$, remain bounded as $\varepsilon \to 0$. We denote by Γ_ε' that part of Γ obtained by removing from Γ the small arc with end points t_1' and t_2'. If $u \in L_p(\Gamma)$, where $1/p + 1/p' = 1$, then

$$\int_\Gamma \frac{u(\zeta)}{\zeta - t}\, d\zeta = \lim_{\varepsilon \to 0} \int_{\Gamma_\varepsilon'} \frac{u(\zeta)}{\zeta - t}\, d\zeta. \tag{6.23}$$

The equality (6.23) is valid in the sense that, if, for some t, one of its sides has sense then the other has sense and they are equal. From Theorem 2.1, it follows that (6.23) is valid almost everywhere on Γ.

2.6 *Change of variable*

Let the function $\zeta = \omega(\tau)$ define a one-to-one and continuous mapping of the contour Γ onto a contour C in the τ-plane, with the derivatives $\omega'(\tau)$, $\omega''(\tau)$ continuous and $\omega'(\tau)$ non-zero on C. Then, if $u \in L_2(\Gamma)$, it follows that

$$\int_\Gamma \frac{u(\zeta)}{\zeta - t}\, d\zeta = \int_C \frac{u(\omega(\tau))\, \omega'(\tau)}{\omega(\tau) - t}\, d\tau. \tag{6.24}$$

§3 Singular operators in functional spaces

3.1 *The singular operator*

From Theorem 2.1, it follows that the singular Cauchy integral (6.8) can be considered as an operator which maps functions $u(t)$, summable on Γ, into correspondence with new functions

$$v(t) = \frac{1}{\pi i} \int_{\Gamma} \frac{u(\zeta)}{\zeta - t} d\zeta, \tag{6.25}$$

defined almost everywhere on Γ. We denote this operator by $\boldsymbol{S}$:

$$(\boldsymbol{S}u)(t) = \frac{1}{\pi i} \int_{\Gamma} \frac{u(\zeta)}{\zeta - t} d\zeta. \tag{6.26}$$

3.2 *Invariant spaces*

In general, we can say nothing about the properties of the function $v(t)$: if the function $u(\zeta)$ is only summable, then we cannot affirm the summability of the function $v(\zeta)$ defined by (6.25). But, if the function $u(\zeta)$ belongs to some appropriate narrower class, then it is possible to establish certain analytical properties of $v(\zeta)$. The problem is usually formulated as following: if u belongs to a given functional space, then what space does v belong to? A more specific but more important question is: what functional spaces are such that u and v belong to one and the same space? Such spaces are called *invariant* with respect to the operator $\boldsymbol{S}$. Below, we cite basic facts about these problems. In this chapter, p will always mean a fixed number such that $1 < p < \infty$; by p' we denote the number which satisfies $1/p + 1/p' = 1$.

THEOREM 3.1. *The space $L_p(\Gamma)$ is invariant for the operator $\boldsymbol{S}$. $\boldsymbol{S}$ is bounded in this space.*

For the case $p = 2$, Theorem 3.1 was proved by Lusin; for arbitrary $p \neq 1$, by M. Riesz. For $p = 1$ and $p = \infty$ the theorem is not valid.

For details concerning Theorem 3.1 see, for instance, [13].

Theorem 3.1 can be generalized for spaces of functions which are summable with respect to given exponent and weight. Thus, the space $L_p(\Gamma; |t - t_0|^\beta)$ of functions, for which the integral

$$\|u\|^p = \int_{\Gamma} |u(t)|^p \, |t - t_0|^\beta \, |dt|,$$

is finite, where $t_0 \in \Gamma$ and the constant β lies within the bounds

$$-1 < \beta < p - 1,$$

is invariant with respect to $\boldsymbol{S}$. In this space, $\boldsymbol{S}$ is bounded [3a].

THEOREM 3.2. *If Γ is a closed Liapunov contour, then the space* $\mathrm{Lip}_\alpha(\Gamma)$ *of functions satisfying a Lipschitz condition with a fixed exponent α, $0 < \alpha < 1$, on Γ, is invariant with respect to $\boldsymbol{S}$. The operator $\boldsymbol{S}$ is bounded in this space.*

Theorem 3.2 was proved by Plemely and Privalov. For further details see [25].

We cite some additional results about the properties of singular integrals on Lipschitz functions. These properties were also studied by Plemely and Privalov.

The space $\mathrm{Lip}_1(\Gamma)$ is not invariant for the Cauchy operator: if Γ is a closed Liapunov contour, $u \in \mathrm{Lip}_1(\Gamma)$ so that

$$|u(t_1) - u(t_2)| \leqslant A\,|t_1 - t_2|, \qquad A = \text{const},$$

and $v = \boldsymbol{S}u$, then for sufficiently close t_1 and t_2

$$|v(t_1) - v(t_2)| \leqslant B\,|t_1 - t_2| \ln \frac{1}{|t_1 + t_2|}, \qquad B = \text{const}. \tag{6.27}$$

If Γ is an open Liapunov contour, then for any α the space $\mathrm{Lip}_\alpha(\Gamma)$ is not invariant for the Cauchy operator: if $u \in \mathrm{Lip}_\alpha(\Gamma)$, $0 < \alpha < 1$, and $v = \boldsymbol{S}u$, then we can only affirm that $v \in \mathrm{Lip}_\alpha(\Gamma')$, where Γ' is any interior part of Γ. Analogously, if $\alpha = 1$, then the function $v = \boldsymbol{S}u$ satisfies the inequality (6.27) not on the whole contour, but only on any of its interior parts Γ'.

We also note that the space of continuous functions is not invariant for the Cauchy operator: there exist continuous functions which $\boldsymbol{S}$ transforms into discontinuous ones. Examples of this kind are given in [13].

3.3 *Non-Liapunov contours*

We present a generalization of Theorem 3.2 for the case of a non-Liapunov contour [3]. Let Γ be a smooth closed contour (i.e., with continuously changing tangent). By $\varphi(\tau)$, we denote a function defined for sufficiently small

non-negative values of τ and having the following properties:
1) $\varphi(\tau)$ is a continuous and increasing function;
2) $\varphi(0) = 0$;
3) $\varphi(\tau) > 0, \tau > 0$;
4) the ratio $\varphi(\tau)/\tau$ is a decreasing function.

Let $u(t)$ be a function defined on Γ. By $\omega(u, \tau)$ we denote its modulus of continuity

$$\omega(u, \tau) = \sup_{|t_1 - t_2| < \tau} |u(t) - u(t_2)|.$$

We denote by $H_\varphi(\Gamma)$ the space of functions which are defined on Γ and satisfy the condition

$$\|u\|_\varphi = \max_{t\in\Gamma} |u(t)| + \sup_\tau \frac{\omega(u, \tau)}{\varphi(\tau)} < \infty. \tag{6.28}$$

It is not difficult to see that $H_\varphi(\Gamma)$ is a Banach space.

3.4 *Further theorems about invariant spaces*

THEOREM 3.3. *If there exists a constant $c > 1$ such that*

$$\underline{\lim}_{\tau\to 0} \frac{\varphi(c\tau)}{\varphi(\tau)} > 1, \qquad \overline{\lim}_{\tau\to 0} \frac{\varphi(c\tau)}{\varphi(\tau)} < c,$$

then the space H_φ is invariant for the Cauchy operator and this operator is bounded in H_φ.

We return now to the case of a closed contour with continuous curvature. We denote by I_q, $q \geqslant 0$, the class of functions which are continuous on Γ and satisfy the condition

$$\int_0^l \frac{\omega(u, \tau)}{\tau}\left(\ln \frac{l}{\tau}\right)^q d\tau < \infty \tag{6.29}$$

where l is the length of the contour Γ. The following theorems are valid ([17]).

THEOREM 3.4. *The Cauchy operator maps any function of the class I_{q+1}, $q > 0$, into a function of the class I_q.*

THEOREM 3.5. *The class* $I_\infty = \bigcap_{0 \leqslant q < \infty} I_q$ *is invariant for the Cauchy operator.*

Theorem 3.5 is clearly a consequence of Theorem 3.4.

Some general spaces are considered in [19].

We also note the following result [8]: if D_α is the class of functions the modulus of continuity of which satisfies the inequality

$$\omega(u, \tau) \leqslant A\left(\ln \frac{a}{\tau}\right)^{-\alpha},$$

then the Cauchy operator transforms the functions of D_α into functions of $D_{\alpha-1}$.

§4 Differentiation and integration formulas involving singular integrals

4.1 *Differentiation formulas*

In the present section, it is supposed that the contour satisfies the assumptions of §2.1. The following differentiation formulas are valid:

1) If $u \in L_p(\Gamma)$, $1 < p < \infty$, then

$$\frac{d}{dt}\int_\Gamma u(\zeta) \ln \frac{1}{\zeta - t}\, d\zeta = \pi i u(t) + \int_\Gamma \frac{u(\zeta)}{\zeta - t}\, d\zeta. \tag{6.30}$$

2) If $u(\zeta)$ is absolutely continuous on Γ and $u'(\zeta) \in L_p(\Gamma)$, then

$$\frac{d}{dt}\int_\Gamma \frac{u(\zeta)}{\zeta - t}\, d\zeta = \int_\Gamma \frac{u'(\zeta)}{\zeta - t}\, d\zeta. \tag{6.31}$$

3) Analogous formulas are valid for Hilbert integrals

$$\frac{d}{ds}\int_{-\pi}^{\pi} u(\sigma) \ln \left|\sin \frac{\sigma - s}{2}\right| d\sigma = -\int_{-\pi}^{\pi} u(\sigma) \operatorname{ctg} \frac{\sigma - s}{2}\, d\sigma, \tag{6.32}$$

$$\frac{d}{ds}\int_{-\pi}^{\pi} u(\sigma) \operatorname{ctg} \frac{\sigma - s}{2}\, d\sigma = \int_{-\pi}^{\pi} u'(\sigma) \operatorname{ctg} \frac{\sigma - s}{2}\, d\sigma. \tag{6.33}$$

The formula (6.33) is valid, if $u(\sigma)$ is absolutely continuous and $u'(\sigma)$ is summable with degree p on the interval $[-\pi, \pi]$ and if $u(-\pi) = u(\pi)$. The formula (6.32) is valid, if $u \in L_p(-\pi, \pi)$, $1 < p < \infty$.

4.2 *Integration formulas*

1) If $u \in L_p(\Gamma)$, $v \in L_{p'}(\Gamma)$, then

$$\int_\Gamma v(t) \left\{ \int_\Gamma \frac{u(\zeta)}{\zeta - t} \, d\zeta \right\} dt = \int_\Gamma u(\zeta) \left\{ \int_\Gamma \frac{v(t)}{\zeta - t} \, dt \right\} d\zeta. \tag{6.34}$$

2) If for some fixed p, $1 < p < \infty$,

$$\int_\Gamma \int_\Gamma |\varphi(\zeta, \tau)|^p \, |d\zeta| \cdot |d\tau| < \infty,$$

then

$$\int_\Gamma \frac{1}{\zeta - t} \left\{ \int_\Gamma \varphi(\zeta, \tau) \, d\tau \right\} d\zeta = \int_\Gamma \left\{ \int_\Gamma \frac{\varphi(\zeta, \tau)}{\zeta - t} \, d\zeta \right\} d\tau. \tag{6.35}$$

3) *Change of Order of Integration for Double Singular Integrals.*
If $u \in L_p(\Gamma)$ and Γ is a closed Liapunov contour, then

$$\frac{1}{(\pi i)^2} \int_\Gamma \frac{1}{\tau - t} \left\{ \frac{u(\zeta)}{\zeta - \tau} \, d\zeta \right\} d\tau = u(t). \tag{6.36}$$

The formula (6.36) is one of the most important in the theory of one-dimensional singular integral equations. It is known as the formulas of Poincaré-Bertrand. The formula (6.36) can be rewritten in the form

$$\boldsymbol{S}^2 = \boldsymbol{I}, \tag{6.37}$$

where $\boldsymbol{I}$ is the identity operator.

The formula for Hilbert integrals, analogous to (6.36), has the form

$$\frac{1}{4\pi^2} \int_{-\pi}^{\pi} \operatorname{ctg} \frac{\omega - s}{2} \left\{ \int_{-\pi}^{\pi} u(\sigma) \operatorname{ctg} \frac{\sigma - \omega}{2} \, d\sigma \right\} d\omega = -u(s) + \frac{1}{2\pi} \int_{-\pi}^{\pi} u(\sigma) \, d\sigma. \tag{6.38}$$

Let $\varphi(\zeta, \tau)$ satisfy a Lipschitz condition with a positive exponent with respect to both variables and let Γ be a closed or open contour satisfying the assumptions of §2.1. Then

$$\frac{1}{(\pi i)^2}\int_\Gamma \frac{1}{\tau - t}\left\{\int_\Gamma \frac{\varphi(\zeta, \tau)}{\zeta - \tau}\,d\zeta\right\} d\tau$$
$$= \frac{1}{(\pi i)^2}\int_\Gamma \left\{\int_\Gamma \frac{\varphi(\zeta, \tau)}{(\zeta - \tau)(\tau - t)}\,d\tau\right\} d\zeta + \varphi(t, t). \quad (6.39)$$

§5 Regularization

The notions and theorems of the present section apply not only to one-dimensional singular operators, but also to general linear operators in Banach spaces. The results of this section will be used not only in this chapter, but also in Chapters VIII and IX. For proofs and bibliography see [23].

5.1 *Right and left regularization*

Let X and X_1 be Banach spaces (which may coincide), and $\boldsymbol{A}$ be a linear closed operator [16] mapping from X into X_1. We say that $\boldsymbol{A}$ admits a *left regularization*, if there exists a bounded operator $\boldsymbol{B}$ mapping from X_1 into X such that

$$\boldsymbol{BA} = \boldsymbol{I} + \boldsymbol{T}, \quad (6.40)$$

where $\boldsymbol{I}$ is the identity and $\boldsymbol{T}$ is a compact operator in X. Analogously, $\boldsymbol{A}$ admits a *right regularization*, if there exists a bounded operator $\boldsymbol{C}$ mapping from X into X_1 such that

$$\boldsymbol{AC} = \boldsymbol{I}_1 + \boldsymbol{T}_1, \quad (6.41)$$

where $\boldsymbol{I}_1$ is the identity and $\boldsymbol{T}_1$ is a compact operator in X_1.

The operators $\boldsymbol{B}$ and $\boldsymbol{C}$ are called, respectively, the *left and right regularizators* of $\boldsymbol{A}$.

An operator $\boldsymbol{A}$ admits a *left equivalent regularization*, if there exists a left regularizator $\boldsymbol{B}$ such that

$$\boldsymbol{A}u = f \quad (6.42)$$

and

$$\boldsymbol{BA}u = \boldsymbol{B}f \tag{6.43}$$

are equivalent for all $f \in X_1$.

An operator $\boldsymbol{A}$ admits a *right equivalent regularization*, if there exists a right regularizator $\boldsymbol{C}$ such that, for any $f \in X_1$, the equation (6.42) and

$$\boldsymbol{AC}v = f \tag{6.44}$$

are equivalent in the following sense: (6.42) and (6.44) are either simultaneously solvable or not solvable. To any solution u of (6.42) there corresponds a solution v of (6.44) such that $\boldsymbol{C}v = u$.

Below, we denote by a star the adjoint equation and operator.

If $\boldsymbol{A}$ admits a left regularization, then the adjoint operator A^* admits a right regularization, and conversely.

5.2 *The index of an operator*

The solutions of the homogeneous equation

$$\boldsymbol{A}u = 0 \tag{6.45}$$

are called the *zeros* of the operator $\boldsymbol{A}$. If $\boldsymbol{A}$ is a closed linear operator then its zeros form a subspace. The dimension of this subspace is given by the number of zeros of $\boldsymbol{A}$, and is denoted by $\alpha(\boldsymbol{A})$. If at least one of the numbers $\alpha(\boldsymbol{A})$ and $\alpha(\boldsymbol{A}^*)$ is finite, then the difference

$$\operatorname{Ind} \boldsymbol{A} = \alpha(\boldsymbol{A}) - \alpha(\boldsymbol{A}^*) \tag{6.46}$$

is called *the index* of $\boldsymbol{A}$. Obviously,

$$\operatorname{Ind} \boldsymbol{A} = -\operatorname{Ind} \boldsymbol{A}^*. \tag{6.47}$$

An operator $\boldsymbol{A}$ is called *normally solvable*, if, for the solvability of (6.42), it is sufficient (the necessity of this condition is trivial) that its r.h.s. be orthogonal to the zeros of the adjoint $\boldsymbol{A}^*$. In other words, that the Fredholm Theorem 1.4 is valid for (6.42) (see Chapter II, §1). The following theorem of Hausdorf is known: an operator $\boldsymbol{A}$ is normally solvable if and only if the range of this operator is closed. In other words, if the range of its values form a subspace in X_1.

THEOREM 5.1. *If* $\boldsymbol{A}$ *admits a left regularization, then the number* $\alpha(\boldsymbol{A})$ *is finite.*

COROLLARY 1. *If* $\boldsymbol{A}$ *admits a right regularization, then the number* $\alpha(\boldsymbol{A}^*)$ *is finite.*

COROLLARY 2. *If* $\boldsymbol{A}$ *admits a double-side regularization (i.e. both left and right), then* Ind $\boldsymbol{A}$ *is finite.*

THEOREM 5.2. *If a closed operator admits a left regularization, then it is normally solvable.*

THEOREM 5.3. *Let a bounded operator* $\boldsymbol{A}$ *mapping from* X *into* X_1 *admit a left regularization. Then, for any compact operator* $\boldsymbol{T}$*, mapping from* X *into* X_1*, the following identity is valid*

$$\text{Ind}\,(\boldsymbol{A} + \boldsymbol{T}) = \text{Ind}\,\boldsymbol{T}. \tag{6.48}$$

THEOREM 5.4. *Let* $\boldsymbol{A}$ *admit a left regularization and* $\boldsymbol{B}$ *be the left regularizator of the operator* $\boldsymbol{A}$*. Let* $\boldsymbol{K}$ *be an operator mapping from* X *into* X_1 *such that* $\|\boldsymbol{B}\| \cdot \|\boldsymbol{K}\| < 1$*. Then* $\text{Ind}\,(\boldsymbol{A} + \boldsymbol{K}) = \text{Ind}\,\boldsymbol{A}$.

We introduce another concept: Bounded operators $\boldsymbol{A}_0$ and $\boldsymbol{A}_1$ mapping from X into X_1 are called *homotopic*, if there exists a set of operators $\boldsymbol{A}_\lambda$, depending on the real parameter $\lambda \in [0, 1]$, with the following properties:

1) For $\lambda = 0$, $\boldsymbol{A}_\lambda$ coincides with A_0 and, for $\lambda = 1$, with $\boldsymbol{A}_1$.
2) For any λ, the $\boldsymbol{A}_\lambda$ are bounded.
3) The $\boldsymbol{A}_\lambda$ depend continuously on λ in the sense that

$$\lim_{\mu \to 0} \|\boldsymbol{A}_{\lambda+\mu} - \boldsymbol{A}_\lambda\| = 0.$$

4) For any λ, the $\boldsymbol{A}_\lambda$ admit a left regularization where the regularizator $\boldsymbol{B}_\lambda$ is bounded independently of λ:

$$\|\boldsymbol{B}_\lambda\| \leqslant M = \text{const.}$$

THEOREM 5.5. (Atkinson [1]). *The indices of homotopic operators are equal.*

THEOREM 5.6. *If* $\boldsymbol{A}_1$ *and* $\boldsymbol{A}_2$ *are bounded operators with finite indices, then*

$$\text{Ind}\,(\boldsymbol{A}_1\boldsymbol{A}_2) = \text{Ind}\,(\boldsymbol{A}_2\boldsymbol{A}_1) = \text{Ind}\,\boldsymbol{A}_1 + \text{Ind}\,\boldsymbol{A}_2. \tag{6.49}$$

THEOREM 5.7. *In order that an operator admit a left equivalent regularization, it is necessary and sufficient that it be normally solvable and that its index be finite and non-negative.*

§6 Closed contours; symbols; Nöther theorems

6.1 *The general singular operator*

We consider the singular integral equation of the form

$$\boldsymbol{A}u = a(t)\,u(t) + b(t)(\boldsymbol{S}u)(t) + (\boldsymbol{T}u)(t) = f(t) \tag{6.50}$$

in the space $L_p(\Gamma)$. The contour Γ will be assumed to be closed with continuous curvature. As above, we denote by $\boldsymbol{S}$ the Cauchy operator (6.8). $\boldsymbol{T}$ is a compact operator in $L_p(\Gamma)$. The functions $a(t)$ and $b(t)$ are defined and continuous on Γ, and $f(t)$ is a function from $L_p(\Gamma)$. The function to be determined is required to belong to $L_p(\Gamma)$. We note at this stage that all the results formulated below in this present section are valid for the spaces $\mathrm{Lip}_\alpha(\Gamma)$, $0 < \alpha < 1$, if it is assumed that $a(t)$ and $b(t)$ belong to this space and that the operator $\boldsymbol{T}$ is compact in the same space. For the case of the $L_2(\Gamma)$ space, the subject-matter of the present section is given in [22]. Similar arguments are used for the $p \neq 2$ case in [41]. A detailed study of one-dimensional singular equations in Lipschitz spaces is given in [25].

The operator $\boldsymbol{A}$ defined by the l.h.s. of (6.50) will be called *a general singular operator* or, simply, *a singular operator*. If $T \equiv 0$, then $\boldsymbol{A}$ will be called the *simplest singular operator* (such operators are called *characteristic operators* in [25]).

The following lemma plays an important role in the theory of one-dimensional singular equations.

LEMMA 6.1. *If the function $c(t)$ is continuous on Γ, then the operator*

$$\boldsymbol{S}(cu) - c\boldsymbol{S}u = \frac{1}{\pi i}\int_\Gamma \frac{c(\zeta) - c(t)}{\zeta - t}\,u(\zeta)\,d\zeta$$

is compact in $L_p(\Gamma)$.

Lemma 6.1 enables one to introduce the concept of the *symbol* of a singular operator.

6.2 *The symbol of a singular operator*

We introduce the independent variable θ which only takes the two values, $+1$ and -1. The function of two variables t and θ

$$\Phi_A(t, \theta) = a(t) + b(t)\theta \tag{6.51}$$

is called the *symbol of the singular operator* $\boldsymbol{A}$ {defined by the l.h.s. of the equation (6.50)}, and also the *symbol of the integral equation* (6.50).

Simple properties of the symbol are an immediate consequence of the definition:

1) The symbol of the unit operator is equal to unity.
2) The symbol of any compact operator is equal to zero.
3) The symbol of the sum of two singular operators is equal to the sum of their symbols.
4) Given the symbol, then the corresponding singular operator can be defined up to a compact term.

The basic property of the symbol follows from Lemma 6.1 and the Poincaré-Bertrand formula (6.37).

5) The symbol of the product of two singular operators is equal to the product of their symbols.

It follows immediately from properties 4) and 5) that the multiplication of singular operators is commutative up to a compact operator: If $\boldsymbol{A}_1$ and $\boldsymbol{A}_2$ are two singular operators, then the difference $\boldsymbol{A}_1\boldsymbol{A}_2 - \boldsymbol{A}_2\boldsymbol{A}_1$ is a compact operator. We say that the symbol is *degenerate*, if, for at least one pair of values $t \in \Gamma$ and $\theta = \pm 1$, the value of the symbol is equal to zero. If, for all $t \in \Gamma$ and $\theta = \pm 1$, the symbol differs from zero, then we say that it is *non-degenerate*. Obviously, the symbol (6.51) is non-degenerate, if and only if

$$a^2(t) - b^2(t) \neq 0, \qquad t \in \Gamma. \tag{6.52}$$

Below, in this section, we consider only operators with non-degenerate symbols.

THEOREM 6.1. *If the symbol of the operator* (6.50) *is non-degenerate, then this operator admits a double-sided regularization. The operator* $\boldsymbol{B}$, *defined by*

$$\boldsymbol{B}u = \frac{1}{a^2(t) - b^2(t)} [a(t)\, u(t) - b(t)\, \boldsymbol{S}u] \tag{6.53}$$

is both the left and the right regularizator.

Thus, the requirement of non-degeneracy of the symbol (under the assumption of continuity) is sufficient for a one-dimensional singular operator to admit a regularization.

In [9], it is proved that this requirement is also necessary: if a one-dimensional singular operator admits a regularization, then its symbol is non-degenerate.

6.3 *Nöther theorems*

From Theorem 6.1 and the results of §5, one can derive theorems established for the first time by Nöther [51]. We formulate them in the form of one theorem.

THEOREM 6.2. *Let the contour Γ be closed with continuous curvature and the symbol of the singular operator be continuous and non-degenerate. Then this operator is normally solvable and has a finite index which is independent of the compact term in the operator.*

An important addition to Theorem 6.2 is the formula for the calculation of the index which was obtained for the first time by Nöther in [51]: if the singular operator $\boldsymbol{A}$ has the form (6.50), then

$$\operatorname{Ind} A = \frac{1}{2\pi} \int_{\Gamma} d \arg \frac{a(t) - b(t)}{a(t) + b(t)}. \tag{6.54}$$

If $\operatorname{Ind} \boldsymbol{A} \geqslant 0$, then as a consequence of Theorem 5.7, $\boldsymbol{A}$ admits a left equivalent regularization. The equivalent regularizator is, in this case, the operator (6.53).

If the coefficients a and b are constant and $a^2 - b^2 \neq 0$, then the existence of a regularization leads at once to the solution of the singular equation. Let this equation have the form

$$au(t) + b(\boldsymbol{S}u)(t) = f(t).$$

Applying the regularizator

$$\frac{1}{a^2 - b^2}(a\boldsymbol{I} - b\boldsymbol{S}),$$

to both sides of this equation, we obtain the solution

$$u(t) = \frac{a}{a^2 - b^2} f(t) - \frac{b}{a^2 - b^2} (\boldsymbol{S}f)(t).$$

If a and b are constants and $a^2 - b^2 = 0$, then the homogeneous singular equation has infinitely many linearly independent solutions. In fact, after the cancellation of a (which is, certainly, required to be non-zero), the homogeneous equation takes one of the following forms:

$$u(t) - (\boldsymbol{S}u)(t) = 0, \qquad u(t) + (\boldsymbol{S}u)(t) = 0.$$

The solution of the first is an arbitrary function, which is holomorphic in D and satisfies a Lipschitz condition in $\bar{D} = D + \Gamma$. The solution of the second equation is an arbitrary function which is regular in $C\bar{D}$[6]), satisfies a Lipschitz condition in $CD = C\bar{D} \cup \Gamma$ and equals zero at infinity.

6.4 *Singular equations with Hilbert kernels*

In the general case {see Nöther [51]}, such equations take the form

$$(\boldsymbol{A}_0 u)(s) = a(s)\, u(s) + \frac{b(s)}{2\pi} \int_{-\pi}^{\pi} u(\sigma) \operatorname{ctg} \frac{\sigma - s}{2}\, d\sigma + (\boldsymbol{T}u)(s) = f(s), \qquad (6.55)$$

where $\boldsymbol{T}$ is a compact operator in $L_p(-\pi, \pi)$. For this equation, the condition (6.52) becomes

$$a^2(s) + b^2(s) \neq 0, \qquad s \in [-\pi, \pi].$$

If this condition is satisfied, then the double-sided regularizator for (6.55) is defined by

$$(\boldsymbol{B}_0 u)(s) = \frac{1}{a^2(s) + b^2(s)} \left[a(s)\, u(s) - \frac{b(s)}{2\pi} \int_{-\pi}^{\pi} u(\sigma) \operatorname{ctg} \frac{\sigma - s}{2}\, d\sigma \right]. \qquad (6.56)$$

[6]) CD is the complement of the region D with respect to the complex plane.

The index of A_0 is given by

$$\operatorname{Ind} A_0 = \frac{1}{2\pi}\int_{-\pi}^{\pi} d \arg \frac{a(s) - ib(s)}{a(s) + ib(s)}. \tag{6.57}$$

6.5 *Supplementary notes*

There is a question of practical importance: under what conditions are the solutions of (6.50) not only summable with given degree but also satisfy a Lipschitz condition with positive exponent.

Let the functions $a(t)$, $b(t)$ and $f(t)$ satisfy a Lipschitz condition with exponent α, $0 < \alpha < 1$, on the contour Γ, and the operator $\boldsymbol{T}$ have the form

$$(\boldsymbol{T}u)(t) = \int_{\Gamma} K(t, \zeta)\, u(\zeta)\, d\zeta.$$

Further, let the kernel $K(t, \zeta)$ be bounded and satisfy the inequality

$$|K(t_1, \zeta) - K(t_2, \zeta)| \leqslant N\,|t_1 - t_2|^{\alpha},$$

where N is independent of t_1, t_2 and ζ. Then, any solution of (6.50), which is summable on Γ with degree greater than unity, satisfies a Lipschitz condition with exponent α on Γ.

In the context of the present section, the fact that the Cauchy operator $\boldsymbol{S}$ satisfies the algebraic equation $\boldsymbol{S}^2 - \boldsymbol{I} = 0$ is of central importance.

Halilov [38] has considered operators of a general form (not necessarily singular) which satisfy the same equation:

$$\boldsymbol{S}^2 - \boldsymbol{I} = 0.$$

Operators satisfying the algebraic equation

$$p_0\boldsymbol{I} + p_1\boldsymbol{S} + p_2\boldsymbol{S}^2 + \ldots + p_n\boldsymbol{S}^n = 0$$

have been studied by Przeworska-Rolewicz [52], where $p_0, p_1, \ldots, p_n$ are constants and $p_n \neq 0$.

It is appropriate to pose one more problem: is Theorem 6.2 valid, if the coefficients of (6.50) are not continuous? In [14], the case is examined when Γ is the circumference of the unit circle, $\boldsymbol{T} = 0$, $a \in L_{p_1}(\Gamma)$, $b \in L_{p_1}(\Gamma)$,

$p_1 > 1$. It was shown, under these conditions, that $\boldsymbol{A}$ (equation (6.50)) maps from $L_p(\Gamma)$ into $L_{p_2}(\Gamma)$ with $p > \frac{p_1}{p_1 - 1}$ and $p_2 = \frac{pp_1}{p + p_1}$; and that $\boldsymbol{A}$ is bounded in the following sense

$$\|\boldsymbol{A}\| \leqslant \|a\|_{L_{p_1}} + \|b\|_{L_{p_1}} \cdot \|\boldsymbol{S}\|_{L_p}.$$

If the symbol of $\boldsymbol{A}$ is non-degenerate such that

$$|a(t) \pm b(t)| \geqslant M = \text{const} > 0, \tag{6.58}$$

then this operator is not normally solvable and its index can be infinite.

In Simonenko [34], a class of discontinuous coefficients for which Theorem 6.2 remains valid is examined: let the coefficients $a(t)$ and $b(t)$ be measurable and bounded and let the inequality (6.58) be satisfied. Further, for $\delta > 0$ and any point $t_0 \in \Gamma$, assume that there exists a neighbourhood of t_0 on Γ such that the variation of the function

$$\arg \frac{a(t) - b(t)}{a(t) + b(t)}$$

in this neighbourhood is less than $\pi - \delta$. If $a(t)$ and $b(t)$ satisfy all these conditions, then Theorem 6.2 remains valid for (6.50).

§7 The Carleman method for a closed contour

A remarkable method for the solution of one-dimensional singular integral equations was proposed by Carleman [47]. Carleman only considered the case when the contour Γ is an open interval on the real axis. But the method can be extended. Carleman's method was widely used and generalized in the papers of Gahov [5], Muskhelishvili, I. N. Vekua and N. P. Vekua, Kveselawa, Khvedelidze (see [4], [25], [41]) and many others. In the present section, we describe briefly the essence of the method for the case of a closed contour.

7.1 *Reduction of a singular equation to a boundary value problem*

Consider the simplest singular equation

$$a(t)\, u(t) + b(t)(\boldsymbol{S}u)(t) = f(t). \tag{6.59}$$

Suppose that its symbol is non-degenerate so that the condition (6.50) is satisfied and that the coefficients $a(t)$ and $b(t)$ satisfy a Lipschitz condition with positive exponent. We require that the r.h.s. $f \in L_p(\Gamma)$. For simplicity, we let Γ be a simply connected closed contour. Let this contour bound the region D^+ externally and the region D^- internally (see Figure 3).

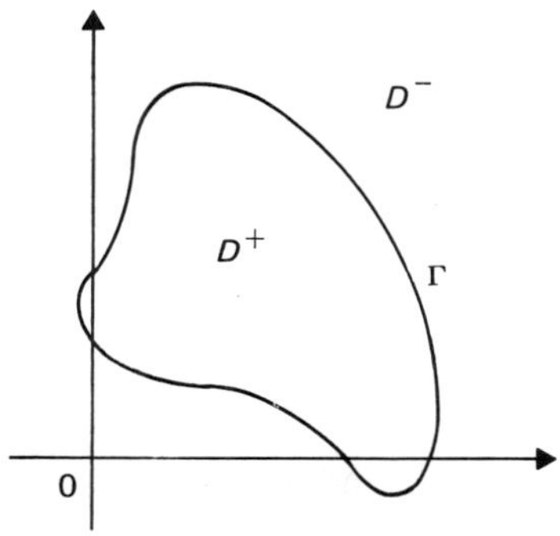

Figure 3

Note. The full solution of (6.59) by Carleman's method for the case of a simply connected closed contour was given by Gahov (see [5]). For the case of a contour bounding a multiply-connected region see [40]. The more general case when Γ is the set of a finite number of closed curves bounding no region is considered in [25]. Magnaradze [18] showed that Carleman's method is also applicable when $a(t)$ and $b(t)$ belong to the class I_∞ (see §3).

We introduce into the examination the integral of Cauchy type

$$\Phi(z) = \frac{1}{2\pi i}\int_\Gamma \frac{u(\zeta)}{\zeta - z}\,d\zeta, \qquad z \notin \Gamma. \tag{6.60}$$

Equation (6.60) defines two holomorphic functions of the complex variable z depending on whether z is in D^+ or in D^-. We denote these functions by $\Phi^+(z)$ and $\Phi^-(z)$ and note that $\Phi^-(\infty) = 0$. It follows from (6.16) that (6.59) can be rewritten as

$$\Phi^+(t) - \frac{a(t) - b(t)}{a(t) + b(t)}\Phi^-(t) = \frac{f(t)}{a(t) + b(t)}, \qquad t \in \Gamma. \tag{6.61}$$

Thus, the singular equation (6.59) is reduced to the following boundary value problem in the complex plane: find holomorphic functions $\Phi^+(z)$ and

$\Phi^-(z)$ in the regions D^+ and D^-, respectively, with their boundary values connected by a linear equation. Different authors call this problem the Riemann, Hilbert, and Riemann-Hilbert problem. In [25], it is called 'the problem of adjoints'.

7.2 *The solution of the boundary value problem*

Let us denote by m the index of (6.59) {i.e. the index of the operator defined by the l.h.s. of this equation} which is defined by (6.54). Putting the origin inside D^+, we can write

$$\ln \frac{a(t) - b(t)}{a(t) + b(t)} = \mu(t) + m \ln t, \tag{6.62}$$

where $\mu(t)$ is continuous on Γ and satisfies a Lipschitz condition with a positive exponent.

We introduce the following function of a complex variable z

$$F(z) = \Phi(z) e^{-\varphi(z)}, \qquad \varphi(z) = \frac{1}{2\pi i} \int_\Gamma \frac{\mu(\zeta)}{\zeta - z}\, d\zeta. \tag{6.63}$$

The value of $F(z)$ on D^+ and D^- will be denoted by $F^+(z)$ and $F^-(z)$. Obviously, $F^-(\infty) = 0$. Equation (6.61) may be rewritten in the form

$$\left.\begin{aligned} &F^+(t) - t^m F^-(t) = \frac{e^{-\omega(t)} t^{m/2} f(t)}{\sqrt{[a^2(t) - b^2(t)]}}, \\ &\text{where} \\ &\omega(t) = \tfrac{1}{2}(S\mu)(t) = \frac{1}{2\pi i} \int_\Gamma \frac{\mu(\zeta)}{\zeta - t}\, d\zeta. \end{aligned}\right\} \tag{6.64}$$

We now examine some properties of the index m.

1) *If* $m = 0$, *then* (6.64) *has the unique solution*

$$F(z) = \frac{1}{2\pi i} \int_\Gamma \frac{e^{-\omega(\zeta)} f(\zeta)}{\sqrt{[a^2(\zeta) - b^2(\zeta)]}} \frac{d\zeta}{\zeta - z},$$

This yields the following solution of (6.59):

$$u(t) = \frac{a(t)}{a^2(t) - b^2(t)} f(t) - \frac{b(t)e^{\omega(t)}}{\pi i \sqrt{[a^2(t) - b^2(t)]}} \int_\Gamma \frac{e^{-\omega(\zeta)} f(\zeta)}{\sqrt{[a^2(\zeta) - b^2(\zeta)]}} \frac{d\zeta}{\zeta - t}. \tag{6.65}$$

In this case, the solution of (6.59) exists and is unique.

2) *If* $m > 0$, *then a solution exists but is not unique*. We can prove this as following: we seek the solution of (6.64) such that $F^-(z) = 0(|z|^{-m-1})$ as $z \to \infty$. Putting

$$F_1(z) = \begin{cases} F^+(z), & z \text{ inside } \Gamma, \\ z^m F^-(z), & z \text{ outside } \Gamma, \end{cases}$$

we find that $F_1(z)$ satisfies the equation

$$F_1^+(t) - F_1^-(t) = \frac{e^{-\omega(t)}\, t^{m/2} f(t)}{\sqrt{[a^2(t) - b^2(t)]}} \tag{6.66}$$

and $F_1^-(\infty) = 0$. The unique solution of this problem has the form

$$F_1(z) = \frac{1}{2\pi i} \int_\Gamma \frac{e^{-\omega(\zeta)}\, \zeta^{m/2} f(\zeta)}{\sqrt{[a^2(\zeta) - b^2(\zeta)]}} \frac{d\zeta}{\zeta - z} \tag{6.66$_1$}$$

and yields a particular solution of (6.64)

$$F^+(z) = F_1^+(z), \qquad F^-(z) = z^{-m} F_1^-(z).$$

We can add to it the general solution of the corresponding homogeneous problem

$$F_0^+(t) - t^m F_0^-(t) = 0. \tag{6.67}$$

The function $F_0^-(z)$ must vanish at infinity and it is easy to show that the general solution of (6.67) has the form

$$F_0^+(z) = \sum_{k=0}^{m-1} \gamma_k z^k, \qquad F_0^-(z) = \sum_{k=0}^{m-1} \frac{\gamma_k}{z^{m-k}}, \tag{6.68}$$

where $\gamma_0, \gamma_1, \ldots, \gamma_{m-1}$ are arbitrary constants. A particular solution of (6.59) can be constructed using

$$u_0(t) = \frac{a(t)}{a^2(t) - b^2(t)} f(t)$$
$$- \frac{b(t)e^{\omega(t)}}{\pi i t^{m/2} \sqrt{[a^2(t) - b^2(t)]}} \int_\Gamma \frac{e^{-\omega(\zeta)} \zeta^{m/2} f(\zeta)}{\sqrt{[a^2(\zeta) - b^2(\zeta)]}} \frac{d\zeta}{\zeta - t}. \tag{6.69}$$

We obtain the general solution of (6.59) by adding to $u_0(t)$ an arbitrary solution of the homogeneous equation

$$a(t)\, \varphi(t) + b(t)\, (S\varphi)(t) = 0. \tag{6.70}$$

This equation has exactly m linearly independent solutions which can be obtained from (6.68). For details about their construction see [5], [25]. Denoting these solutions by $\varphi_1(t), \varphi_2(t), \ldots, \varphi_m(t)$, we obtain the general solution of (6.59) in the form

$$u(t) = u_0(t) + \sum_{k=1}^{m} c_k\, \varphi_k(t), \tag{6.71}$$

where the c_k are arbitrary constants and $u_0(t)$ is defined by (6.69).

3) *If* $m < 0$, *then the homogeneous equation* (6.70) *has only the trivial solution* $\varphi(t) \equiv 0$. *Its adjoint homogeneous equation has exactly* $|m|$ *linearly independent solutions. The equation* (6.59) *has no more than one solution, which exists if and only if the function* $f(t)$ *satisfies the following conditions:*

$$\int_\Gamma \frac{e^{-\omega(\zeta)} \zeta^{-m/2-k}}{\sqrt{[a^2(\zeta) - b^2(\zeta)]}} f(\zeta)\, d\zeta = 0, \qquad k = 1, 2, \ldots, |m|. \tag{6.72}$$

If the conditions (6.72) are satisfied then the r.h.s. of (6.68) yields the solution of (6.59). We can derive these results in the following way. Let $m < 0$ and assume a solution exists. In (6.64), we set

$$F_1^+(z) = F^+(z), \qquad F_1^-(z) = z^m F^-(z). \tag{6.73}$$

Then $F_1^-(\infty) = 0$ and (6.64) reduces to the form (6.66). This last equation has the unique solution (6.66_1). From (6.73), it follows that at infinity $F_1(z) = 0(|z|^{-|m|-1})$, and this holds if and only if the conditions (6.72) are satisfied.

The conditions (6.72) are orthogonality conditions on $f(t)$ with respect to the zeros of the operator which is adjoint to the operator of (6.59). Hence, it is easy to find an expression for these zeros. If we denote them by $\psi_k(t)$, we obtain

$$\overline{\psi_k(t)} = \frac{e^{-\omega(t)}\, t^{-m/2-k}}{\sqrt{[a^2(t) - b^2(t)]}} \frac{dt}{ds}, \qquad ds = |dt|, \qquad k = 1, 2, \ldots, |m|. \tag{6.74}$$

On the basis of the above results we obtain in particular the following result. Let the coefficients of the operator

$$Au = a(t)\, u(t) + b(t)(Su)(t)$$

with non-degenerating symbol satisfy a Lipschitz condition with a positive exponent and let Ind $A = m$. Then A has m zeros, if $m > 0$; and has no non-trivial zeros, if $m \leqslant 0$. In [11], this result is extended to the case when the coefficients $a(t)$ and $b(t)$ are only continuous.

Example 7.1. Let Γ be the circumference of $|\zeta| = 1$. Consider the equation

$$\cos \vartheta\, u(t) + \frac{\sin \vartheta}{\pi} \int_\Gamma \frac{u(\zeta)}{\zeta - t}\, d\zeta = f(t), \qquad e^{i\vartheta} = t. \tag{6.75}$$

Here

$$a(t) = \cos \vartheta = \frac{1}{2}\left(t + \frac{1}{t}\right), \qquad b(t) = i \sin \vartheta = \frac{1}{2}\left(t - \frac{1}{t}\right).$$

Applying Carleman's method, we introduce the function $\Phi(z)$ defined by (6.60). Then

$$u(t) = \Phi^+(t) - \Phi^-(t),$$

$$(Su)(t) = \frac{1}{\pi i} \int_\Gamma \frac{u(\zeta)}{\zeta - t}\, d\zeta = \Phi^+(t) + \Phi^-(t),$$

and the given equation becomes

$$t\Phi^+(t) - \frac{1}{t}\Phi^-(t) = f(t),$$

or

$$\Phi^+(t) - \frac{1}{t^2}\Phi^-(t) = \frac{1}{t}f(t). \tag{6.76}$$

The index of (6.75) is equal to

$$\frac{1}{2\pi}\int_\Gamma d\arg\frac{\cos\vartheta - i\sin\vartheta}{\cos\vartheta + i\sin\vartheta} = \frac{1}{2\pi}\int_\Gamma d\arg\frac{1}{t^2} = -\frac{1}{\pi}\int_{-\pi}^{\pi} d\vartheta = -2.$$

From the results of the present section, it follows that (6.75) has no more than one solution which exists, if and only if $f(t)$ satisfies two orthogonality conditions of the form (6.72). We verify this immediately.

In our example

$$\frac{a(t) - b(t)}{a(t) + b(t)} = \frac{1}{t^2}.$$

Formula (6.62) yields $\mu(t) = 0$ and hence $F(z) = \Phi(z)$. Further, $a^2(t) - b^2(t) = 1$ and (6.64) coincides with (6.76). We introduce the function $\Phi_1(z)$. Putting

$$\Phi_1^+(z) = \Phi^+(z), \qquad \Phi_1^-(z) = \frac{1}{z^2}\Phi^-(z),$$

we obtain from (6.76)

$$\Phi_1^+(t) - \Phi_1^-(t) = \frac{1}{t}f(t).$$

The only solution of this equation which equals zero at infinity is

$$\Phi_1(z) = \frac{1}{2\pi i}\int_\Gamma \frac{f(\zeta)}{\zeta(\zeta - z)}\,d\zeta.$$

Hence

$$\Phi^+(z) = \frac{1}{2\pi i}\int_\Gamma \frac{f(\zeta)}{\zeta(\zeta - z)}\,d\zeta, \qquad |z| < 1,$$

$$\Phi^-(z) = \frac{z^2}{2\pi i}\int_\Gamma \frac{f(\zeta)}{\zeta(\zeta - z)}\,d\zeta, \qquad |z| > 1. \tag{6.77}$$

From (6.60), it follows that $\Phi(\infty) = 0$. *A* necessary and sufficient condition for this is that the Laurent series of the function

$$\int_\Gamma \frac{f(\zeta)}{\zeta(\zeta - z)}\, d\zeta$$

in the neighbourhood of the point $z = \infty$ should begin with a term containing z^{-3}. For sufficiently large z

$$\frac{1}{\zeta - z} = -\sum_{n=0}^{\infty} \frac{\zeta^n}{z^{n+1}}.$$

For the solvability of (6.75), it is necessary and sufficient that the two conditions

$$\int_\Gamma \frac{f(\zeta)}{\zeta}\, d\zeta = 0, \qquad \int_\Gamma f(\zeta)\, d\zeta = 0 \tag{6.78}$$

be satisfied.

We shall show that these two conditions imply the orthogonality of f with respect to the zeros of the operator which is adjoint to the operator of (6.75). These zeros are the solutions of

$$\cos \vartheta\, v(t) + \frac{1}{\pi} \int_\Gamma \frac{\sin \sigma\, v(\zeta)}{\bar{t} - \bar{\zeta}} \frac{d\bar{t}}{d\vartheta}\, d\sigma = 0 \tag{6.79}$$

with $\zeta = e^{i\sigma}$. Taking complex conjugates of (6.79) and dividing by $\dfrac{dt}{d\vartheta}$, we transform (6.79) to

$$\cos \vartheta\, \overline{v(t)}\, \frac{d\vartheta}{dt} - \frac{1}{\pi} \int_\Gamma \frac{\sin \sigma\, \overline{v(\zeta)}}{\zeta - t} \frac{d\sigma}{d\zeta}\, d\zeta = 0.$$

We introduce the new unknown function

$$w(t) = \sin \zeta\, \overline{v(t)}\, \frac{d\vartheta}{dt}. \tag{6.80}$$

It follows that

$$\cos\vartheta\, w(t) - \frac{\sin\vartheta}{\pi}\int_\Gamma \frac{w(\zeta)}{\zeta - t}\,d\zeta = 0. \tag{6.81}$$

The index of (6.81) is equal to

$$\frac{1}{2\pi}\int_\Gamma d\arg\frac{\cos\vartheta + i\sin\vartheta}{\cos\vartheta - i\sin\vartheta} = \frac{1}{2\pi}\int_\Gamma d\arg e^{2i\vartheta} = 2.$$

Following Carleman's method, we put

$$\Omega(z) = \frac{1}{2\pi i}\int_\Gamma \frac{w(\zeta)}{\zeta - z}\,d\zeta = 0.$$

Equation (6.81) transforms into

$$\Omega^+(t) - t^2\Omega^-(t) = 0$$

which shows that the functions $\Omega^+(z)$ and $t^2\Omega^-(t)$ analytically continue one another through the contour Γ, and that the function $\Omega^+(z)$ can have a pole of not more than first order at infinity. In such a case {compare with (6.68)}

$$\Omega^+(z) = \gamma_0 + \gamma_1 z, \qquad \Omega^-(z) = \frac{\gamma_0}{z^2} + \frac{\gamma_1}{z},$$

where γ_0 and γ_1 are arbitrary constants. This gives

$$w(t) = \Omega^+(t) - \Omega^-(t) = \gamma_0\left(1 - \frac{1}{t^2}\right) + \gamma_1\left(t - \frac{1}{t}\right)$$

$$= 2i\gamma_0\frac{\sin\vartheta}{t} + 2i\gamma_1\sin\vartheta.$$

Thus, equation (6.81) has two linearly independent solutions

$$w_1(t) = \frac{\sin\vartheta}{t}, \qquad w_2(t) = \sin\vartheta.$$

Using (6.80), we obtain two solutions of (6.79) which is homogeneous and adjoint to (6.75):

$$v_1(t) = \frac{1}{\bar{t}}\frac{d\bar{t}}{d\vartheta}, \qquad v_2(t) = \frac{d\bar{t}}{d\vartheta},$$

It is clear that (6.78) can be rewritten in the form

$$(f, v_1) = \int_\Gamma f(\zeta)\,\overline{v_1(\zeta)}\,d\sigma = 0, \qquad (f, v_2) = \int_\Gamma f(\zeta)\,\overline{v_2(\zeta)}\,d\sigma = 0.$$

In conclusion, we derive a formula giving the solution of (6.75), if this solution exists. We have $u(t) = \Phi^+(t) - \Phi^-(t)$. According to (6.77)

$$\Phi^+(t) = \frac{f(t)}{2t} + \frac{1}{2\pi i}\int_\Gamma \frac{f(\zeta)}{\zeta(\zeta - t)}\,d\zeta,$$

$$\Phi^-(t) = -\frac{tf(t)}{2} + \frac{t^2}{2\pi i}\int_\Gamma \frac{f(\zeta)}{\zeta(\zeta - t)}\,d\zeta.$$

Hence

$$u(t) = f(t)\cos\vartheta - \frac{t\sin\vartheta}{\pi}\int_\Gamma \frac{f(\zeta)}{\zeta(\zeta - t)}\,d\zeta,$$

which is in agreement with the general formula (6.69).

Many papers are concerned with the application of Carleman's method to singular integral equations 'with shift'. In these equations, the usual singular Cauchy integral contains singular integrals of the form

$$\frac{1}{\pi i}\int_\Gamma \frac{u(\zeta)}{\zeta - \alpha(t)}\,d\zeta,$$

where the function $\alpha(t)$ is a one-to-one mapping of the contour Γ onto itself. Results concerning this are presented in detail in [5] {second edition}.

§8 Systems of singular equations defined on a closed contour

8.1 *The singular operator of a system*

We shall consider a singular system of the following form

$$\sum_{k=1}^{n} \{a_{jk}(t)\, u_k(t) + b_{jk}(t)(\boldsymbol{S}u_k)(t) + (\boldsymbol{T}_{jk}u_k)(t)\} = f_j(t), \tag{6.82}$$

$(j = 1, 2, \ldots, n)$.

Here, the $a_{jk}(t)$ and $b_{jk}(t)$ are continuous on the contour Γ. We require that the contour Γ be closed and satisfy the conditions of §2.1. As usual, we denote by $\boldsymbol{S}$ the singular Cauchy operator, and by $\boldsymbol{T}_{jk}$ a compact operator in $L_p(\Gamma)$. We assume both the functions $f_j(t)$ and $u_j(t)$ belong to the same space $L_p(\Gamma)$.

The system (6.82) can be rewritten as one equation which coincides with (6.50) in form. We introduce n-th order matrices $a(t)$, $b(t)$ and $\boldsymbol{T}$ with elements $a_{jk}(t)$, $b_{jk}(t)$ and $\boldsymbol{T}_{jk}$ $(j, k = 1, 2, \ldots, n)$, respectively, and vectors $u(t)$ and $f(t)$ with components $u_1(t), u_2(t), \ldots, u_n(t)$ and $f_1(t), f_2(t), \ldots, f_n(t)$. Now the system (6.82) can be rewritten as

$$a(t)\, u(t) + b(t)(\boldsymbol{S}u)(t) + (\boldsymbol{T}u)(t) = f(t). \tag{6.83}$$

The vectors $u(t)$ and $f(t)$ can be considered as elements of the Banach space $L_p(\Gamma)$, which consists of n-component vectors defined almost everywhere on Γ, and having components which are summable with p-th degree. The norm in $L_p(\Gamma)$ can be defined, for instance, by

$$\|u\|_{L_p(\Gamma)} = \sum_{k=1}^{n} \|u_k\|_{L_p(\Gamma)}. \tag{6.84}$$

The operator $\boldsymbol{S}$ is bounded and the operator $\boldsymbol{T}$ is compact in $L_p(\Gamma)$.

8.2 *The symbol*

We shall denote the operator, defined by the l.h.s. of (6.83), by $\boldsymbol{A}$. The matrix

$$\Phi_A(t, \theta) = a(t) + b(t)\theta, \tag{6.85}$$

where, as in §6, the independent variable θ can have only two values $+1$ and

-1, is called the *symbol matrix* or, simply, the *symbol* of A (and also the *symbol* of the equation (6.83)). The symbol matrix (6.85) retains all the properties (1)–(5) of §6, with unity (property 1) and zero (property 2) replaced by the unit and zero matrices. The multiplication of matrices is, generally speaking, not commutative, hence the assertion in §6 about the compactness of the operator $A_1A_2 - A_2A_1$ (A_1 and A_2 are singular operators) cannot be transferred to the case of matrix singular operators.

We say that the symbol (6.85) is degenerate, if at least for one pair of values $t \in \Gamma$ and $\theta = \pm 1$ the determinant of the matrix $\Phi_A(t, \theta)$ vanishes. If, for all $t \in \Gamma$ and $\theta = \pm 1$, this determinant differs from zero, then we say that the symbol (6.85) is non-degenerate.

The variable θ has only two values, so any matrix-function $\Psi(t, \theta)$ can be represented in terms of a linear function of θ

$$\Psi(t, \theta) = \alpha(t) + \beta(t)\theta.$$

If $\Psi(t, \theta)$ is a continuous function of t, then $\alpha(t)$ and $\beta(t)$ are also continuous and the matrix $\Psi(t, \theta)$ can be regarded as the symbol of some matrix singular operator.

8.3 *Nöther theorems*

If the symbol (6.85) is non-degenerate, then the operator (6.83) admits a two-sided regularization: the regularizator {simultaneously left and right} is the singular operator, the symbol matrix of which is $[\Phi_A(t, \theta)]^{-1}$. Hence, for the system (6.82) with continuous and non-degenerate symbol matrix, the Theorem of Nöther {viz., Theorem 6.2} is valid. The index of this system {i.e. the index of A of (6.83)} is calculated using

$$\begin{aligned}\operatorname{Ind} A &= \frac{1}{2\pi}\int_\Gamma d \arg \det \{[\Phi(t, +1)]^{-1} [\Phi(t, -1)]\} \\ &= \frac{1}{2\pi}\int_\Gamma d \arg \{\det [a(t) + b(t)]^{-1} \det [a(t) - b(t)]\}. \qquad (6.86)\end{aligned}$$

The Theorem of Nöther for singular systems was established for the first time by Giraud [49] and a little later by Muskhelishvili and N. Vekua [26] {see also [4] and [25]}. These authors used different arguments than that used above. In particular, in [26], an extension of Carleman's method for systems of singular equations is given. We should like to note that, in contrast with one equation, a system of the form

$$a(t)\,u(t) + b(t)(Su)(t) = f(t)$$

is solvable by quadrature methods only in exceptional cases. This is the case when the matrices $a(t)$ and $b(t)$ are rational functions of t {see [4], pp. 44–49}.

For the space $L_2(\Gamma)$, it has been proved (see [10], [12]) that the condition of the non-degeneracy of the symbol matrix is not only sufficient but also necessary in order that the system of one-dimensional singular equations should admit a regularization. The formula (6.86) was first obtained by Muskhelishvili and N. Vekua in [26]; their argument is also presented in [4] and [25]. The formula (6.86) can be deduced by simpler tools using certain topological facts. One such argument is given in [24]. The following is valid:

Let the functions $a_{jk}(t)$, $b_{jk}(t)$ and $f_j(t)$ satisfy a Lipschitz condition with exponent α, $0 < \alpha < 1$, on Γ and let $\boldsymbol{T}_{jk}$ be an integral operator of the form

$$(\boldsymbol{T}_{jk}u)(t) = \int_\Gamma K_{jk}(t, \zeta)\,d\zeta,$$

the kernels $K_{jk}(t, \zeta)$ of which are bounded and satisfy the inequality

$$|K_{jk}(t_1, \zeta) - K_{jk}(t_2, \zeta)| \leqslant N\,|t_1 - t_2|^\alpha,$$

where N is independent of t_1, t_2 and ζ. Then any solution of the system (6.82), *which is summable on Γ with degree greater than unity, satisfies a Lipschitz condition with exponent α on Γ.*

Example 8.1. We illustrate the application of Carleman's method for systems of singular equations with rational matrices $a(t)$ and $b(t)$. Let Γ be the circumference of $|\zeta| = 1$ and consider the system

$$\frac{1 + \cos\vartheta}{2}u_1(t) + \tfrac{1}{2}\sin\vartheta\,u_2(t)$$

$$+ \frac{1 - \cos\vartheta}{2\pi i}\int_\Gamma \frac{u_1(\zeta)}{\zeta - t}\,d\zeta - \frac{\sin\vartheta}{2\pi i}\int_\Gamma \frac{u_2(\zeta)}{\zeta - t}\,d\zeta = f_1(t), \quad (6.87_1)$$

$$-\frac{\sin\vartheta}{2}u_1(t)+\frac{1+\cos\vartheta}{2}u_2(t)$$
$$+\frac{\sin\vartheta}{2\pi i}\int_\Gamma\frac{u_1(\zeta)}{\zeta-t}\,d\zeta+\frac{1-\cos\vartheta}{2\pi i}\int_\Gamma\frac{u_2(\zeta)}{\zeta-t}\,d\zeta=f_2(t). \quad (6.87_2)$$

Here, as in Example 7.1, $t = e^{i\vartheta}$. For this system, we have

$$a(t)=\begin{pmatrix}\dfrac{1+\cos\vartheta}{2} & \dfrac{\sin\vartheta}{2}\\[2ex] -\dfrac{\sin\vartheta}{2} & \dfrac{1+\cos\vartheta}{2}\end{pmatrix},\qquad b(t)=\begin{pmatrix}\dfrac{1-\cos\vartheta}{2} & -\dfrac{\sin\vartheta}{2}\\[2ex] \dfrac{\sin\vartheta}{2} & \dfrac{1-\cos\vartheta}{2}\end{pmatrix}$$

and, hence,

$$a(t)+b(t)=I,\qquad a(t)-b(t)=\begin{pmatrix}\cos\vartheta & \sin\vartheta\\ -\sin\vartheta & \cos\vartheta\end{pmatrix}.$$

It is clear that the symbol matrix of the system (6.87) is non-degenerate and that for this system the Nöther theorem (viz., Theorem 6.2) is valid. In addition, (6.86) shows that the index of this system is zero and that the Fredholm theorems (viz., Theorems 1.2 and 1.4 of Chapter II) are valid.

Let us construct the solution of (6.87). We denote by u and f the vectors with components u_1, u_2, and f_1, f_2, respectively. Let us now introduce the vector valued function

$$\Phi(z)=\frac{1}{2\pi i}\int_\Gamma\frac{u(\zeta)}{\zeta-z}\,d\zeta.$$

From (6.87), it follows that $\Phi(z)$ satisfies

$$\Phi^+(t)-\delta(t)\,\Phi^-(t)=f(t),\qquad t\in\Gamma, \quad (6.88)$$

in which, for the sake of brevity, we have set

$$\delta(t)=a(t)-b(t)=\begin{pmatrix}\dfrac{1+t^2}{2t} & -\dfrac{1-t^2}{2it}\\[2ex] \dfrac{1-t^2}{2it} & \dfrac{1+t^2}{2t}\end{pmatrix}.$$

Consider, initially, the homogeneous problem when $f(t) \equiv 0$. In this case

$$\Phi^+(t) - \delta(t)\,\Phi^-(t) = 0.$$

This shows that $\Phi^+(z)$ can be extended analytically to the whole plane. It is bounded at infinity, for, if $z = \infty$, then the factor $\delta(z)$ has a pole of order one and the factor $\Phi^-(z)$ has a zero of the same order. But then, $\Phi^+(z) = \alpha =$ const. and $\Phi^-(z) = \delta^{-1}(z)\alpha$. Let us denote by α_1 and α_2 the components of the vector α. It is easy to check that

$$\delta^{-1}(z) = \begin{pmatrix} \dfrac{z^2+1}{2z} & -\dfrac{z^2-1}{2iz} \\ \dfrac{z^2-1}{2iz} & \dfrac{z^2+1}{2z} \end{pmatrix}$$

and, therefore,

$$\Phi^-(z) = \left(\frac{z^2-1}{2z}\alpha_1 - \frac{z^2-1}{2iz}\alpha_2, \quad \frac{z^2-1}{2iz}\alpha_1 + \frac{z^2-1}{2z}\alpha_2\right).$$

From the condition $\Phi^-(\infty) = 0$, it follows that

$$\alpha_1 + i\alpha_2 = 0.$$

Thus, the homogeneous system has only one non-trivial solution corresponding to the analytic vector-valued functions

$$\Phi^+(z) = (1, i), \qquad \Phi^-(z) = (1, i)\frac{1}{z}.$$

The index of the system is equal to zero and the adjoint homogeneous system has only one non-trivial solution. Hence, the non-homogeneous system (6.87) is solvable if and only if the vector $f(t)$ is orthogonal to the solution of the adjoint equation mentioned above.

We multiply both sides of (6.88) by the Cauchy kernel $\dfrac{1}{2\pi i}\dfrac{dt}{t-z}$, where $z \in D^+$, and integrate over Γ to obtain

$$\Phi^+(z) - \frac{1}{2\pi i}\int_\Gamma \frac{\delta(t)\,\Phi^-(t)}{t-z}\,dt = \frac{1}{2\pi i}\int_\Gamma \frac{f(t)}{t-z}\,dt.$$

The function $\delta(z)\,\Phi^-(z)$ is regular outside Γ, including the point at infinity, with the left integral equal to the value of this function at $z = \infty$. Hence,

$$\Phi^+(z) - \beta = \frac{1}{2\pi i}\int_\Gamma \frac{f(t)}{t - z}\,dt, \qquad z \in D^+, \tag{6.89}$$

where

$$\beta = [\delta(z)\,\Phi^-(z)]_{z=\infty}.$$

Now, we integrate (6.88), after first multiplying it by the Cauchy kernel with $z \in D^-$. This yields the new relationship

$$-\delta(z)\,\Phi^-(z) + \beta = \frac{1}{2\pi i}\int_\Gamma \frac{f(t)}{t - z}\,dt, \qquad z \in D^-.$$

Multiplying this relationship on the left by the matrix $\delta^{-1}(z)$, we find that

$$\Phi^-(z) = \delta^{-1}(z)\left[\beta - \frac{1}{2\pi i}\int_\Gamma \frac{f(t)}{t - z}\,dt\right], \qquad z \in D^-. \tag{6.90}$$

The r.h.s. of (6.90) has, in the general case, a pole of order one at infinity. The condition $\Phi^-(\infty) = 0$ forces us to require that linear terms in z should vanish. These requirements yield the relationships

$$\beta_2 = i\beta_1, \qquad \int_\Gamma [f_2(t) - if_1(t)]\,dt = 0. \tag{6.91}$$

Here, β_1 and β_2 are compon·nts of the vector β. The first of the relationships (6.91) implies that the vector β defines a solution of the corresponding homogeneous problem, the second is the orthogonality condition mentioned above to which the vector $f(t)$ must be subjected. It is not difficult to find an expression for the solution of the adjoint homogeneous equation, and also a formula defining the unknown vector $u(t)$. We leave this to the reader.

§9 The open contour case

9.1 *Example*

Consider the equation

$$\frac{1}{\pi i}\int_{-1}^{+1}\frac{u(\zeta)}{\zeta - t}\,d\zeta = f(t), \qquad -1 < t < 1. \tag{6.92}$$

We suppose that the function $f(t)$ satisfies a Lipschitz condition on the interval $[-1, 1]$. We shall require that the unknown function $u(\zeta)$ satisfy a Lipschitz condition with one and the same exponent on any interval $[-1 + \varepsilon, 1 - \varepsilon]$, where ε is positive, and be summable on the interval $-1 < t < 1$.

We solve (6.92) by Carleman's method. We set

$$\Phi(z) = \frac{1}{2\pi i}\int_{-1}^{+1}\frac{u(\zeta)}{\zeta - z}\,d\zeta, \qquad z \notin [-1, 1].$$

The function $\Phi(z)$ is holomorphic in the plane cut along the segment $[-1, +1]$ and is equal to zero at infinity. It has limit values (generally speaking, different) on the interval $(-1, +1)$ for $z \to t$ $\{t \in (-1, +1)\}$ from above and below. We denote these limit values by $\Phi^+(t)$ and $\Phi^-(t)$, respectively. The limit formulas (6.16) are valid for them. In this case, (6.92) becomes

$$\Phi^+(t) + \Phi^-(t) = f(t), \qquad -1 < t < 1.$$

We choose a branch of the function $\sqrt{(1 - z^2)}$ (for example, the one for which $\sqrt{(1 - z^2)} > 0$ on the upper side of the slit $-1 < z < 1$) and put $F(z) = \Phi(z)\sqrt{(1 - z^2)}$. Then

$$F^+(t) - F^-(t) = \sqrt{(1 - t^2)}\,f(t), \qquad -1 < t < 1.$$

The function $F(z)$ is bounded at infinity, so the general solution of the last equation is

$$F(z) = \frac{1}{2\pi i}\int_{-1}^{+1}\frac{\sqrt{(1 - \zeta^2)}\,f(\zeta)}{\zeta - z}\,d\zeta + c,$$

where c is an arbitrary constant. Hence

$$\Phi(z) = \frac{1}{2\pi i\sqrt{(1 - z^2)}} \int_{-1}^{+1} \frac{\sqrt{(1 - \zeta^2)}\, f(\zeta)}{\zeta - z} + \frac{c}{\sqrt{(1 - z^2)}}.$$

We determine the value of $u(t)$ using the formula $u(t) = \Phi^+(t) - \Phi^-(t)$:

$$u(t) = \frac{1}{\pi i\sqrt{(1 - t^2)}} \int_{-1}^{+1} \frac{\sqrt{(1 - \zeta^2)}\, f(\zeta)}{\zeta - t}\, d\zeta + \frac{C}{\sqrt{(1 - t^2)}}, \quad C = 2c. \tag{6.93}$$

Using the properties of Cauchy integrals (for details, see [25]), it is not difficult to prove that, for any value of the arbitrary constant C, the function (6.93) satisfies all the conditions imposed on it. The constant C can be chosen so that the solution satisfies either supplementary condition. For instance, if we require that $u(t)$ be bounded as $t \to 1$, then it is necessary that

$$C = \frac{1}{\pi i} \int_{-1}^{+1} \sqrt{\left(\frac{1 + \zeta}{1 - \zeta}\right)}\, f(\zeta)\, d\zeta$$

and hence,

$$u(t) = \frac{1}{\pi i} \sqrt{\left(\frac{1 - t}{1 + t}\right)} \int_{-1}^{+1} \sqrt{\left(\frac{1 + \zeta}{1 - \zeta}\right)} \frac{f(\zeta)}{\zeta - t}\, d\zeta. \tag{6.94}$$

It can be proved that the last function is actually bounded as $t \to 1$ {see [25]}. The requirement of boundedness for $t \to -1$ yields the value

$$C = -\frac{1}{\pi i} \int_{-1}^{+1} \sqrt{\left(\frac{1 - \zeta}{1 + \zeta}\right)}\, f(\zeta)\, d\zeta$$

and the solution

$$u(t) = \frac{1}{\pi i} \sqrt{\left(\frac{1 + t}{1 - t}\right)} \int_{-1}^{+1} \sqrt{\left(\frac{1 - \zeta}{1 + \zeta}\right)} \frac{f(\zeta)}{\zeta - t}\, d\zeta. \tag{6.95}$$

The Problem: find the solution of (6.92) which is bounded at both end-points $t = -1$ and $t = +1$, is solvable only when $f(t)$ satisfies certain orthogonality conditions. In this case, it is possible to use (6.94) and to require in addition that the integral in it vanish for $t = -1$:

$$\int_{-1}^{1} \frac{f(\zeta)}{\sqrt{(1 - \zeta^2)}}\, d\zeta = 0. \tag{6.96}$$

It is possible to prove that the condition (6.96) is sufficient for any of the formulas (6.93) and (6.94) to yield a solution which is bounded at both endpoints $t = +1$ and $t = -1$.

9.2 *The general case*
{see Muskhelishvili and Kveselana [27] as well as [25]}

Let Γ be a finite set of n simple open Liapunov curves. We denote the ends of these arcs, enumerated in arbitrary order, by $c_1, c_2, \ldots, c_{2n}$. Consider

$$a(t)\, u(t) + \frac{b(t)}{\pi i} \int_{\Gamma} \frac{u(\zeta)}{\zeta - t}\, d\zeta = f(t), \tag{6.97}$$

and assume that the functions $a(t)$, $b(t)$ and $f(t)$ satisfy a Lipschitz condition on Γ, with exponent $\alpha > 0$. It is also assumed that the expression $a^2(t) - b^2(t)$ does not vanish on Γ. Regarding the unknown, we require firstly that it satisfies a Lipschitz condition with positive exponent on any closed internal section of Γ and, in the neighbourhood of any of the ends c_k, it satisfies the inequality

$$|u(t)| \leqslant \frac{\text{const}}{|t - c_k|^{\beta_k}}, \qquad 0 \leqslant \beta_k < 1.$$

A function satisfying such conditions is obviously summable on Γ. More definite restrictions on the solution will be imposed below.

On the basis of Carleman's method, we introduce the function

$$\Phi(z) = \frac{1}{2\pi i} \int_{\Gamma} \frac{u(\zeta)}{\zeta - z}\, d\zeta.$$

Then (6.97) becomes

$$\Phi^+(t) - G(t)\,\Phi^-(t) = g(t), \tag{6.98}$$

where

$$G(t) = \frac{a(t) - b(t)}{a(t) + b(t)}, \qquad g(t) = \frac{f(t)}{a(t) + b(t)}.$$

Obviously,

$$G, g \in \mathrm{Lip}_\alpha(\Gamma) \quad \text{and} \quad G(t) \neq 0 \quad \text{on} \quad \Gamma.$$

A classification of the ends c_k is required for the subsequent analysis. We choose a continuous branch on Γ of the multi-valued function $\ln G(t)$ and put

$$\gamma_k = \pm \frac{1}{2\pi i} \ln G(c_k), \tag{6.99}$$

where we take a plus sign, if c_k is at the end of an arc, and a minus sign, if c_k is at the beginning, and choose the integers μ_k so that

$$-1 < \mathrm{Re}(\gamma_k) + \mu_k < 1. \tag{6.100}$$

If $\mathrm{Re}(\gamma_k)$ is an integer then μ_k is uniquely defined: $\mu_k = \mathrm{Re}(\gamma_k)$. In this case, the end c_k will be called *special*. In the other case, the end c_k will be called *non-special*. On the non-special ends, it is possible to choose μ_k so that either $-1 < \mathrm{Re}(\gamma_k) + \mu_k < 0$ or $0 < \mathrm{Re}(\gamma_k) + \mu_k < 1$. We enumerate the ends c_k so that the ends $c_1, c_2, \ldots, c_m$ are non-special and the ends $c_{m+1}, \ldots, c_{2n}$ are special. Now, we impose the following condition on the solution of (6.97): it must be bounded on a set of non-special ends $c_1, c_2, \ldots, c_q$ for $q \leqslant m$. For the problem formulated in such a manner, the following assertions are valid:

1) *Any solution of this problem {and hence, any solution of* (6.92)} *is bounded on all special ends.*
2) *When solving our problem, we must choose the μ_k so that*

$$0 < \mathrm{Re}(\gamma_k) + \mu_k < 1 \qquad (k = 1, 2, \ldots, q),$$
$$-1 < \mathrm{Re}(\gamma_k) + \mu_k < 0 \qquad (k = q + 1, q + 2, \ldots, m),$$

at special ends, since we have seen that $\mu_k = -\mathrm{Re}\,(\gamma_k)$. *Let us put*

$$\kappa = -\sum_{k=1}^{2n} \mu_k. \tag{6.101}$$

The problem formulated for the homogeneous equation

$$a(t)\, v(t) + \frac{b(t)}{\pi i} \int_\Gamma \frac{v(\zeta)}{\zeta - t}\, d\zeta = 0, \tag{6.102}$$

has κ *linearly independent solutions, if* $\kappa > 0$, *and only the trivial solution, if* $\kappa \leqslant 0$.

3) *The problem is solvable for the non-homogeneous equation* (6.97), *if and only if* $f(t)$ *satisfies certain orthogonality conditions. If these conditions are written in the form*

$$\int_\Gamma f(t)\, w_j(t)\, dt = 0, \tag{6.103}$$

then the $w_j(t)$ *are the solutions of the homogeneous equation*

$$a(t)\, w(t) - \frac{b(t)}{\pi i} \int_\Gamma \frac{w(t)}{\zeta - t}\, d\zeta = 0, \tag{6.104}$$

bounded at the non-special ends $c_{q+1}, c_{q+2}, \ldots, c_m$.

4) *The index of our problem is* κ.

The solution can be constructed in the following way: We put

$$\chi(z) = \prod_{k=1}^{2n} (z - c_k)^{\mu_k} \exp \left\{ \frac{1}{2\pi i} \int_\Gamma \frac{\ln G(\zeta)}{\zeta - z}\, d\zeta \right\}. \tag{6.105}$$

The function $\chi(z)$ is called the *canonical function of the problem*. It is bounded at the ends $c_1, c_2, \ldots, c_q$ as well as the special ends. Its limit values as well as the limit values of the function $[\chi(z)]^{-1}$ are summable on Γ. It satisfies the relationship

$$\chi^+(t) = G(t)\, \chi^-(t), \qquad t \in \Gamma.$$

Dividing (6.98) by $\chi^+(t)$, we obtain

$$\frac{\Phi^+(t)}{\chi^-(t)} - \frac{\Phi^-(t)}{\chi^-(t)} = \frac{g(t)}{\chi^+(t)},$$

and hence

$$\Phi(z) = \frac{\chi(z)}{2\pi i}\int_\Gamma \frac{g(\zeta)}{\chi^+(\zeta)}\frac{d\zeta}{\zeta + t} + \chi(z)\,Q(z), \tag{6.106}$$

where $Q(z)$ is a holomorphic function on the whole finite plane. The condition $\Phi(\infty) = 0$ leads to the conclusion that, if $\kappa > 0$, then $Q(z)$ is a polynomial of degree $\leqslant \kappa - 1$ and, if $\kappa \leqslant 0$, then necessarily $Q(z) \equiv 0$. If $Q(z)$ is defined in this manner, then, for $\kappa \geqslant 0$, (6.106) solves the problem. For $\kappa < 0$, this formula solves the problem, if and only if

$$\int_\Gamma \frac{g(\zeta)\zeta^k}{\chi^+(\zeta)}\,d\zeta = 0 \qquad (k = 0, 1, \ldots, -\kappa + 1). \tag{6.107}$$

9.3 *Supplementary notes*

We now consider the more general equation

$$a(t)\,u(t) + \frac{b(t)}{\pi i}\int_\Gamma \frac{u(\zeta)}{\zeta - t}\,d\zeta + \int_\Gamma K(t, \zeta)\,u(\zeta)\,d\zeta = f(t), \tag{6.108}$$

where $K(t, \zeta)$ satisfies a Lipschitz condition with exponent α with respect to both variables on Γ. The functions $a(t), b(t), f(t) \in \mathrm{Lip}_\alpha(\Gamma)$. As above we seek the solution which is bounded at the prescribed non-special ends $c_1, c_2, \ldots, c_q$. Transferring the second integral to the r.h.s. and applying the method presented above, we carry out the regularization of (6.108). In this case, the problem formulated for (6.108) is normally solvable and has finite index as before. It can be proved that the index of this new problem is defined by (6.101). If necessary and sufficient conditions for solvability are written in the form (6.103), then, on this occasion the functions $w_j(t)$ are the solutions of the homogeneous equation

$$a(t)\,w(t) - \frac{b(t)}{\pi i}\int_{\Gamma} \frac{w(\zeta)}{\zeta - t}\,d\zeta + \int_{\Gamma} K(\zeta, t)\,w(\zeta)\,d\zeta = 0, \tag{6.109}$$

which are bounded at the ends $c_{q+1}, \ldots, c_m$.

In Khvedelidze [41], theory analogous to that presented above is developed. In this work, the solution is constructed in the space of functions which are summable with degree $p > 1$ and with weight defined by the choice of non-special ends.

In Akhiezer [2] and Freidkin [36], cases when the contour Γ consists of a countable set of intervals of the real axes are considered.

Systems of singular integral equations with integration along an open contour are studied in the book of N. Vekua [4]. This author uses a slightly different method: after the use of Carleman's method to derive the adjoint problems, he completes Γ in an arbitrary manner to form a closed contour and on the added section of this contour, he requires that an analytic continuation for the vector $\Phi(x)$ exists. This condition takes the form

$$\Phi^{+}(t) - \Phi^{-}(t) = 0.$$

Now, the problem is transformed to the problem of conjugation with discontinuous coefficient for $\Phi^{-}(t)$ and with discontinuous free term. This last problem is the one studied by Vekua. In general, the results obtained for systems are similar to those obtained for one equation.

§10 Tricomi and Gellerstedt equations

10.1 *The formulation of the problem*

Boundary value problems for partial differential equations of mixed elliptic-hyperbolic type in gass-dynamics can sometimes be reduced to integral equations which contain, besides the usual singular integral, integrals the kernels of which are non-summable with respect to both variables and are non-singular.

The Tricomi integral equation [35] can be reduced to the form

$$u(x) - \lambda \int_{0}^{1} \left(\frac{1}{t - x} - \frac{1}{x + t - 2tx} \right) u(t)\,dt = f(x). \tag{6.110}$$

The second term $(x + t - 2tx)^{-1}$ in the kernel is continuous, if at least one of the variables x or t remains strictly inside the interval $(0, 1)$. If, however,

$x = t = 0$ or $x = t = 1$, then this term becomes infinite. Hence, it is not summable in the square $0 \leqslant x, t \leqslant 1$. The Gellerstedt equation [48] is a little more complex

$$u(x) - \lambda \int_{-1}^{1} K(x, t)\, u(t)\, dt = f(x), \tag{6.111}$$

with

$$K(x, t) = \begin{cases} \dfrac{1}{t - x} + \left(\dfrac{1 - tx_0}{x_0 - t}\right)^{\alpha} \dfrac{1}{1 - xt}, & 1 \leqslant t \leqslant x_0, \\[2ex] -\dfrac{1}{t - x} + \left(\dfrac{1 - tx_0}{t - x_0}\right)^{\alpha} \dfrac{1}{1 - xt}, & x_0 \leqslant t \leqslant 1. \end{cases} \tag{6.112}$$

Here, x_0 and α are constants such that $-1 \leqslant x_0 \leqslant 1$, $0 < \alpha < 1$. For $x_0 = 1$ the Gellerstedt equation becomes the Tricomi equation: it is sufficient to make the following change of variables

$$x_1 = \frac{x + 1}{2}, \qquad t_1 = \frac{t + 1}{2}.$$

in (6.111). Carleman's method enables (see [48]) one to solve (6.111) on a closed interval. Here, we consider the simpler equation (6.110) for which the application of Carleman's method can be simplified {[21], see also [22]}. The reader can find an examination of the general equation (6.111) in Gellerstedt [48].

We assume that $f(x)$ satisfies a Lipschitz condition with positive exponent on the interval $0 \leqslant x \leqslant 1$. The unknown function $u(x)$ will be sought in the class of functions satisfying the following conditions:
1) on any interval $a \leqslant x \leqslant b$, where $0 < a < b < 1$, $u(x)$ satisfies a Lipschitz condition with positive exponent;
2) the products $u(x) \ln x$ and $u(x) \ln (1 - x)$ are summable on the interval $0 \leqslant x \leqslant 1$.

10.2 *Reduction to a boundary value problem*

Let z be an arbitrary point of the complex plane. We put

$$\Phi(z) = \frac{1}{2\pi i} \int_{0}^{1} \left(\frac{1}{t - z} - \frac{1}{z + t - 2zt}\right) u(t)\, dt. \tag{6.113}$$

Obviously, $\Phi(z)$ is holomorphic in both the upper and lower half-planes.

We denote by $\Phi^+(x)$ and $\Phi^-(x)$ the limit value of $\Phi(z)$ as z tends to the point x on the real axes from the upper or lower half-plane, respectively. Equation (6.110) becomes

$$(1 - i\pi\lambda)\, \Phi^+(x) - (1 + i\pi\lambda)\, \Phi^-(x) = f(x), \qquad 0 < x < 1. \tag{6.114}$$

It is easily seen from (6.113) that

$$\Phi\left(\frac{z}{2z - 1}\right) = (2z - 1)\, \Phi(z). \tag{6.115}$$

The linear-fractional transformation $\zeta = \dfrac{z}{2z - 1}$ maps the upper half-plane into the lower and conversely. The interval (0, 1) transforms into the two rays $(0, -\infty)$ and $(\infty, 1)$. Replacing x by $\dfrac{x}{2x - 1}$ in (6.114) and using the relationship (6.115), we obtain

$$(1 - i\pi\lambda)\, \Phi^-(x) + (1 + i\pi\lambda)\, \Phi^+(x) = \frac{1}{2x - 1} f\left(\frac{x}{2x - 1}\right),$$
$$x < 0 \quad \text{or} \quad x > 1. \tag{6.116}$$

Now we put

$$G(x) = \begin{cases} \dfrac{1 + i\pi\lambda}{1 - i\pi\lambda}, & 0 < x < 1, \\[2ex] \dfrac{1 - i\pi\lambda}{1 + i\pi\lambda}, & x < 0 \quad \text{or} \quad x > 0; \end{cases}$$

$$h(x) = \begin{cases} \dfrac{f(x)}{1 - i\pi\lambda}, & 0 < x < 1, \\[2ex] \dfrac{f\left(\dfrac{x}{2x - 1}\right)}{(1 + i\pi\lambda)(2x - 1)}, & x < 0 \quad \text{or} \quad x > 1. \end{cases}$$

Equations (6.114) and (6.116) can be combined to form:

$$\Phi^+(x) - G(x)\, \Phi^-(x) = h(x),$$
$$-\infty < x < \infty, \qquad x \neq 0, \qquad x \neq 1. \tag{6.117}$$

10.3 *Solution of the homogeneous problem*

Initially, we solve the corresponding homogeneous problem. We seek the function $\chi(z)$ which is holomorphic in both the upper and lower half-planes, is bounded at infinity, and satisfies the equation

$$\chi^+(x) = G(x)\,\chi^-(x)$$

on the real axis, or, equivalently,

$$\ln \chi^+(x) - \ln \chi^-(x) = \ln G(x).$$

A solution of the last equation is

$$\ln \chi(x) = \frac{1}{2\pi i} \lim_{N\to\infty} \int_{-N}^{N} \frac{\ln G(x)}{t - z}\, dt.$$

The last integral can be easily evaluated: writing $\theta = 1/\pi \operatorname{arc\,tg} \lambda\pi$, $-\frac{1}{2} < \theta < \frac{1}{2}$, we obtain

$$\ln \chi^+(z) = 2\theta \ln \frac{1 - z}{z} + i\pi\theta, \qquad \ln \chi^-(z) = 2\theta \ln \frac{1 - z}{z} - i\pi\theta.$$

Here, $\ln \{(1 - z)/z\}$ means that branch of the logarithm which is holomorphic in the plane cut along the rays $(-\infty, 0)$ and $(1, \infty)$ and which has real values on the interval $(0, 1)$. Consequently

$$\chi^+(z) = e^{i\pi\theta}\left(\frac{1 - z}{z}\right)^{2\theta}, \qquad \chi^-(z) = e^{-i\pi\theta}\left(\frac{1 - z}{z}\right)^{2\theta}.$$

10.4 *Solution of the non-homogeneous problem*

Dividing (6.117) by $\chi^+(z)$, we transform it to

$$\frac{\Phi^+(z)}{\chi^+(z)} - \frac{\Phi^-(z)}{\chi^-(z)} = \frac{h(z)}{\chi^+(z)}.$$

A solution of the last equation is

$$\Phi(z) = \frac{\chi(z)}{2\pi i} \int_{-\infty}^{+\infty} \frac{h(t)}{\chi(t)} \frac{dt}{t - z}.$$

In order to find the general solution, we consider the homogeneous equation

$$\frac{\Phi_0^+(x)}{\chi^+(x)} - \frac{\Phi_0^-(x)}{\chi^-(x)} = 0.$$

This equation shows that the function $F(z) = \Phi_0(z)/\chi(z)$ is holomorphic on the whole plane, except possibly for the points $z = 0$ and $z = 1$. Assuming that the products $\varphi(x) \ln x$ and $\varphi(x) \ln (1 - x)$ are summable, we easily find that $F(z) = a/(1 - z)$, for $\theta > 0$, and $F(z) = a/z$, for $\theta < 0$, where a is an arbitrary constant.

Now, the general solution of (6.117) takes the form

$$\begin{aligned} \Phi(z) &= \frac{\chi(z)}{2\pi i}\int_{-\infty}^{+\infty}\frac{h(t)}{\chi^+(t)}\frac{dt}{t-z} + a\frac{\chi(z)}{z}, \qquad \theta < 0, \\ \Phi(z) &= \frac{\chi(z)}{2\pi i}\int_{-\infty}^{+\infty}\frac{h(t)}{\chi^+(t)}\frac{dt}{t-z} + a\frac{\chi(t)}{1-z}, \qquad \theta > 0. \end{aligned} \tag{6.118}$$

We replace the integral in (6.118) by three integrals on the intervals $(-\infty, 0)$, $(1, \infty)$ and $(0, 1)$. In the first two, we replace t by $t/(2t - 1)$, and define $u(x)$ by the formula

$$u(x) = \Phi^+(x) - \Phi^-(x), \qquad 0 < x < 1.$$

After certain essential calculations, we obtain the solution of Tricomi's equation

$$u(t) = \frac{1}{1+\lambda^2\pi^2}\left\{f(x) + \lambda\left(\frac{1-x}{x}\right)^{2\theta}\int_0^1\left(\frac{t}{1-t}\right)^{2\theta} \times \left(\frac{1}{t-x} - \frac{1}{x+t-2xt}\right)f(t)\,dt\right\} + \frac{A}{x^{1+2\theta^*}(1-x)^{-2\theta^*}},$$

where tg $\theta^*\pi = \lambda\pi$, $-1 < \theta^* < 0$, and A is an arbitrary constant.

§11 Equations with degenerate symbol

11.1 *Unbounded regularization*

One-dimensional singular equations with their symbol degenerate at a finite number of points have been examined in detail recently. The first results in this direction were obtained by Sherman [44], [45]. Using Riemann's problem as a base, this problem was examined by Chikin [43], (see also [4]) and Khvedelidze [41]. These results were essentially improved, presented from a sharper operator point of view and extended to systems of singular equations with a degenerate symbol matrix by Prößdorf [28]. In [28], it was established that singular systems of this kind can be regularized with the help of an unbounded regularizator. The properties of operators admitting an unbounded regularization were also studied there. Kosulin [15] investigated equations with degenerate symbols in spaces of generalized functions. This topic is examined in the next section. This section contains the results of [28] and [29].

Initially, we note some general results concerning operators (or equations) admitting an unbounded regularization.

We say that an operator $\boldsymbol{A}$ admits an unbounded regularization, if there exists an unbounded operator $\boldsymbol{B}$ the domain of which includes the range of $\boldsymbol{A}$ $\{$i.e. $D(\boldsymbol{B}) \supset R(\boldsymbol{A})\}$ and such that

$$\boldsymbol{B}\boldsymbol{A}u = u + \boldsymbol{T}u, \qquad u \in D(\boldsymbol{A}), \tag{6.119}$$

where $\boldsymbol{T}$ is a compact operator.

The following statements are valid:

1) *In order that a closed operator admit a bounded or unbounded regularization, it is necessary and sufficient that this operator have a finite number of zeros.*

2) *If a closed operator admits an unbounded regularization, then it is not normally solvable.*

3) *If a closed operator admits a bounded regularization, then it does not admit an unbounded regularization and conversely.*

11.2 *The general singular equation*

We now consider the singular integral equation

$$\boldsymbol{A}u = a(t)\, u(t) + b(t)\, \boldsymbol{S}u + \int_{\Gamma} K(t, \zeta)\, u(\zeta)\, d\zeta = f(t). \tag{6.120}$$

As above, we set

$$Su = \int_{\Gamma} \frac{u(\zeta)}{\zeta - t}\, d\zeta,$$

and assume that Γ is a sufficiently smooth contour dividing the complex variable plane into two regions: D^+ (internal) and D^- (external). We also assume that the functions $a(t)$, $b(t)$ and $K(t, \zeta)$ satisfy a Lipschitz condition with positive exponent on the whole contour Γ. Further, let $f \in L_p(\Gamma)$, $1 < p < \infty$. A solution u is sought in the same space. We write

$$\delta(t) = a(t) - b(t), \qquad \sigma(t) = a(t) + b(t)$$

and consider the following two cases:

I The function $\delta(t)$ has one root $t = \alpha$ on Γ with multiplicity m. The function $\sigma(t)$ does not vanish anywhere on Γ.

II The function $\delta(t)$ does not vanish anywhere on Γ. The function $\sigma(t)$ has one root β on Γ with multiplicity m. We require that, in some neighbourhood of the root $t = \alpha$ (or $t = \beta$ otherwise), the functions $a(t)$, $b(t)$ and $K(t, \zeta)$ should have m derivatives with respect to t and that the m-th derivatives should satisfy a Lipschitz condition with exponent $\lambda > (p - 1)/p$.

In I, it can be proved that there exists an unbounded regularizator

$$\boldsymbol{B}^{(1)} = \boldsymbol{B}_4^{(1)} \boldsymbol{B}_3^{(1)} \boldsymbol{B}_2^{(1)} \boldsymbol{B}_1^{(1)}, \tag{6.121}$$

where

$$\left.\begin{aligned} \boldsymbol{B}_1^{(1)} u &= a(t)\, u(t) - b(t)\, \boldsymbol{S}u, \\ \boldsymbol{B}_2^{(1)} u &= \frac{(t - \alpha)^m}{\delta(t)}\, u(t), \\ \boldsymbol{B}_3^{(1)} u &= u(t) - \sum_{k=1}^{m} \frac{1}{k!}\, u^{(k)}(\alpha)(t - \alpha)^k, \\ \boldsymbol{B}_4^{(1)} u &= \frac{u(t)}{(t - \alpha)^m\, \sigma(t)}. \end{aligned}\right\} \tag{6.122}$$

In II, there also exists an unbounded regularization $\boldsymbol{B}^{(2)}$ obtained from $\boldsymbol{B}^{(1)}$ by replacing α with β and interchanging the functions $\delta(t)$ and $\sigma(t)$:

$$\boldsymbol{B}^{(2)} = \boldsymbol{B}_4^{(2)} \boldsymbol{B}_3^{(2)} \boldsymbol{B}_2^{(2)} \boldsymbol{B}_1^{(2)}, \tag{6.123}$$

where

$$\left.\begin{aligned} \boldsymbol{B}_1^{(2)} &= \boldsymbol{B}_1^{(1)}, \qquad \boldsymbol{B}_2^{(2)} = \frac{(t-\beta)^m}{\sigma(t)}\boldsymbol{I}, \\ \boldsymbol{B}_3^{(2)}u &= u(t) - \sum_{k=1}^{m} \frac{1}{k!} u^{(k)}(\beta)(t-\beta)^k, \\ \boldsymbol{B}_4^{(2)}u &= \frac{u(t)}{(t-\beta)^m\,\delta(t)}. \end{aligned}\right\} \tag{6.124}$$

Since an unbounded regularizator exists, the homogeneous equation for both I and II {obtained from (6.120) with $f(t) \equiv 0$} will only have a finite number of linearly independent solutions. This assertion is also valid for the adjoint equations. For the non-homogeneous equation (6.120), the orthogonality of $f(t)$ to the solutions of the homogeneous adjoint equation ceases to be a sufficient condition for solvability, but remains a necessary condition.

11.3 *Systems of singular equations*

Let $a(t)$, $b(t)$ and $K(t, \zeta)$ in (6.120) be quadratic matrices of order n, and write

$$\left.\begin{aligned} c(t) &= a(t) + b(t), \qquad & d(t) &= a(t) - b(t), \\ \sigma(t) &= \det c(t), & \delta(t) &= \det d(t). \end{aligned}\right\} \tag{6.125}$$

As before we consider the two cases I and II which can be formulated in the notation of (6.125) in the same manner as for one equation. We assume also that the matrices $a(t)$, $b(t)$ and $K(t, \zeta)$ satisfy the smoothness conditions formulated above for the corresponding coefficients. Then, in I and II, there exists an unbounded regularizator $\boldsymbol{B}^{(1)}$ and $\boldsymbol{B}^{(2)}$, respectively: as above, they are defined by (6.121) and (6.123) where we now have

$$\left.\begin{aligned} \boldsymbol{B}_1^{(1)}u &= a(t)\,\tilde{d}(t)\,u(t) - b(t)\,\tilde{d}(t)(\boldsymbol{S}u)(t), \\ \boldsymbol{B}_2^{(1)}u &= \frac{(t-\alpha)^m}{\delta(t)}\,u(t), \\ \boldsymbol{B}_3^{(1)}u &= u(t) - \sum_{k=1}^{m} \frac{1}{k!} u^{(k)}(\alpha)(t-\alpha)^k, \\ \boldsymbol{B}_4^{(1)}u &= \frac{c^{-1}(t)}{(t-\alpha)^m}\,u(t). \end{aligned}\right\} \tag{6.126}$$

The sign '$\tilde{}$' denotes the adjoint matrix, i.e. the transposed matrix of cofactors of the elements of the given matrix. In addition,

$$\left.\begin{aligned} &\boldsymbol{B}_1^{(2)}u = a(t)\,\tilde{c}(t)\,u(t) - b(t)\,\tilde{c}(t)(\boldsymbol{S}u)(t), \\ &\boldsymbol{B}_2^{(2)}u = \frac{(t-\beta)^m}{\sigma(t)}\,u(t), \qquad \boldsymbol{B}_3^{(2)}u = u(t) - \boldsymbol{P}u, \\ &\boldsymbol{B}_4^{(2)}u = \frac{d^{-1}(t)}{(t-\beta)^m}\,u(t). \end{aligned}\right\} \tag{6.127}$$

The operator $\boldsymbol{P}$ in the expression for $\boldsymbol{B}_3^{(2)}$ is defined in the following way: we denote by γ_k the coefficients defined by the relations

$$\left[\frac{d^\nu}{dt^\nu}\sum_{k=1}^{m}\frac{\gamma_k}{t^k}\right]_{t=a} = u^{(\nu)}(\alpha), \qquad \nu = 1, 2, \ldots, m.$$

Then

$$(\boldsymbol{P}u)(t) = \sum_{k=1}^{m}\frac{\gamma_k}{t^k}.$$

The qualitative results formulated above for one equation, which follow from the existence of an unbounded regularizator, remain valid for systems.

By applying the unbounded regularizator $\boldsymbol{B}^{(1)}$ {or $\boldsymbol{B}^{(2)}$} to both sides of (6.120), we reduce this equation to an equation of Fredholm types. The problem of the equivalence of both equations is studied in [25] and [45].

§12 Singular equations in generalized function spaces {for information on the theory of generalized functions see [7], [8], [42], [46], [53], and [54]}

In Kosulin [15], Rogozhin [32], [33] and others, one-dimensional singular equations in the spaces of distributions [6] are studied. It is worth noting that recently, not one-dimensional, but multi-dimensional singular equations were examined in the spaces of generalized functions. The first work of this kind appears to have been done by Horváth [50]. In the present section, we present the results of [15] for they are the most complete.

§12 *Singular equations in generalized function spaces*

12.1 *Equations with a non-degenerate symbol*

We consider the singular equation of the form

$$a(t)\, u(t) + b(t)(Su)(t) = f(t) \tag{6.128}$$

under the supposition that Γ is a closed infinitely differentiable contour dividing the plane into two regions: D^+ (internal) and D^- (external). As the basic space, we take the countably-normalized space Φ of infinitely differentiable functions on Γ. A system of norms is introduced into Φ as follows: if $u \in \Phi$ then

$$\|u\|_k = \max_{t \in \Gamma} \{|\varphi(t)|, |\varphi'(t)|, \ldots, |\varphi^{(k)}(t)|\} \qquad (k = 0, 1, 2, \ldots). \tag{6.129}$$

Below we shall assume that $a, b \in \Phi$.

We initially consider the case when $f \in \Phi$ and the symbol of the equation is non-degenerate. Then it is natural to consider the equation (6.128) in the space of basis functions. In this space, the Nöther theorem (see §6) is valid. What is more, the homogeneous equation

$$a(t)\, u(t) + b(t)(Su)(t) = 0 \tag{6.130}$$

has m linearly independent solutions in Φ, if $m > 0$, and only the trivial solution, if $m \leqslant 0$. Here, m is the index of the equation (6.128) defined by (6.54). The solutions of (6.128) and (6.130), belonging to the space Φ, will be referred to as classical.

We denote by Φ' the space of generalized functions, i.e. the space of linear functionals on Φ, which are continuous in the topology of (6.129). Suppose now that $f \in \Phi'$. Then the unknown function $u(t)$ also belongs to the space Φ'. The fact that the generalized function $u(t)$ satisfies the equation (6.128) implies: whatever be the function $\varphi \in \Phi$, the following identity is valid

$$\left(u, \overline{a(t)}\, \varphi(t) - \frac{1}{\pi i} \int_\Gamma \frac{\varphi(\zeta)}{\bar{t} - \bar{\zeta}}\, d\bar{\zeta} \right) = (f, \varphi). \tag{6.131}$$

The round brackets denote the value of the given functional on the given function: thus, (f, φ) is the value of the functional f on the function φ.

We retain for the time being the assumption that *the symbol of the*

equation is non-degenerate. Then the following statements are valid:
1) *The equation* (6.130) *as well as its adjoint homogeneous equation has only classical solutions in the space* Φ'.
2) *In order that the equation* (6.128) *be solvable in* Φ', *it is necessary and sufficient that* $f(t)$ *should be orthogonal to all the solutions of the homogeneous equation conjugate to* (6.130).

12.2 *Equations with a degenerate symbol*

We now assume that the symbol of the equation is degenerate and that cases I or II of the preceding section hold. Equation (6.130) as well as its adjoint homogeneous equation can have both classical, and generalized function solutions, which are functionals from Φ'. The number of both types of solutions is finite.

Let the Fredholm equation obtained by the application of an unbounded regularizator $\boldsymbol{B}$ to (6.130) have the form

$$u(t) + \int_{\Gamma} P(t, \zeta)\, u(\zeta)\, d\zeta = 0. \tag{6.132}$$

Consider the equation

$$v(t) + \int_{\Gamma} \overline{P(\zeta, t)}\, v(\zeta)\, d\zeta = 0, \tag{6.133}$$

which is adjoint to the equation (6.132). This equation has $p > 0$ solutions $\tilde{v}_j(t)$ which do not belong to the range of definition of the operator $\boldsymbol{B}^*$ adjoint to the regularization $\boldsymbol{B}$. We denote by $v_j(t)$, $j = 1, 2, \ldots, s$, the classical solutions of the adjoint homogeneous equation of (6.130).

The following statements are valid:
1) Let $f \in \Phi$. In order that (6.128) be solvable, it is necessary and sufficient that

$$\left.\begin{aligned} (f, v_j) &= 0, \qquad j = 1, 2, \ldots, s,\\ (\boldsymbol{B}f, \tilde{v}_j) &= 0, \qquad j = 1, 2, \ldots, p. \end{aligned}\right\} \tag{6.134}$$

2) Let $f \in \Phi'$. In order that (6.128) be solvable in Φ', it is necessary and sufficient that

$$(f, v_j) = 0, \qquad j = 1, 2, \ldots, s. \tag{6.135}$$

Chapter VII

THE INTEGRAL EQUATIONS OF MATHEMATICAL PHYSICS[1])

§1 The integral equations of potential theory

1.1 *The integral equations of the Dirichlet and Neumann problems, for simply connected boundaries*

This section deals with the integral equations of the commonest problems of potential theory (viz., the Dirichlet and Neumann problems), when the bounded or unbounded domain, in which the problem is solved, lies in three-dimensional Euclidean space and is bounded by a simply connected closed surface S satisfying the Liapunov conditions. These conditions are:

1) At every point of S, there exists a well-defined tangent plane and normal.

2) If θ is the angle between the normals at the points M_1 and M_2 of S, and r is the distance between these points, then there exist two numbers $A > 0$ and α $(0 < \alpha \leqslant 1)$ such that the following inequality holds

$$|\theta| \leqslant Ar^{\alpha}.$$

3) There exists a number d, which is the same for all points of S and possesses the property: parallels to the normal at any point M of S cut that part of the surface, which lies inside the sphere with radius d and centre M, not more than once.

We denote by Ω_i the finite region lying inside S and by Ω_e its complement. We recall that the *Dirichlet problem* is: Find the harmonic function which takes given specified values on the boundary:

$$u|_S = f(M). \tag{7.1}$$

[1]) The application of integral equations to the problems of mathematical physics, hydrodynamics and the theory of elasticity are presented in the books [2]–[14] which differ in their completeness and the extent of their treatment of this topic. In this chapter, we rely mostly on the books [4], [6] and [7].

If the problem is solved in the region Ω_i, then it is called *the interior Dirichlet problem*, and if in Ω_e, then *the exterior Dirichlet problem*.

If, on the boundary S, the normal derivative of a function S, viz.

$$\left.\frac{\partial u}{\partial n}\right|_S = \varphi(M), \tag{7.2}$$

is given, and the harmonic function u is sought in the region Ω_i which satisfies it, then this problem is called the *internal Neumann problem*. When such a harmonic function u is required in Ω_e, this problem is referred to as the *external Neumann problem*.

We also recall that a function $u(M)$ is harmonic in the infinite region Ω_e, if it is harmonic in any finite region contained in Ω_e, and $u(M) \to 0$ as M tends to infinity.

We denote by $V(M)$ the *simple layer potential*

$$V(M) = \int_S \frac{\rho(M_1)}{r} \, dS_1, \tag{7.3}$$

where r is the distance between the points M and M_1, dS_1 is an element of the surface S, and $\rho(M_1)$ is a continuous density distributed over the surface S. We always require that the vector $\boldsymbol{r} = \overline{MM_1}$ is directed from the point M to the point M_1.

By $W(M)$, we denote the *double layer potential*

$$W(M) = \int_S \sigma(M_1) \frac{\partial(1/r)}{\partial n_1} \, dS_1, \tag{7.4}$$

where n_1 is the outward normal to the surface S at the point M_1 and $\sigma(M)$ is a continuous density distributed over the surface S.

The potentials (7.3) and (7.4) are continuous harmonic functions both in Ω_i and Ω_e. The potential of a simple layer (7.3) is continuous across the surface, but the potential of a double layer is discontinuous. We denote the limiting values of $W(M)$ on the surface S from inside and outside of S by W_i and W_e, respectively. Then the following relationships hold

$$W_i(M) = -2\pi\sigma(M) - \int_S \sigma(M_1) \frac{\cos(r, n_1)}{r^2} \, dS_1, \tag{7.5}$$

$$W_e(M) = 2\pi\sigma(M) - \int_S \sigma(M_1)\frac{\cos(r, n_1)}{r^2}\,dS_1, \tag{7.6}$$

where M is an arbitrary point of S.

The term

$$\overline{W(M)} = -\int_S \sigma(M_1)\frac{\cos(r, n_1)}{r^2}\,dS_1 \tag{7.7}$$

in (7.5) and (7.6) is called the *principal value of the potential* $W(M)$. It is obtained by the direct substitution of the coordinates of M on S into (7.4).

Let $\partial V_i/\partial n$ and $\partial V_e/\partial n$ be the normal derivatives of the potential of a simple layer calculated across the boundary S from inside and outside of S, respectively. These values are defined by

$$\frac{\partial V_i}{\partial n} = 2\pi\rho(M) + \int_S \rho(M_1)\frac{\cos(r, n)}{r^2}\,dS_1, \tag{7.8}$$

$$\frac{\partial V_e}{\partial n} = -2\pi\rho(M) + \int_S \rho(M_1)\frac{\cos(r, n)}{r^2}\,dS_1. \tag{7.9}$$

The expression

$$\frac{\partial V}{\partial n} = \int_S \rho(M_1)\frac{\cos(r, n)}{r^2}\,dS_1 \tag{7.10}$$

is called the *principal value of the normal derivative* of the potential of a simple layer.

With the help of (7.5)–(7.9) the Dirichlet and Neumann problems can be reduced to Fredholm integral equations of the second kind.

The solution of the Dirichlet problem is sought in the form of the potential $W(M)$ of a double layer. In the case of the internal problem, the boundary condition takes the form

$$W_i(M) = f(M),$$

where $f(M)$ is a given continuous function of M on S. Using (7.5), we obtain the following integral equation for the density $\sigma(M)$:

$$\sigma(M) + \frac{1}{2\pi}\int_S \sigma(M_1)\frac{\cos(r, n_1)}{r^2}\,dS_1 = -\frac{1}{2\pi}f(M). \tag{7.11}$$

Solving this equation, we find the potential $W(M)$ from (7.4).

In the case of the external problem, the boundary condition yields

$$\sigma(M) - \frac{1}{2\pi}\int_S \sigma(M_1)\frac{\cos(r, n_1)}{r^2}\,dS_1 = \frac{1}{2\pi}f(M). \tag{7.12}$$

We could also consider the more general equation

$$\sigma(M) - \frac{\lambda}{2\pi}\int_S \sigma(M_1)\frac{\cos(r, n_1)}{r^2}\,dS_1 = \pm\frac{1}{2\pi}f(M), \tag{7.13}$$

where λ is a parameter.

We now examine the Neumann problem. The solution of this problem is sought in the form of the potential of a simple layer (7.3). In the case of the internal problem, the following condition must be satisfied:

$$\frac{\partial V_i}{\partial n} = g,$$

where g is a given continuous function. In the case of the external problem, the boundary condition will take the form

$$\frac{\partial V_e}{\partial n} = g.$$

Using (7.8) and (7.9), we obtain that the density $\rho(M)$ of the potential has to satisfy the integral equation

$$\rho(M) + \frac{1}{2\pi}\int_S \rho(M_1)\frac{\cos(r, n)}{r^2}\,dS_1 = \frac{1}{2\pi}g(M) \tag{7.14}$$

in the case of the internal problem; and has to satisfy

$$\rho(M) - \frac{1}{2\pi}\int_S \rho(M_1)\frac{\cos(r,n)}{r^2}\,dS_1 = -\frac{1}{2\pi}\,g(M) \tag{7.15}$$

in the case of the external problem (n is the normal to S at the point M and r is the distance between M and M_1).

We could also consider the more general equation with a parameter λ:

$$\rho(M) + \frac{\lambda}{2\pi}\int_S \rho(M_1)\frac{\cos(r,n)}{r^2}\,dS_1 = \pm\frac{1}{2\pi}\,g(M). \tag{7.16}$$

It can be shown that, if M and M_1 belong to S, then the kernels of (7.13) and (7.16) satisfy the inequalities

$$\left|\frac{\cos(n_1,r)}{r^2}\right| < \frac{c}{r^{2-\alpha}}, \qquad \left|\frac{\cos(r,n)}{r^2}\right| < \frac{c}{r^{2-\alpha}},$$

where $\alpha > 0$ is the constant in the Liapunov conditions. The last inequalities show that the kernels (7.13) and (7.16) have a weak singularity and hence that the Fredholm theory is applicable {see Chapter II, §6}.

The equation (7.16) is adjoint to (7.13). Thus, the external Neumann problem and the internal Dirichlet problem can be described as adjoint integral equations. An analogous assertion holds for the internal Neumann problem and the external Dirichlet problem.

It can be proved that $\lambda = -1$ is not a characteristic value of (7.13) and (7.16); i.e., the homogeneous equations

$$\sigma(M) + \frac{1}{2\pi}\int_S \sigma(M_1)\frac{\cos(r,n_1)}{r^2}\,dS_1 = 0 \tag{7.11$_1$}$$

and

$$\rho(M) - \frac{1}{2\pi}\int_S \rho(M_1)\frac{\cos(r,n)}{r^2}\,dS_1 = 0 \tag{7.15$_1$}$$

have only trivial (zero) solutions.

Thus, the integral equations of the internal Dirichlet problem and the external Neumann problem are solvable for any continuous boundary functions $f(M)$ and $g(M)$.

In the case when $\lambda = 1$ (the internal Neumann problem and the external Dirichlet problem), the homogeneous equations

$$\sigma(M) - \frac{1}{2\pi}\int_S \sigma(M_1)\frac{\cos(r, n_1)}{r^2}\,dS_1 = 0, \tag{7.12$_1$}$$

$$\rho(M) + \frac{1}{2\pi}\int_S \rho(M_1)\frac{\cos(r, n)}{r^2}\,dS_1 = 0, \tag{7.14$_1$}$$

have non-trivial solutions, i.e. the value $\lambda = 1$ will be a characteristic value of (7.13) and (7.16). The only non-trivial solution of the homogeneous equation (7.12$_1$) is a constant. According to the Fredholm theory for the solvability of the adjoint homogeneous equation (7.14), it is necessary and sufficient that

$$\int_S g(M_1)\,dS_1 = 0. \tag{7.17}$$

Thus, for the solvability of the internal Neumann problem, it is necessary and sufficient that the boundary function $g(M)$ satisfy (7.17).

The question of the solvability of the external Dirichlet problem will be considered in §1.3.

1.2 *The Robin problem*

We consider in detail the equation

$$\rho(M) + \frac{1}{2\pi}\int_S \rho(M_1)\frac{\cos(r, n)}{r^2}\,dS_1 = 0. \tag{7.14$_1$}$$

As shown above, this equation has non-trivial solutions.

The problem of their determination is called the *Robin problem*. We assume as above that the region Ω is bounded by a closed surface satisfying the Liapunov condition.

Except for a multiplicative constant, the equation (7.14$_1$) has only one non-trivial solution $\rho_0(M)$.

The function $\rho_0(M)$ can be constructed by the method of successive approximations.

Let $f(M)$ be an arbitrary continuous function defined on S. We set $\rho_0^{(0)}(M) = f(M)$ and define $\rho_0^{(1)}(M)$ by

$$\rho_0^{(1)}(M) = -\frac{1}{2\pi}\int_S \rho_0^{(0)}(M_1)\frac{\cos(r, n)}{r^2}\, dS_1.$$

Analogously,

$$\rho_0^{(n)}(M) = -\frac{1}{2\pi}\int_S \rho_0^{(n-1)}(M_1)\frac{\cos(r, n)}{r^2}\, dS_1 \qquad (n = 1, 2, \ldots).$$

The sequence $\{\rho_0^{(K)}(M)\}$ converges uniformly to the continuous function $\rho_0(M)$ satisfying (7.14_1), with

$$\int_S \rho_0(M)\, dS = \int f(M)\, dS.$$

1.3 *The external Dirichlet problem*

We now return to the integral equation of the external Dirichlet problem

$$\sigma(M) - \frac{1}{2\pi}\int_S \sigma(M_1)\frac{\cos(r, n_1)}{r^2}\, dS_1 = \frac{1}{2\pi} f(M), \tag{7.12}$$

which is a special case of

$$\sigma(M) - \frac{\lambda}{2\pi}\int_S \sigma(M_1)\frac{\cos(r, n_1)}{r^2}\, dS_1 = \frac{1}{2\pi} f(M) \tag{7.13}$$

with $\lambda = 1$.

Since the homogeneous equation corresponding to (7.12) has a non-trivial solution $\sigma(M) = c$, it is necessary and sufficient, for the solvability of (7.12), that its r.h.s. be orthogonal to the eigenfunctions of (7.14_1).

In §1.2, a method for determining the eigenfunctions of (7.14_1) has been described. Let $\rho_0(M)$ be such a function. Then, for the solvability of (7.12), it is necessary and sufficient that

$$\int_S f(M)\, \rho_0(M)\, dS = 0 \tag{7.18}$$

be valid.

It is not difficult to understand the reason why (7.12) is not always solvable. The solution of the external Dirichlet problem in the form of the potential of a double layer (7.4) decreases near infinity like $1/R^2$, where R is the distance from the origin to the point M, whereas it is known that an arbitrary harmonic function decreases near infinity like $1/R$.

It is possible to obtain an integral equation which has a solution for arbitrary $f(M)$. In this case, we seek the solution as

$$w^*(M) = W(M) + \frac{1}{R}\int_S \sigma(M_1)\, dS_1, \tag{7.19}$$

where R is the distance from the origin inside S to the point M.

The boundary condition

$$w_e^* = f(M)$$

is satisfied, if $\sigma(M)$ is the solution of

$$\sigma(M) - \frac{1}{2\pi}\int_S \sigma(M_1)\frac{\cos(r, n_1)}{r^2}\, dS_1 + \frac{1}{2\pi R}\int_S \sigma(M_1)\, dS_1 = \frac{1}{2\pi} f(M).$$

This equation appears to have a solution for any continuous $f(M)$.

Thus, a solution of the external Dirichlet problem can be found in the extended form (7.19).

1.4 *The case of a disconnected boundary*

Let the boundary of Ω consist of several closed surfaces satisfying Liapunov conditions (see §1.1).

Consider firstly the case of a finite region outwardly bounded by the surface S_0 and inwardly by the surfaces $S_1, S_2, \ldots, S_l$.

For this type of region, the external problems of Dirichlet and Neumann reduce to l internal problems for the regions bounded by the surfaces $S_1, S_2, \ldots, S_l$ and an external problem for the surface S_0. As before, the whole boundary of Ω is denoted by S. The solution of the internal Neumann problem is sought in the form of the potential of a simple layer (7.3). Since (7.8) and (7.9) do not depend on the fact that S is simply or multiply connected, the integral equation for the density

$$\rho(M) + \frac{1}{2\pi}\int_S \rho(M_1)\frac{\cos(r, n)}{r^2}\, dS_1 = \frac{1}{2\pi}\, g(M) \tag{7.14}$$

must also be satisfied for a multi-connected region. For the solvability of this equation, it is necessary and sufficient that the condition

$$\int_S g(M)\, dS = 0$$

be satisfied. This condition does not differ from (7.17). For the case of a multi-connected region, we note that for the last result we have to integrate over the totality of surfaces $S_0, S_1, \ldots, S_l$.

We now turn to the internal Dirichlet problem. If the solution is sought in the form of the potential of a double layer (7.4), then we obtain the integral equation

$$\sigma(M) + \frac{1}{2\pi}\int_S \sigma(M_1)\frac{\cos(r, n_1)}{r^2}\, dS_1 = -\frac{1}{2\pi}\, f(M), \tag{7.11}$$

which was considered for the case of a simply connected region.

This equation does not have a solution for arbitrary continuous $f(M)$. Since the adjoint homogeneous equation

$$\rho(M) - \frac{1}{2\pi}\int_S \rho(M_1)\frac{\cos(r, n)}{r^2}\, dS_1 = 0 \tag{7.15$_1$}$$

has non-trivial solutions, it is necessary and sufficient, for the solvability of (7.11), that the r.h.s. of (7.11) be orthogonal to the eigenfunctions of (7.14).

We denote by $\rho_k(M)$ $(k = 1, \ldots, l)$ the particular non-zero solution of

$$\rho(M) - \frac{1}{2\pi}\int_{S_k} \rho(M_1)\frac{\cos(r, n)}{r^2}\, dS_1 = 0 \qquad (k = 1, \ldots, l).$$

More concisely, we denote by $\rho_k(M)$ $(k = 1, \ldots, l)$ the solutions of the Robin problem for the contour S_k.

For the solvability of (7.11) and, hence, the representation of the solution of the Dirichlet problem for the region in question in the form of the potential of a double layer (7.4), it is necessary and sufficient that the conditions

$$\int_{S_k} f(M)\,\rho_k(M)\,dS = 0 \qquad (k = 1, \ldots, l) \tag{7.20}$$

be satisfied. If the contour S_0 of the region Ω is absent (i.e., Ω is infinite), then (7.20) is also a necessary and sufficient condition for (7.11) to have a solution.

1.5 *Mixed problems in potential theory*

We seek the function $u(x)$ which is harmonic inside Ω and satisfies on S the boundary condition

$$\left.\frac{\partial u}{\partial n} + p(M)\,u\right|_S = f(M), \tag{7.21}$$

where $p(M)$ and $f(M)$ are given continuous functions on the boundary S. This problem is called the *mixed problem* of potential theory.

Let

$$p(M) > 0 \tag{7.22}$$

and S be a closed surface satisfying Liapunov conditions (see §1.1).

The solution of the problem is sought in the form of the potential of a simple layer (7.3). By means of (7.8) and the condition (7.21), we obtain for the density $\rho(M)$ the following integral equation

$$\rho(M) + \frac{1}{2\pi}\int_S \rho(M_1)\left[\frac{\cos(r, n)}{r^2} + p(M)\frac{1}{r}\right] dS_1 = \frac{1}{2\pi} f(M), \tag{7.23}$$

which appears to have a solution for arbitrary r.h.s., if the inequality (7.22) is satisfied. As before, we suppose that the vector $\boldsymbol{r}$ is directed from the point M_1 to the point M. If $p(M)$ does not satisfy the inequality (7.22), then generally speaking (7.23) does not have a solution.

1.6 *The distribution of the characteristic values of the integral equations of potential theory*

Consider the equations

$$\sigma(M) - \frac{\lambda}{2\pi} \int_S \sigma(M_1) \frac{\cos(r, n_1)}{r^2} dS_1 = 0, \tag{7.24}$$

$$\rho(M) + \frac{\lambda}{2\pi} \int_S \rho(M_1) \frac{\cos(r, n)}{r^2} dS_1 = 0, \tag{7.25}$$

which are the homogeneous versions of (7.13) and (7.16).

It can be proved that the characteristic values of these equations are real and lie on the rays $\lambda \geqslant 1$ and $\lambda < -1$ for regions bounded by simply connected surfaces of Liapunov type. For multi-connected regions, $\lambda = -1$ can be a characteristic value.

In the case of (7.24), an eigenfunction equal to a constant corresponds to the characteristic value $\lambda = 1$ for regions bounded by a simply-connected surface. In the case of (7.25) eigenfunctions, corresponding to the value $\lambda = 1$, will be the solutions of the Robin problem.

1.7 *Extension to multi-dimensional spaces*

The notions and methods of potential theory {presented in §§1.1–1.6} can be generalized to the case of m-dimensional spaces, $m > 3$. The function $u(M)$ is called *harmonic* in an infinite region of m-dimensional space, if it satisfies the Laplace equation and decreases like $1/R^{m-2}$ near infinity (R is the distance from the origin to the point M). Let S be a surface satisfying the Liapunov conditions, M an arbitrary point of the space, M_1 a point on S, and r be the distance between M and M_1.

The integrals

$$V(P) = \int_S \rho(M_1) \frac{dS_1}{r^{m-2}}$$

and

$$W(P) = \int_S \sigma(M_1) \frac{\partial(1/r^{m-2})}{\partial n_1} dS_1,$$

where n_1 is the outward normal at the point M_1, are called the *potentials of a simple and double layer*, respectively. Their densities $\rho(M_1)$ and $\sigma(M_1)$ are required to be continuous.

If the solution of the Dirichlet problem is sought in the form of the potential of a double layer, then we obtain the integral equations

$$\sigma(M) \pm \frac{\Gamma(m/2)}{\pi^{m/2}} \int_S \sigma(M_1) \frac{\cos(r, n_1)}{r^{m-1}} dS_1 = \mp \frac{\Gamma(m/2)}{\pi^{m/2}} f(M),$$

where M, S and Γ is the Euler integral of the second kind. The upper sign corresponds to the internal problem and the lower to the external.

The solution of the Neumann problem is sought in the form of the potential of a simple layer. Then the integral equation for the density will be

$$\rho(M) \pm \frac{\Gamma(m/2)}{\pi^{m/2}} \int_S \rho(M_1) \frac{\cos(r, n)}{r^{m-1}} dS_1 = \pm \frac{\Gamma(m/2)}{\pi^{m/2}} \varphi(M),$$

where n is the direction of the inward normal at M. The upper sign corresponds to the internal problem, the lower sign to the external one.

An examination of these equations shows that the internal Dirichlet problem and the external Neumann problem always have a solution. For the solvability of the internal Neumann problem, a necessary and sufficient condition is

$$\int_S \varphi(M_1)\, dS_1 = 0.$$

The solution of the external Dirichlet problem can be sought in the form

$$u(M) = \int_S \sigma(M_1) \frac{\partial(1/r^{m-2})}{\partial n_1} dS_1 + \frac{1}{R^{m-2}} \int_S \sigma(M_1)\, dS_1,$$

where R is the distance from some fixed point M_0 inside S to the point M. This expression leads to the integral equation

$$\sigma(M) - \frac{\Gamma(m/2)}{\pi^{m/2}} \int_S \sigma(M_1) \left[\frac{\cos(r, n_1)}{r^{m-1}} - \frac{1}{R^{m-2}}\right] dS_1 = \frac{\Gamma(m/2)}{\pi^{m/2}} f(M),$$

which has a solution for any continuous $f(M)$.

The basic problems of potential theory for the case of two variables will be considered in the next section.

§2 The application of complex variable to the problems of potential theory in plane regions

2.1 *Dirichlet's problem for a simply connected plane region*

Let D be a simply connected finite plane region bounded by a smooth contour L with continuous curvature. The harmonic function will be denoted by $U(x, y)$ and its value on the contour L by $u(t)$, where t is the complex coordinate of a point on L. The function $U(x, y)$ is the real part of a certain analytic function $\varphi(z)$. We seek the function $\varphi(z)$ in the form of a Cauchy-type integral

$$\varphi(z) = \frac{1}{2\pi i} \int_L \frac{\mu(\zeta)}{\zeta - z} \, d\zeta \tag{7.26}$$

with real density $\mu(\zeta)$, where ζ is a point of the contour L and z is any point inside D. Letting z tend to a certain point t of L, we obtain from the properties of Cauchy-type integrals that

$$\varphi(t) = \tfrac{1}{2}\mu(t) + \frac{1}{2\pi i} \int_L \frac{\mu(\zeta)}{\zeta - t} \, d\zeta.$$

Noting that $\operatorname{Re} \varphi(t) = u(t)$, we find that the unknown density $\mu(t)$ satisfies the integral equation

$$\mu(t) + \frac{1}{\pi} \int_L \mu(\zeta) \operatorname{Im} \left(\frac{d\zeta}{\zeta - t} \right) = 2u(t).$$

Since

$$\operatorname{Im} \left(\frac{d\zeta}{\zeta - t} \right) = -\frac{\cos(r, v)}{r} \, d\sigma,$$

where r is the distance from the point ζ to the point t (the direction of r is chosen from ζ to t), $d\sigma$ is an element of length of the contour L, and v is the outward normal to L, we obtain the following integral equation for the density $\mu(t)$

$$\mu(t) - \frac{1}{\pi}\int_L \mu(\zeta)\frac{\cos(r, v)}{r}\,d\sigma = 2u(t), \tag{7.27}$$

the kernel $\cos(v, r)/r$ of which is continuous. It can be proved that this equation has a solution for any r.h.s. Having solved this equation, we find by (7.26) the analytic function $\varphi(z)$, the real part of which is the harmonic function taking on the contour L the given values $u(t)$.

Consider now the infinite region D' lying outside D. In this case, the function $\varphi(z)$ can be sought in the form

$$\varphi(z) = \frac{1}{2\pi i}\int_L \frac{\mu(\zeta)}{\zeta - z}\,d\zeta + \frac{1}{2\pi}\int_L \mu(\zeta)\,d\zeta, \tag{7.28}$$

where $\mu(\zeta)$ is assumed to be real. Then, for the density $\mu(\zeta)$, we obtain

$$\mu(t) + \frac{1}{\pi}\int_L \mu(\zeta)\frac{\cos(r, v)}{r}\,d\sigma - \frac{1}{\pi}\int_L \mu(\zeta)\,d\sigma = -2u(t).$$

This equation has a unique solution for arbitrary r.h.s. Substituting $\mu(\zeta)$ into (7.28) and separating off the real part of $\varphi(z)$, we obtain the solution of the external Dirichlet problem for the given continuous boundary function $u(t)$. Using the solution of the Dirichlet problem, we can determine the function $w = \omega(z)$, which gives the conformal transformation of the simply connected region D onto the circle $|w| < 1$.

Let $z = a$ be the point of D which transforms into the centre of the circle, $w = 0$, and let $\psi(z) = \omega(z)/(z - a)$. Writing $\varphi(z) = \ln \psi(z)$, we obtain that $\mathrm{Re}\{\varphi(t)\} = -\ln|t - a|$ on the contour. Having found $\varphi(t)$, we can easily find $\omega(z)$. Thus, in order to find $\omega(z)$, it is sufficient to solve Dirichlet's problem for the boundary function $u(t) = -\ln|t - a|$.

Example 2.1. Consider the ellipse $x = a\cos\theta$, $y = b\sin\theta$. We require that, under the conformal transformation, the centre of the ellipse corresponds to the centre of the circle. For the function $\varphi(t)$, we obtain on the contour the condition

$$\mathrm{Re}\{\varphi(t)\} = -\ln|t|.$$

Setting

$$\varphi(z) = \frac{1}{2\pi i}\int_L \frac{\mu(\zeta)}{\zeta - t}\, d\zeta,$$

we derive the integral equation

$$\mu(t) + \frac{b}{2\pi a}\int_0^{2\pi} \mu(\zeta)\frac{d\tau}{1 - \varepsilon^2 \cos^2 \dfrac{\theta + \tau}{2}} = -2 \ln |t|, \quad \varepsilon^2 = \frac{a^2 - b^2}{a^2} = \frac{c^2}{a^2}.$$

The solution of this equation can be obtained in the form of a series

$$\mu(\zeta) = -\ln \frac{a + b}{2} - 2\sum_{k=1}^{\infty} \frac{(-1)^{k-1}}{k} \frac{c^{2k}}{(a + b)^{2k} + (a - b)^{2k}} \cos 2k\theta.$$

It follows then, that

$$\omega(z) = \frac{2z}{a + b} \exp\left\{-\sum_{k=1}^{\infty} \frac{(-1)^{k-1}}{k} \frac{c^{2k}}{(a + b)^{2k} + (a - b)^{2k}} F_{2k}(z)\right\},$$

where

$$F_{2k}(z) = (z + \sqrt{(z^2 - c^2)})^{2k} + (z - \sqrt{(z^2 - c^2)})^{2k}.$$

2.2 *Dirichlet's problem for multi-connected regions*

Let D be a finite $(n + 1)$-connected region, the contour L of which consists of closed curves $L_0, L_1, L_2, \ldots, L_n$, with the contour L_0 outwardly bounding the region D. Let $z_1, z_2, \ldots, z_n$, be fixed points inside the regions bounded by the curves $L_1, L_2, \ldots, L_n$, respectively. We require the function $U(x, y)$ which is harmonic inside D, and takes on the contour the values of a given continuous function $u(t)$. Let $\varphi(z)$ be an analytic function inside D the real part of which is equal to $U(x, y)$. The function $\varphi(z)$ can be expressed in the form

$$\varphi(z) = \sum_{k=1}^{n} A_k \ln (z - z_k) + \varphi^*(z), \tag{7.29}$$

where $\varphi^*(z)$ is a single-valued function which is regular in D. It is sufficient to find $\varphi^*(z)$ and the coefficients A_k $(k = 1, \ldots, n)$.

On the contour, $\varphi^*(z)$ satisfies the condition

$$\operatorname{Re}\{\varphi^*(t)\} = u(t) - \sum_{k=1}^{n} A_k \ln|t - z_k|.$$

It is sought in the form of a Cauchy-type integral

$$\varphi^*(z) = \frac{1}{2\pi i}\int_L \frac{\mu(\zeta)}{\zeta - z}\, d\zeta \tag{7.30}$$

with real density $\mu(\zeta)$. The density $\mu(\zeta)$ must satisfy the integral equation

$$\mu(t) - \frac{1}{\pi}\int_L \mu(\zeta)\frac{\cos(r, \nu)}{r}\, d\sigma = 2u(t) - 2\sum_{k=1}^{n} A_k \ln|t - z_k|, \tag{7.31}$$

which does not appear to have a solution for arbitrary r.h.s.

Since $1/\pi$ is a characteristic value of the kernel $(\cos(r, \nu))/r$ with multiplicity n, the constants A_k are defined by the condition that the r.h.s. of (7.31) be orthogonal to the eigenfunctions of the adjoint integral equation. We can give a method in which it is not necessary to find the eigenfunctions of the adjoint equation. We replace (7.31) by

$$\mu(t) - \frac{1}{\pi}\int_L \left(\frac{\cos(r, \nu)}{r} + a(t, \zeta)\right)\mu(\zeta)\, d\sigma = 2u(t) - 2\sum_{k=1}^{n} A_k \ln|t - z_k|, \tag{7.32}$$

where $a(t, \zeta)$ is the function which equals unity, if the points t and ζ belong to one and the same internal curve L_k $(k = 1, 2, \ldots, n)$ and zero in all other cases. The equation (7.32) is uniquely solvable. Having solved this equation, we require that the A_k satisfy the condition

$$\int_{L_k} \mu(\zeta)\, d\sigma = 0 \qquad (k = 1, \ldots, n). \tag{7.33}$$

The system (7.33) is a system of n homogeneous linear equations with respect to the coefficients A_k. It appears to be always solvable. Choosing the A_k in such a way, we obtain that (7.32) and (7.31) coincide and thus that the solution of the problem is given by (7.30) and (7.29).

If the region D is infinite, i.e. if D lies outside the n contours $L_1, L_2, \ldots, L_n$, then (7.29) remains valid and the coefficients $A_1, \ldots, A_n$ satisfy the complementary condition

$$A_1 + \ldots + A_n = 0. \tag{7.34}$$

In this case, the function $\varphi^*(z)$ is sought in the form

$$\varphi^*(z) = \frac{1}{2\pi i} \int_L \frac{\mu(\zeta)}{\zeta - z} \, d\zeta + \frac{1}{2\pi} \int_L \mu(\zeta) \, d\sigma. \tag{7.35}$$

This leads to the equation

$$\mu(t) + \frac{1}{\pi} \int_L \mu(\zeta) \frac{\cos(r, \nu)}{r} \, d\sigma - \frac{1}{\pi} \int_L \mu(\zeta) \, d\sigma = -2u(t) + 2 \sum_{k=1}^{n} A_k \ln |t - z_k|, \tag{7.36}$$

which has $n - 1$ eigenfunctions.

The coefficients A_k can be obtained from (7.34) and the condition that the r.h.s. of (7.36) be orthogonal to the $(n - 1)$ linearly independent eigenfunctions of the adjoint homogeneous equation. In order to avoid a search for eigenfunctions, we replace (7.36) by

$$\mu(t) + \frac{1}{\pi} \int_L \mu(\zeta) \left(\frac{\cos(r, \nu)}{r} + b(t, \zeta) \right) d\sigma - \frac{1}{\pi} \int_L \mu(\zeta) \, d\sigma = -2u(t) + 2 \sum_{k=1}^{n} A_k \ln |t - z_k|, \tag{7.37}$$

where $b(t, \zeta)$ is the function which equals unity when the points t and ζ belong to one and the same curve L_k $(k = 1, \ldots, n - 1)$ and zero in all other cases.

The equation (7.37) is solvable for any r.h.s. Solving this equation, and finding the A_k $(k = 1, \ldots, n)$ from the conditions

$$\int_{L_k} \mu(\zeta) \, d\sigma = 0 \qquad (k = 1, \ldots, n - 1)$$

and the condition (7.34), we obtain $\varphi^*(z)$ as (7.35). Then separating off the real part of the function $\varphi(z)$ (see (7.29)), we find the required solution.

2.3 *The Neumann problem*

The Neumann problem consists in finding the harmonic function $U(x, y)$ in the region D, which has a given normal derivative $F(\zeta)$ on the contour L. We assume that the contour L satisfies the conditions of §2.2.

For the solvability of the Neumann problem, a necessary and sufficient condition is

$$\int_L F(\zeta)\, d\sigma = 0.$$

Let $V(x, t)$ be a harmonic function conjugate to $U(x, y)$, and let $\varphi(z) = U + iV$. Then (7.29) is valid for $\varphi(z)$, where $\varphi^*(z)$ is a single-valued regular function in D and

$$A_k = -\frac{1}{2\pi}\int_{L_k} F(t)\, d\sigma.$$

Let $\varphi^*(z) = U^* + iV^*$. Then, on the contour L,

$$\frac{\partial U^*}{\partial v} = F(t) - \sum_{k=1}^{n} A_k \frac{\partial \ln |t - z_k|}{\partial v} = F^*(t).$$

Choosing on every contour L_k $(k = 0, 1, \ldots, n)$ a point t_k, we obtain on these contours

$$V^* = \int_{t_k}^{t} F^*(\zeta)\, d\sigma + b_k, \tag{7.38}$$

where the b_k are undefined constants.

We now examine the function $-i\varphi^*(z) = V^* - iU^*$, the real part of which is known except for a constant item, which is defined for each contour L_k $(k = 0, 1, \ldots, n)$ by (7.38). The following problem 'Find the analytic function regular in a multi-connected region, if on each contour $L_0, L_1, \ldots, L_n$ the real part of this function is given except for an additive constant', is called the *modified Dirichlet problem*. Thus, the Neumann problem reduces to the modified Dirichlet problem for the function $-i\varphi^*(z)$. It is known that the last problem has a unique solution, if one of the constants $b_0, b_1, \ldots, b_n$ is fixed. The remaining constants are uniquely defined by the condition that the unknown function be single-valued. We show that the solution of the modified Dirichlet problem can be reduced to the solution of an integral equation.

We put $-i\varphi^*(z) = \varphi(z)$. Then

$$\operatorname{Re}\{\varphi(t)\} = f(t) + b_k \qquad (t \in L), \tag{7.39}$$

where $f(t)$ is a given function on the r.h.s. of (7.38). The function $\varphi(z)$ is sought in the form of a Cauchy-type integral (7.30). In the case of a finite region, we obtain the following equation for the density $\mu(\zeta)$:

$$\mu(t) - \frac{1}{\pi}\int_L \left(\frac{\cos(r, v)}{r} + a(t, \zeta)\right)\mu(\zeta)\, d\sigma = 2f(t), \tag{7.32$_1$}$$

the l.h.s. of which coincides with the l.h.s. of (7.32) of the previous section. Solving these equations and putting

$$b_0 = 0,\ b_k = \frac{1}{2\pi}\int_{L_k} \mu(\zeta)\, d\sigma \qquad (k = 1, 2, \ldots, n),$$

we obtain that the solutions of the last equation and (7.37) coincide.

If the region D is infinite, then $\mu(\zeta)$ satisfies

$$\mu(t) + \frac{1}{\pi}\int_L \left(\frac{\cos(r, v)}{r} + b(t, \zeta)\right)\mu(\zeta)\, d\sigma - \frac{1}{\pi}\int_L \mu(\zeta)\, d\sigma = -2f(t). \tag{7.36$_1$}$$

This equation coincides with (7.36) except for its r.h.s. Having solved this equation, we put

$$b_k = \frac{1}{2\pi}\int_{L_k} \mu(\zeta)\, d\sigma - \frac{1}{2\pi}\int_L \mu(\zeta)\, d\sigma \qquad (k = 1, \ldots, n).$$

In this case, the integral (7.30) also gives the solution of the examined problem.

2.4 *The conformal mapping of multi-connected regions*

We consider the problem: Map the ring-shaped region D, bounded on the outside by a closed smooth curve L_0 and on the inside by a closed smooth curve L_1, onto the circular annulus $R < |w| < 1$, so that the curve L_0 is mapped into one-to-one correspondence with the circle $|w| = 1$, and the curve L_1 into correspondence with the circle $|w| = R$. Let $w = \omega(z)$ be the unknown mapping. We put the origin inside L_1. Then the function

$$\varphi(z) = \ln \frac{\omega(z)}{z}$$

is regular in D, and the function $\varphi(z)$ satisfies the boundary conditions

$$\operatorname{Re} \varphi(t) = \begin{cases} -\ln |t| & \text{for} \quad t \in L_0, \\ -\ln |t| + \ln R & \text{for} \quad t \in L_1. \end{cases} \tag{7.40}$$

Hence, in order to find $\varphi(z)$, it is necessary to solve the modified Dirichlet problem with the conditions (7.40) ($b_0 = 0$, $b_1 = \ln R$ and $f(t) = -\ln |t|$). It can be proved that there exists a unique value $R < 1$ for which $w = \varphi(z)$ {obtained by means of the solution of the modified Dirichlet problem} maps the region conformally on to the ring $R < |w| < 1$.

Similarly, we can reduce to a modified Dirichlet problem, the problem of the conformal mapping of a multi-connected region onto a plane with cuts parallel to the imaginary axis, or onto a plane cut along arcs of concentric circles, or along segments of straight lines all passing through a common point.

2.5 *The Green's function and the Schwarz kernel*

Let D be a region of the type considered in §2.2, which is bounded by the contours $L_0, L_1, \ldots, L_n$, and let $z = x + iy$ and $\zeta = \xi + i\eta$ be arbitrary points of D. Let $G(x, y; \xi, \eta)$ be the Green's function for this region {see [1; Chapter III, §6]}.

The function $G(x, y; \xi, \eta)$ is symmetric and harmonic with respect to both (x, y) and (ξ, η). If the point (ξ, η) lies on the contour L, then $G(x, y; \xi, \eta) = 0$. The Green's function can be represented in the form

$$G(x, y; \xi, \eta) = g(x, y; \xi, \eta) - \ln r,$$

where r is the distance from (x, y) to (ξ, η), and $g(x, y; \xi, \eta)$ is a harmonic function in D with respect to (ξ, η) and (x, y) which equals $\ln r$ on L.

We denote $G(x, y; \xi, \eta)$ by $G(z; \zeta)$. The real function $H(z; \zeta)$, conjugate to $G(z; \zeta)$ with respect to (x, y), can be defined by

$$H(z; \zeta) = \int_a^z -\frac{\partial G}{\partial y}\,dx + \frac{\partial G}{\partial x}\,dy,$$

where a is a fixed point of D. The function

$$M(z; \zeta) = G(z; \zeta) + iH(z; \zeta)$$

is called the *complex Green's function*. The function $M(z; \zeta)$ is an analytic function, but non-regular, of z, and a non-analytic function of ζ.

The complex Green's function $M(z; \zeta)$ is multivalued, since it contains the term $-\ln(z - \zeta)$, and besides, in a multi-connected region it can change its value when passing round any of the internal contours $L_1, \ldots, L_n$. The complex Green's function can be expressed in the form

$$M(z; \zeta) = M_0(z; \zeta) + \sum_{k=1}^{n} b_k(\zeta)\ln(z - z_k) - \ln(\zeta - z), \tag{7.41}$$

where the z_k $(k = 1, \ldots, n)$ are fixed points lying inside L_k, $M_0(z; \zeta)$ is a regular function of z in D, and $b_k(\zeta)$ is a harmonic function of ζ in D which equals unity on L_k and zero on L_j $(j \neq k)$.

The functions $b_k(\zeta)$ are defined by

$$b_k(\zeta) = \frac{1}{2\pi}\int_{L_k} \frac{\partial G}{\partial \nu_1}\,d\sigma,$$

where ν_1 is the direction of the inward normal to L_k at the point z. Let $\zeta \in L$, and ν be the direction of the inward normal to the contour L at the point ζ.

Differentiating (7.41) with respect to ν, we obtain

$$T(z; \zeta) = \frac{\partial M_0}{\partial \nu} + \sum_{k=1}^{n} a_k(\zeta)\ln(z - z_k) - \frac{1}{\zeta - z}\frac{\partial \zeta}{\partial \nu}, \tag{7.42}$$

where

$$T(z; \zeta) = \frac{\partial M(z; \zeta)}{\partial \nu} \tag{7.43}$$

and

$$a_k(\zeta) = \frac{\partial b_k(\zeta)}{\partial \nu}.$$

The function $T(z; \zeta)$ is called the *Schwarz kernel*. It is an analytic function of z, and is many valued, if the region D is multi-connected. The Schwarz kernel is a non-analytic and single valued function with respect to ζ.

In addition,

$$\text{Re } T(z; \zeta) = \frac{\partial G}{\partial \nu}$$

and, for one of the branches of the Schwarz kernel,

$$\text{Im } T(a; \zeta) = 0.$$

Let $F(z) = u(z) + iv(z)$ be an analytic function without a singular point inside D, and have a single-valued and continuous real part up to the contour. Then the following formula is valid:

$$\frac{1}{2\pi} \int_L u(\zeta)\, T(z; \zeta)\, d\sigma = F(z) - iv(a),$$

where a is a fixed point inside D.

Let $t = \omega(z)$ map the region D of the plane z conformally onto the region D' of the plane t, and let $M'(t; \tau)$ be the complex Green's function of the region D'. Then the complex Green's function of the region D is defined by

$$M(z; \zeta) = M'(\omega(z), \omega(\zeta)).$$

Under this mapping, the relationship between the Schwarz kernels is

$$T(z; \zeta)\, d\sigma = T'(\omega(z), \omega(\zeta))\, d\sigma', \qquad d\sigma' = |d\tau|.$$

For the circle $|t| < 1$, the Schwarz kernel is

$$T(t; \tau) = \frac{\tau + t}{\tau - t}. \tag{7.44}$$

§3 The biharmonic equation and the plane problem in the theory of elasticity

3.1 *The application of Green's function. The formulation and the study of the plane problem of the theory of elasticity*

Equations of the form

$$\Delta^2 W = \frac{\partial^4 W}{\partial x^4} + 2\frac{\partial^4 W}{\partial x^2\,\partial y^2} + \frac{\partial^4 W}{\partial y^4} = 0 \tag{7.45}$$

are called biharmonic. The function, satisfying (7.45), is called *biharmonic*. Every biharmonic function can be expressed in terms of two analytic functions $\varphi(z)$ and $\chi(z)$ of the complex variable z

$$W(x, y) = \operatorname{Re}\{\bar{z}\varphi(z) + \chi(z)\}. \tag{7.46}$$

The functions $\varphi(z)$ and $\psi(z) = \chi'(z)$ are called *Goursat* functions. The plane problem of the theory of elasticity, when body forces are absent, can be reduced to a boundary value problem for the biharmonic equation.

Let u and v be the components of displacement of a plane deformation parallel to the (x, y)-plane; σ_x, τ_{xy}, σ_y be the components of stress. The equations of the plane theory of elasticity can be satisfied by setting

$$\sigma_x = \frac{\partial^2 W}{\partial x^2} \qquad \tau_{xy} = -\frac{\partial^2 W}{\partial x\,\partial y} \qquad \sigma_y = \frac{\partial^2 W}{\partial y^2},$$

where $W(x, y)$, called the *Airy function*, satisfies (7.45) (i.e. it is biharmonic).

We denote the region in which we solve (7.45) by D and its contour by L. The region D can be multi-connected.

Let L_0 be the external contour and $L_1, \ldots, L_n$ be the internal contours bounding D.

We consider two boundary value problems of the plane theory of elasticity:

1) the displacements u and v are given on the contour L;
2) the external forces, with components X_ν and Y_ν, are given on the contour L.

Using (7.46) and writing $\chi'(z) = \psi(z)$, we can reduce these problems to the problem: find two analytic functions $\varphi(z)$ and $\psi(z)$ satisfying at every point $\zeta \in L$ the boundary condition

$$\kappa\varphi(\zeta) - \overline{\zeta\varphi'(\zeta)} - \overline{\psi(\zeta)} = g(\zeta). \tag{7.47}$$

Here, $g(z) = 2\mu(u + iv)$ is a given point of the contour L, and μ and $\kappa = (\lambda + 3\mu)/(\lambda + \mu)$ are the constants of elasticity ($\kappa > 1$) for the first problem.

In the second problem

$$g(z) = -f(z) - b_k \quad \text{on} \quad L_k\,(b_k = \text{const})\,(k = 0, 1, \ldots, n), \tag{7.48}$$

where $f(z) = i\int_{s_0}^{s}(X_\nu - iY_\nu)\,ds$ and $\kappa = -1$. The constants b_k are defined so that the displacements are single-valued.

The functions $\varphi(z)$ and $\psi(z)$ are regular in the region D, if it is simply-connected. In the case of a multi-connected region, $\varphi(z)$ and $\psi(z)$ take the form

$$\left.\begin{aligned} \varphi(z) &= \sum_{k=1}^{n} B_k \ln(z - z_k) + \varphi^*(z), \\ \psi(z) &= -\kappa \sum_{k=1}^{n} \bar{B}_k \ln(z - z_k) + \psi^*(z), \end{aligned}\right\} \tag{7.49}$$

where $\varphi^*(z)$ and $\psi^*(z)$ are regular analytic functions, the B_k are constants and the z_k are fixed points inside the regions bounded by the L_k ($k = 1, \ldots, n$).

Let $z = \omega(t)$ map the region D conformally onto the region D^* with a sufficiently smooth contour γ, and let the Schwartz kernel $T(t, \tau)$ for the region D^* be known {see (7.42)}, where $t \in D^*$, $\tau \in \gamma$. We write

$$\Phi(t) = \varphi(\omega(t)), \qquad \Psi(t) = \psi(\omega(t)) \quad \text{and} \quad G(\tau) = g(\omega(\tau)). \tag{7.50}$$

The boundary condition (7.47) takes the form

$$\kappa\Phi(\tau) - \frac{\omega(\tau)}{\omega'(\tau)}\overline{\Phi'(\tau)} - \overline{\Psi(\tau)} = G(\tau) \tag{7.51}$$

on the contour γ. This is equivalent to the two equations

$$\kappa\Phi(t) - \frac{1}{4\pi}\int_\gamma \overline{\frac{\omega(\tau)}{\omega'(\tau)}\Phi'(\tau)}\, T(t, \tau)\, d\sigma = A(t), \tag{7.52}$$

$$\Psi(t) + \frac{1}{4\pi}\int_\gamma \frac{\overline{\omega(\tau)}}{\omega'(\tau)}\Phi'(\tau)\, T(t, \tau)\, d\sigma = B(t), \tag{7.53}$$

where $d\sigma$ is an element of length on γ and

$$\left.\begin{aligned} A(t) &= \frac{1}{4\pi}\int_{\gamma} G(\tau)\, T(t, \tau)\, d\sigma, \\ B(t) &= -\frac{1}{4\pi}\int_{\gamma} \overline{G(\tau)}\, T(t, \tau)\, d\sigma. \end{aligned}\right\} \tag{7.54}$$

If $\Phi'(\tau)$ is known, then (7.53) defines $\Psi(t)$.

Let us introduce a new function $\vartheta(t)$ by letting

$$\vartheta(t) = \Phi'(t) - l\omega'(t), \tag{7.55}$$

where the constant l must satisfy

$$l - \frac{1}{2\kappa}\bar{l} = \frac{1}{2\kappa}\frac{\overline{\Phi'(a)}}{\omega'(a)} \tag{7.56}$$

(a is an arbitrary fixed point of D).

By means of (7.52), we obtain the equation for the boundary function $\vartheta(\tau)$:

$$\vartheta(t_0) - \frac{1}{4\pi\kappa}\int_{\gamma}\frac{\partial}{\partial t}\left[\frac{\omega(\tau) - \omega(t_0)}{\overline{\omega'(\tau)}}\, T(\tau_0, \tau)\right]\overline{\vartheta(\tau)}\, d\sigma = \frac{1}{\kappa}A'(t_0), \qquad (t_0 \in \gamma). \tag{7.57}$$

Using (7.42) and introducing in D^* the function $K(t, \tau)$ which is regular with respect to t, viz.

$$K(t, \tau) = \frac{1}{4\pi}\frac{d}{dt}\left[\frac{\omega(\tau) - \omega(t)}{\overline{\omega'(\tau)}}\, T(t, \tau)\right] + \frac{\omega'(t)}{4\pi}\sum_{k=1}^{n} a_k(\tau)\ln(t - t_k),$$

where t_k is a point inside γ_k (γ_k is the image of L_k), we can rewrite (7.57) in the form

$$\vartheta(\tau_0) - \frac{1}{\kappa}\int_{\gamma} K(\tau_0, \tau)\,\overline{\vartheta(t)}\, d\sigma = \frac{1}{\kappa}A'(\tau_0), \qquad (\tau_0 \in \gamma). \tag{7.58}$$

Separating off the real and imaginary parts of this equation and taking $\operatorname{Re}\{\vartheta(\tau)\}$ and $\operatorname{Im}\{\vartheta(\tau)\}$ as unknown functions, we can transform (7.58) into a system of two equations of Fredholm type. We proved that (7.58) has a solution for arbitrary r.h.s. In the case of the first boundary value problem, l can be uniquely derived from (7.56), and then $\Phi(t)$ and $\Psi(t)$ can be obtained from (7.55) and (7.52). Thus, the problem is solved. In the second boundary value problem ($\kappa = -1$), the function $\Phi'(t)$ is obtained except for a term $i\alpha\omega'(t)$, where α is a real valued constant. Hence, we require the additional condition

$$\int_L Y_\nu \, dx - X_\nu \, dy = 0.$$

3.2 *Simply connected regions*

Let a simply connected region D be mapped onto the unit circle $|t| < 1$ by means of a sufficiently smooth function $z = \omega(t)$. Then, (7.58) takes the form

$$\vartheta(t) - \frac{1}{2\pi i\kappa} \int_\gamma \frac{\partial}{\partial t}\left[\frac{\omega(t) - \omega(\tau)}{\overline{\omega'(\tau)}(\tau - t)}\right] \overline{\vartheta(t)} \, d\tau = \frac{1}{\kappa} A'(t), \tag{7.59}$$

where

$$\vartheta(t) = \Phi'(t) - l\omega'(t), \qquad l = \frac{1}{\kappa} \frac{\overline{\Phi'(0)}}{\omega(0)}$$

and

$$A(t) = \frac{1}{2\pi i} \int_\gamma G(\tau) \frac{d\tau}{\tau - t}.$$

This equation was obtained by Muskhelishvili.

If $\omega(t)$ is a rational function, then the kernel of the integral equation (7.59) is degenerate, and hence this equation can be solved in finite form.

3.3 *Application of Cauchy-type integrals. Muskhelishvili's equations*

Let the elastic medium fill a finite simply connected region D with a sufficiently smooth contour L. The problem becomes: find two analytic functions $\varphi(z)$ and $\psi(z)$ which are regular in D and satisfy on L the condition

$$\varphi(\zeta) + \overline{\zeta\varphi'(\zeta)} + \overline{\psi(\zeta)} = f(\zeta), \tag{7.60}$$

where $f(\zeta)$ is a known continuous function of the point ζ on the contour.

For the boundary function $\varphi(\zeta)$, Muskhelishvili obtained the equation

$$\overline{\varphi(t)} - \frac{1}{\pi}\int_L \overline{\varphi(\zeta)}\, d\vartheta - \frac{1}{\pi}\int_L \varphi(\zeta)e^{-2i\vartheta}\, d\vartheta = A(t), \tag{7.61}$$

where t is an arbitrary point of the contour L and ϑ is the polar angle defined by

$$\zeta - t = re^{i\vartheta} \qquad (r = |\zeta - t|).$$

In this equation

$$A(t) = -\tfrac{1}{2}f(t) + \frac{1}{2\pi i}\int_L \frac{f(\zeta)}{\zeta - t}\, d\zeta,$$

where the principal value of the integral is taken.

When deriving (7.61), Muskhelishvili used Cauchy-type integrals. Separating off the real and imaginary parts in (7.61), we obtain a system of two integral equations of Fredholm type for Re $\{\varphi(z)\}$ and Im $\{\varphi(z)\}$. On solving this system, we obtain the boundary values of the functions $\varphi(\zeta)$ and $\psi(\zeta)$. It is proved that any solution of the equation (7.61) can be extended analytically inside D, and hence $\varphi(z)$ and $\psi(z)$ can be expressed in terms of Cauchy formulas.

The equation (7.61) does not have a solution for all right hand sides. If

$$\operatorname{Re}\int_L \overline{f(\zeta)}\, d\zeta \neq 0,$$

then (7.61) has no solutions.

Let the origin be inside D. We replace (7.61) by

$$\overline{\varphi(t)} - \frac{1}{\pi}\int_L \overline{\varphi(\zeta)}\, d\vartheta - \frac{1}{\pi}\int_L \varphi(\zeta)e^{-2i\vartheta}\, d\vartheta + \frac{1}{2\pi i}\int_L \frac{\varphi(\zeta)}{\zeta}\, d\zeta$$
$$+ \frac{1}{2\pi i}\int_L \left(\frac{\varphi(\zeta)}{\zeta^2}\, d\zeta + \overline{\frac{\varphi(\zeta)}{\zeta^2}\, d\zeta}\right) = A(t). \tag{7.62}$$

Every solution of this equation can be extended analytically to the whole region D. This equation has a solution for any r.h.s. If

$$\operatorname{Re}\int_L \overline{f(\zeta)}\, d\zeta = 0, \tag{7.63}$$

then the solutions of (7.61) and (7.62) coincide.

The first boundary value problem for the plane problem of the theory of elasticity which has the boundary conditions

$$\kappa\varphi(\zeta) - \zeta\overline{\varphi'(\zeta)} - \overline{\psi(\zeta)} = g(\zeta),$$

where $\kappa > 1$, can be examined in an analogous way.

The case of a multi-connected region demands a more involved analysis, but the integral equations (7.61) and (7.62) retain the form derived above.

3.4 *Lauricella-Sherman equations*

We assume that the conditions formulated at the beginning of the previous section are satisfied. We seek regular analytic functions $\phi(z)$ and $\psi(z)$, which satisfy the condition (7.60) in the form

$$\varphi(z) = \frac{1}{2\pi i}\int_L \frac{\omega(\zeta)}{\zeta - z}\, d\zeta, \tag{7.64}$$

$$\psi(z) = \frac{1}{2\pi i}\int_L \frac{\overline{\omega(\zeta)}}{\zeta - z}\, d\zeta - \frac{1}{2\pi i}\int_L \frac{\bar{\zeta}\omega(\zeta)}{(\zeta - z)^2}\, d\zeta + \frac{1}{2\pi i}\int_L \frac{\omega(\zeta)\, d\bar{\zeta}}{\zeta - z}, \tag{7.65}$$

where $\omega(\zeta)$ is an unknown function which has to be determined on the contour. Let the condition (7.63) be satisfied.

For $\omega(\zeta)$, we obtained the so-called Lauricella equation

$$\omega(t) + \frac{1}{2\pi i}\int_L \omega(\zeta)\, d\ln\frac{\zeta - t}{\bar{\zeta} - \bar{t}} - \frac{1}{2\pi i}\int_L \overline{\omega(\zeta)}\, d\frac{\zeta - t}{\bar{\zeta} - \bar{t}} = f(t), \tag{7.66}$$

in complex form. In the general case, this equation does not have a solution.

Sherman replaced it by

$$\omega(t) + \frac{1}{\pi}\int_L \omega(\zeta)\, d\vartheta - \frac{1}{\pi}\int_L \overline{\omega(\zeta)} e^{2i\vartheta}\, d\vartheta$$
$$+ \left(\frac{1}{t-a} - \frac{1}{\bar{t}-\bar{a}} + \frac{t-a}{(\bar{t}-\bar{a})^2}\right)\frac{1}{\pi i}\operatorname{Re}\int_L \frac{\omega(\zeta)\, d\zeta}{(\zeta-a)^2} = f(t), \qquad (7.67)$$

where a is an internal point of D and $\zeta - t = re^{i\vartheta}$ $(r = |\zeta - t|)$. This equation is solvable for arbitrary r.h.s. Any solution of (7.67) satisfies (7.66). The function $\omega(t)$ thus obtained must be substituted in (7.64) and (7.65) in order to solve the problem.

3.5 *Periodic problem of the theory of elasticity*

Let the elastic medium fill the entire plane with the exception of an infinite series of regularly spaced identical holes D, bounded by a contour L and subjected to the action of similar external forces. We assume that the resulting vector of external forces acting on the boundary of each individual hole is equal to zero, and that all the known functions are periodic functions of $\bar{x}$ with period a.

The problem reduces to the determination of the functions $\varphi_0(z)$ and $\psi_0(z)$ which are periodic in x, and satisfy the boundary condition

$$\varphi_0(\zeta) - (\zeta - \bar{\zeta})\,\overline{\varphi_0'(\zeta)} + \overline{\psi_0(\zeta)} = f(\zeta) \qquad (\zeta \in L).$$

We set $z = a \ln t/2\pi i$ and introduce the functions

$$\Phi(t) = \varphi_0\left(\frac{a}{2\pi i}\ln t\right) \quad \text{and} \quad \Psi(t) = \psi_0\left(\frac{a}{2\pi i}\ln t\right).$$

Then, for the boundary values of $\Phi(t)$, we obtain the integral equation

$$\overline{\Phi(\tau_0)} + \frac{1}{2\pi i}\int_L \overline{\Phi(\tau)}\, d\ln\frac{\tau - \tau_0}{\bar{\tau} - \bar{\tau}_0}$$
$$+ \frac{1}{\pi i}\int_L \Phi(\tau)\, d\,\frac{\tau\ln|\tau| - \tau_0\ln|\tau_0|}{\tau - \tau_0}$$
$$+ \frac{i}{2\pi}\overline{\ln\tau_0}\operatorname{Re}\left\{\int_L \frac{\Phi(\tau)}{(\tau - c)^2}\, d\tau\right\} + \frac{1}{2\pi i}\int_L \frac{\overline{\Phi(\tau)}}{\bar{\tau} - \bar{c}}\, d\bar{\tau} = A(\tau_0), \qquad (7.68)$$

where c is an arbitrary fixed point inside L and τ_0 is an arbitrary point on the contour L. The r.h.s. $A(\tau_0)$ is the limiting value of the integral

$$\frac{1}{2\pi i}\int_L \frac{F(\tau)}{\tau - t'}\, d\tau$$

for t' a point inside L which tends to τ_0. Here $F(\tau) = f[(a/2\pi i)\ln \tau]$.

The equation (7.68) is equivalent to a system of two equations of Fredholm type.

The equation (7.68) has a solution for arbitrary $F(\tau)$. The function $\Phi(\tau_0)$, satisfying (7.68), can be extended analytically throughout the whole exterior of the curve L. Moreover, it is bounded at infinity. The function

$$\Psi(\tau_0) = \overline{F(\tau_0)} - \overline{\Phi(\tau_0)} + 2\tau_0 \ln|\tau_0|\, \Phi'(\tau_0)$$

can be continued analytically throughout the whole exterior of the curve L and is equal to zero at infinity.

3.6 *The characteristic values of the integral equations of the theory of elasticity*

Consider the equation (7.58). Setting

$$\vartheta(\tau) = p(\tau) + iq(\tau), \qquad K(\tau_0, \tau) = R(\tau_0, \tau) + iS(\tau_0, \tau),$$

$$\frac{1}{\kappa}A'(\tau) = A_1(\tau) + iA_2(\tau) \qquad \frac{1}{\kappa} = \lambda,$$

and separating off the real and imaginary parts in (7.58), we obtain a system of integral equations of Fredholm type:

$$\left.\begin{aligned} p(\tau_0) - \lambda\int_\gamma \{R(\tau_0, \tau)\, p(\tau) + S(\tau_0, \tau)\, q(\tau)\}\, d\sigma &= A_1(\tau_0),\\ q(\tau_0) - \lambda\int_\gamma \{S(\tau_0, \tau)\, p(\tau) - R(\tau_0, \tau)\, q(\tau)\}\, d\sigma &= A_2(\tau_0). \end{aligned}\right\} \tag{7.69}$$

The absolute magnitudes of all the characteristic values of this system are greater than unity. Since, for problems of the theory of elasticity,

$$|\lambda| = |1/\kappa| \leqslant 1,$$

the series of successive approximations for the system (7.69) and for the equation (7.58) are convergent.

§4 Potentials for the heat conduction equation

4.1 *Integral equations for heat conduction*

The heat conduction equation takes the form

$$\frac{\partial^2 U}{\partial x^2} + \frac{\partial^2 U}{\partial y^2} = \frac{1}{a^2}\frac{\partial U}{\partial t}. \tag{7.70}$$

This equation is to be solved for a certain region of the (x, y) variables for all $t > 0$. We assume that the contour L bounding the region D is sufficiently smooth. We seek the solution of (7.70) satisfying:

1) the initial condition

$$U|_{t=0} = F(x, y) \tag{7.71}$$

where $F(x, y)$ is a known continuous function;

2) the boundary condition

$$U|_L = f(s, t), \tag{7.72}$$

where $f(s, t)$ is a known function of s on the contour L and time t.

The problem (7.70)–(7.72) is called *the first boundary value problem for the heat conduction equation.* This problem can be reduced using heat potentials to an integral equation. Let σ be a point on L, ν be the external normal to L at the point σ, and r be the distance from the point σ to the point (x, y). {The vector r is taken to be directed from the point σ towards (x, y)}.

The integral

$$W(x, y, t) = \frac{1}{2\pi}\int_0^t d\tau \int_L \frac{\mu(\sigma, \tau)}{t - \tau}\frac{\partial}{\partial \nu} e^{-[r^2/4a^2(t-\tau)]}\, d\sigma \tag{7.73}$$

is called *the heat potential of a double layer*, where $\mu(\sigma, t)$ is a continuous function (the density), defined for $t \geqslant 0$ and $\sigma \in L$. It is assumed that $\mu(\sigma, 0) = 0$.

The potential (7.73) has the following properties:

1) For all (x, y) lying both inside and outside L, $W(x, y, t)$ is differentiable infinitely often and satisfies (7.70).

2) $W(x, y, 0) = 0$.

3) The limiting values W_i and W_e of the potential (7.73) on the contour L, obtained from inside and outside, respectively, are determined by

$$W_i = -\mu(s, t) + \frac{1}{4\pi a^2}\int_0^t d\tau \int_L \frac{\mu(\sigma, \tau)}{(t-\tau)^2} e^{-[r^2/4a^2(t-\tau)]} r \cos(\nu, r)\, d\sigma, \tag{7.74}$$

$$W_e = \mu(s, t) + \frac{1}{4\pi a^2}\int_0^t d\tau \int_L \frac{\mu(\sigma, \tau)}{(t-\tau)^2} e^{-[r^2/4a^2(t-\tau)]} r \cos(\nu, r)\, d\sigma, \tag{7.75}$$

where s is an arbitrary point of L.

4) The normal derivative of the potential (7.73) is continuous across the contour L.

We seek the solution of (7.70)–(7.72) in the form

$$U(x, y, t) = U_0(x, y, t) + W(x, y, t), \tag{7.76}$$

where

$$U_0(x, y, t) = \frac{1}{2a\sqrt{(\pi t)}}\iint_D F(\xi, \eta) e^{-[((x-\xi)^2+(y-\eta)^2)/4a^2 t]}\, d\xi\, d\eta \tag{7.77}$$

and $W(x, y, t)$ is the heat potential (7.73).

For the unknown density $\mu(s, t)$, we obtain the integral equation

$$\mu(s, t) - \frac{1}{4\pi a^2}\int_0^t d\tau \int_L \frac{\mu(\sigma, \tau)}{(t-\tau)^2} e^{-[r^2/4a^2(t-\tau)]} r \cos(\nu, r)\, d\sigma = -g(s, t), \tag{7.78}$$

where r is the distance between the points s and σ, and

$$g(s, t) = f(s, t) - \lim_{(x,y)\to s} U_0(x, y, t).$$

For the exterior problem (the region D lies outside the contour L) the solution of (7.70)–(7.72) is also sought in the form (7.76). In this case, the integral equation for the density $\mu(s, t)$ takes the form

$$\mu(s, t) + \frac{1}{4\pi a^2} \int_0^t d\tau \int_L \frac{\mu(\sigma, \tau)}{(t - \tau)^2} e^{-r^2/[4a^2(t-\tau)]} r \cos(\nu, r)\, d\sigma = g(s, t). \tag{7.79}$$

Often, for (7.70), the condition (7.72) is replaced by the boundary condition

$$\left. \frac{\partial U}{\partial \nu} + h(\sigma)U \right|_L = f(\sigma, \tau). \tag{7.80}$$

Here $h(\sigma)$ is a known continuous positive function.

With the aid of *the heat potential of a simple layer*, which has the form

$$V(x, y, t) = \frac{1}{2\pi} \int_0^t d\tau \int_L \frac{\rho(\sigma, \tau)}{t - \tau} e^{-r^2/[4a^2(t-\tau)]}\, d\sigma, \tag{7.81}$$

where the density $\rho(\sigma, t)$ is sufficiently smooth and vanishes when $t = 0$, this problem is reduced to an integral equation.

When $t > 0$ and $(x, y) \notin L$, the potential (7.81) is a continuously differentiable function of the variables x, y, t and satisfies (7.70). It is continuous across the boundary, and the limiting values of its normal derivative on the contour L, obtained by taking limits from inside and outside, respectively, have the form

$$\frac{\partial V}{\partial n_i} = \rho(s, t) - \frac{1}{4\pi a^2 t} \int_0^t d\tau \int_L \frac{\rho(\sigma, \tau)}{(t - \tau)^2} e^{-r^2/[4a^2(t-\tau)]} r \cos(r, n)\, d\sigma, \tag{7.82}$$

$$\frac{\partial V}{\partial n_e} = -\rho(s, t) - \frac{1}{4\pi a^2 t} \int_0^t d\tau \int_L \frac{\rho(\sigma, \tau)}{(t - \tau)^2} e^{-r^2/[4a^2(t-\tau)]} r \cos(r, n)\, d\sigma, \tag{7.83}$$

where n is the direction of the outward normal at the point $s \in L$.

We seek the solution of (7.70)–(7.71) with the boundary condition

(7.80) in the form

$$U(x, y, t) = U_0(x, y, t) + V(x, y, t), \tag{7.84}$$

where $U_0(x, y, t)$ is defined by (7.77).

The integral equation for the density $\rho(\sigma, t)$ is

$$\pm\rho(s, t) - \frac{1}{4\pi a^2}\int_0^t d\tau \int_L \frac{\rho(\sigma, \tau)}{(t-\tau)^2}\, e^{-r^2/[4a^2(t-\tau)]} r \cos(r, n)\, d\sigma$$

$$+ \frac{h(s)}{2\pi}\int_0^t d\tau \int_L \frac{\rho(\sigma, \tau)}{t-\tau}\, e^{-r^2/[4a^2(t-\tau)]}\, d\sigma$$

$$= f(s, t) - \frac{\partial U_0}{\partial n} - h(s)U_0, \tag{7.85}$$

where the plus sign corresponds to the interior region case and the minus to the exterior.

The integral equations (7.78), (7.79) and (7.85) retain their form when the region D is multi-connected.

The heat potentials for a three-dimensional region D bounded by a surface S are defined by: the potential of a simple layer

$$V(x, y, z, t) = \int_0^t d\tau \iint_S \frac{\rho(\sigma, \tau)}{[4\pi(t-\tau)]^{3/2}}\, e^{-r^2/[4a^2(t-\tau)]}\, d\sigma; \tag{7.86}$$

the potential of a double layer

$$W(x, y, z, t) = \int_0^t d\tau \iint_S \frac{\mu(\sigma, \tau)}{[4\pi(t-\tau)]^{3/2}} \frac{\partial}{\partial \nu}\, e^{-r^2/[4a^2(t-\tau)]}\, d\sigma. \tag{7.87}$$

With the help of these potentials, the fundamental boundary problems for the three-dimensional heat equation can be reduced to integral equations.

4.2 *An examination of the integral equations of heat potentials and the convergence of the method of successive approximations*

Consider the integral equations (7.78) and (7.79) under the assumption that the contour L is convex but not necessarily smooth. We set

$$\frac{\cos(r, \nu)}{r}\, d\sigma = d\alpha,$$

where $d\alpha$ is the angle formed by two infinitely close radius vectors drawn from the point s to the extremities of an arc $d\sigma$ and assume that the integral $\int |d\alpha|$ is finite. This is valid for example, for non-smooth convex contours. Under this assumption, (7.78) and (7.79) can be rewritten in different ways. We consider a more general equation, for which (7.78) and (7.79) are particular cases when $\lambda = \pm 1$:

$$\mu(s, t) - \frac{\lambda}{4\pi a^2} \int_0^t d\tau \int_L \frac{\mu(\sigma, \tau)}{(t - \tau)^2}\, e^{-r^2/[4a^2(t-\tau)]} r^2\, d\alpha = \mp g(s, t). \tag{7.88}$$

The method of successive approximations for this equation converges, if $|\lambda| \leqslant 1$, and the solution becomes

$$\mu(s, t) = \sum_{n=0}^{\infty} \lambda^n \mu_n(s, t), \tag{7.89}$$

where $\mu_0(s, t) = g(s, t)$,

$$\mu_n(s, t) = \frac{1}{4\pi a^2} \int_0^t d\tau \int_L \frac{\mu_{n-1}(s, t)}{(\tau - t)^2}\, e^{-r^2/[4a^2(t-\tau)]} r^2\, d\alpha.$$

The series (7.89) converges uniformly for $0 \leqslant t \leqslant t_0$, where t_0 is an arbitrary fixed number. Let the modulus of the function $g(s, t)$ be bounded for all s and t by some constant A.

Then the series (7.89) converges with the rate of a geometrical progression with ratio

$$q(t_0) = \max_{0 \leqslant t \leqslant t_0} \frac{1}{\pi} \int_L e^{-r^2/(4a^2 t)}\, d\alpha < 1.$$

More precisely, the following estimate holds:

$$|\mu_n(s, t)| < Aq^n(t_0).$$

If L is smooth and the curvature is continuous, then an upper bound for $|\mu_n(s, t)|$ is given by

$$|\mu_n(s, t)| < \frac{A(pt)^{n/2}}{\Gamma(n/2 + 1)},$$

where Γ is Euler's function and the constant p depends on the shape of L.

§5 The generalized Schwarz algorithm

5.1 *The general formulation and the convergence of the algorithm for the plane problem of potential theory*

Let D be a multi-connected region bounded by the contours $L_0, L_1, \ldots, L_n$ (see §2.2), and consider the Dirichlet problem for this region. Let $U(x, y)$ be an unknown harmonic function and $f_k(\zeta)$ $(k = 0, 1, \ldots, n)$ be its known values on the contours L_k. Proceeding from (7.29), the function $U(x, y)$ can be expressed in the form

$$U(x, y) = \sum_{k=0}^{n} U_k(x, y) + \sum_{k=1}^{n} A_k \ln |z - z_k|, \tag{7.90}$$

where the $U_k(x, y)$ are harmonic outside L_k $(k = 0, 1, \ldots, n)$, the z_k are arbitrary fixed complex points inside L_k $(k = 0, 1, \ldots, n)$, the A_k are constants, and $z = x + iy$.

The representation (7.90) is not single-valued, since we can add a constant a_k to every $U_k(x, y)$, if

$$a_0 + a_1 + \ldots + a_n = 0.$$

We denote by $u_k(\zeta)$ the value of $U_k(x, y)$ on the contour L_k and by D_k $(k = 0, 1, \ldots, n)$ the region lying outside L_k. Let $G_k(z, \zeta)$ be the Green's function of the region D_k. Then the functions $u_k(\zeta)$ $(k = 0, 1, \ldots, n)$ must satisfy the system of integral equations

$$u_m(z) + \frac{1}{2\pi} \sum_{k \neq m} \int_{L_k} u_k(\zeta) \frac{\partial G_k(z; \zeta)}{\partial \nu} d\sigma = f_m(z) - \sum_{k=1}^{n} A_k \ln |z - z_k|, \qquad (7.91)$$

where $z \in L_m$, $\zeta \in L_k$ and ν is the outward normal to L_k ($k = 0, 1, \ldots, n$).

The system (7.91) does not have a solution for arbitrary r.h.s.

Let $l_k(\zeta)$ be an arbitrary continuous function except that it satisfies the condition

$$\frac{1}{2\pi} \int_{L_k} l_k(\zeta)\, d\sigma = 1.$$

The system (7.91) can be replaced by

$$u_m(z) + \frac{1}{2\pi} \sum_{k \neq m} \int_{L_k} u_k(\zeta) \left[\frac{\partial G_k(z; \zeta)}{\partial \nu} - l_k(\zeta) \right] d\sigma$$

$$= f_m(z) - \sum_{k=1}^{n} A_k \ln |z - z_k|, \qquad (m = 0, 1, \ldots, n). \quad (7.92)$$

This system does not have a solution for arbitrary r.h.s.

Let $W_m(z)$ be the solution of the system (7.92) when its r.h.s. equals $f_m(z)$, and $W_{km}(z)$ the solution when the r.h.s. equals $\ln |z - z_k|$. Then the solution $U_m(z)$ of the system can be written in the form

$$U_m(z) = W_m(z) - \sum_{k=1}^{n} A_k W_{km}(z).$$

The constants A_k can be derived from the conditions that the value

$$A = \frac{1}{2\pi} \sum_{k \neq m} \int_{L_k} U_k(\zeta)\, l_k(\zeta)\, d\sigma \qquad (7.93)$$

must be one and the same for all $m = 0, 1, \ldots, n$. The conditions (7.93) defines a system of $(n + 1)$ equations in the $(n + 1)$ unknowns $A, A_1, \ldots, A_n$.

Solving this system, we obtain the solution of the Dirichlet problem in the form

$$U(x, y) = \sum_{k=0}^{n} U_k(x, y) + \sum_{k=1}^{n} A_k \ln |z - z_k| - A.$$

For the modified Dirichlet problem (see §2.3), we have $A_k = 0$ and the functions $U_k(x, y)$ can be determined from the system

$$U_m(z) + \frac{1}{2\pi} \sum_{k \neq m} \int_{L_k} U_k(\zeta) \left[\frac{\partial G_k(z; \zeta)}{\partial \nu} - l_k(\zeta) \right] d\sigma = f_m(z).$$

The constants b_m are defined by

$$b_m = -\frac{1}{2\pi} \sum_{k \neq m} \int_{L_k} U_k(\zeta)\, l_k(\zeta)\, d\sigma.$$

$(m = 1, 2, \ldots, n;\ z \in L_m)$.

Let us assume now that the contour L_0 is absent. In the system (7.94), we put

$$l_k(\zeta) = \left. \frac{\partial G_k(z; \zeta)}{\partial \nu} \right|_{z=\infty}$$

and introduce a parameter λ. Then this system takes the form

$$u_m(z) + \frac{\lambda}{2\pi} \sum_{k \neq m} \int_{L_k} u_k(\zeta) \left[\frac{\partial G_k(z; \zeta)}{\partial \nu} - \frac{\partial G_k(\infty; \zeta)}{\partial \nu} \right] d\sigma = f_m(z) \tag{7.95}$$

The solution of this system is sought in the form

$$u_m(z) = \sum_{r=0}^{\infty} (-1)^r \lambda^r u_{mr}(z). \tag{7.96}$$

The functions $u_{mr}(z)$ satisfy the recurrence relations

$$\left. \begin{aligned} u_{m0}(z) &= f_m(z), \\ u_{mr}(z) &= \frac{1}{2\pi} \sum_{k \neq m} \int_{L} u_{k,r-1}(\zeta) \left[\frac{\partial G_k(z; \zeta)}{\partial \nu} - \frac{\partial G_k(\infty; \zeta)}{\partial \nu} \right] d\sigma. \end{aligned} \right\} \tag{7.97}$$

If the contours $L_1, \ldots, L_n$ are sufficiently far from each other, then the series (7.96) converges for $\lambda = 1$.

Let $U_{mr}(z)$ be a harmonic function in D_m which takes the value defined by (7.97) on L_m. Then, on the contour L_m, we have the relations

$$U_{m0}(z) = f_m(z), \qquad U_{mr}(z) = \sum_{k \neq m} [U_{k,r-1}(z) - U_{k,r-1}(\infty)]. \tag{7.98}$$

Thus, for the solution of the Dirichlet problem, we can apply the following method, which is called the *generalized Schwarz algorithm*.

As the initial approximation, we construct the harmonic function $U_{m0}(z)$ in D_m, which takes the value $f_m(z)$ on the contour. Knowing $U_{m0}(z), \ldots, U_{mr-1}(z)$, we construct $U_{mr}(z)$ in the following manner: we subtract from $U_{k,r-1}(z)$ its value at infinity and then sum them for all $k \neq m$. The functions $U_{mr}(z)$ can be obtained in D_m from the value of $U_{m,r}(z)$ on the contour with the aid of the Green's function for the region D_m. The unknown harmonic function $U(z)$ is the sum of the series

$$U(z) = \sum_{r=0}^{\infty} (-1)^r \sum_{m=1} U_{mr}(z),$$

which converges absolutely, if the contours L_k $(k = 0, 1, \ldots, n)$ are sufficiently far from each other.

The process is particularly simple in the case of a doubly-connected region. Let $f_1(z) = f(z)$ and $f_2(z) = 0$. The formulas (7.98) give: on L_1, $U_{10}(z) = f(z)$, $U_{1r}(z) = U_{2,r-1}(\infty)$; on L_2, $U_{20}(z) = 0$,

$$U_{2r}(z) = U_{1,r-1}(z) - U_{1,r-1}(\infty).$$

From these, it follows that $U_{1r}(z)$, with r odd, and $U_{2r}(z)$, with r even, are identically equal to zero. Writing

$$U_{1r}(z) = U_r(z), \qquad U_{2r}(z) = V_r(z),$$

we obtain

$$U(z) = U_0(z) - V_1(z) + U_2(z) - V_3(z) + \ldots$$

The generalized Schwarz algorithm can also be applied, without changes, to the case of a finite multi-connected region. It is sufficient to replace $l_0(\zeta)$ by zero.

5.2 *Application to three-dimensional problems*

The generalized Schwarz algorithm can be applied in a three-dimensional region. Let the region D be bounded by an external surface S_0 and internal surfaces $S_1, S_2, \ldots, S_n$. Let D_m be the region bounded by a surface S_m containing D.

The function $U(M)$ harmonic in D can be uniquely represented as the sum of functions $U_m(z)$ harmonic in D_m:

$$U(M) = \sum_{m=1}^{n} U_m(M),$$

where

$$U_m(M) = \frac{1}{4\pi} \iint_{S_m} \left(\frac{1}{r} \frac{\partial U}{\partial \nu} - U \frac{\partial(1/r)}{\partial \nu} \right) dS$$

(ν is the outward normal to S_m).

Let $f_m(M)$ be the value of $U(M)$ on the surface S_m ($m = 0, 1, \ldots, n$) and let $G_m(M; M)$ be the Green's function for the region D_m. Then, for the values $U_m(M)$ on the contour L_m, we obtain the system

$$U_m(M) + \frac{1}{4\pi} \sum_{k \neq m} \iint_{S_k} U_k(M_1) \frac{\partial G_k(M; M_1)}{\partial \nu} dS = f_m(M), \tag{7.99}$$

where $M \in S_m$ and $m = 0, 1, \ldots, n$.

This system always has a unique solution. After introducing a parameter λ the system (7.99) becomes

$$u_m(M) + \frac{\lambda}{4\pi} \sum_{k \neq m} \iint_{S_k} u_k(M_1) \frac{\partial G_k(M; M_1)}{\partial \nu} dS = f_m(M) \quad \text{on} \quad L_m.$$

We seek its solution in the form

$$u_m(M) = \sum_{r=0}^{+\infty} (-1)^r \lambda^r u_{mr}(M).$$

The functions $u_{mr}(M)$ satisfy the recurrence relations

$$u_{m0}(M) = f(M),$$

$$u_{mr}(M) = \frac{1}{4\pi} \sum_{k \neq m} \iint_{S_k} u_{k,r-1}(M_1) \frac{\partial G(M; M_1)}{\partial \nu} dS.$$

Let the $U_{mr}(M)$ be harmonic in D with values on S_m defined by the last relations. On S_m the functions $U_{mr}(M)$ satisfy

$$U_{m0}(M) = f_m(M), \qquad U_{mr}(M) = \sum_{k \neq m} U_{k,r-1}(M).$$

The successive approximations of the generalized Schwarz algorithm for the Dirichlet problem in the case of a three-dimensional region can be constructed in the following way: for the initial approximation, we construct the harmonic function $U_{m0}(M)$ in D_m which equals $f_m(M)$ on S_m. If the harmonic functions $U_{m0}(M)$, $U_{m1}(M)$, . . . , $U_{m,r-1}(z)$ have been constructed, then the function $U_{mr}(M)$ harmonic in D_m is equal to the sum $\sum_{k \neq m} U_{k,r-1}(M)$ on S_m. If the region D is doubly-connected, then the generalized Schwarz algorithm converges.

The convergence of the generalized Schwarz algorithm can be proved with the aid of methods of functional analysis for a wide class of boundary problems and regions for which the contours bounding them are situated arbitrarily close to each other.

5.3 *Applications to the theory of elasticity*

Let us consider the problem of determining the stresses present in a multi-connected region D bounded on the inside by the contours $L_1, L_2, \ldots, L_n$ and on the outside by L_0. We assume that the components of the external forces, distributed along the boundary L and acting on the elastic medium, are known. The problem reduces to finding the biharmonic function $W(x, y)$ in D, with the value of its first derivative on L known up to an arbitrary constant (see §3.1),

$$\left.\frac{\partial W}{\partial x} + i\frac{\partial W}{\partial y}\right|_{L_k} = f_k(z) + B_k \qquad (k = 0, 1, \ldots, n), \tag{7.100}$$

where z is the complex coordinate of a point on the contour, the $f_k(z)$ are given boundary functions, and the B_k are constants which are chosen so that the displacements are single-valued. (One of the constants may be chosen arbitrarily.) We assume that the resultant vector and the resultant moment of the forces acting on each of the contours L_k vanish. We denote by D_0 the region lying inside L_0 and by D_k the region lying outside L_k, $k = 1, 2, \ldots, n$.

The biharmonic function $W(x, y)$ can be expressed in the form

$$W(x, y) = W_0(x, y) + W_1(x, y) + \ldots + W_n(x, y), \tag{7.101}$$

where $W_0, W_1, \ldots, W_n$ are biharmonic functions which are regular in the corresponding regions $D_0, D_1, \ldots, D_n$.

We assume that the problem can be solved for every region D_k ($k = 0, 1, \ldots, n$). Thus, the problem reduces to the construction of the quantities

$$\frac{\partial W_k}{\partial x} + i\frac{\partial W_k}{\partial y} = g_k(z) \tag{7.102}$$

on the contour L_k ($k = 0, 1, \ldots, n$).

It is known for simply connected regions that the solution of the biharmonic equation with the boundary condition

$$\frac{\partial U}{\partial x} + i\frac{\partial U}{\partial y} = g(z)$$

can be determined at any internal point $z = x + iy$ by

$$\frac{\partial U}{\partial x} + i\frac{\partial U}{\partial y} = \int_L [M_1(x, y; \zeta)\, g(\zeta) + M_2(x, y; \zeta)\, \overline{g(\zeta)}]\, d\zeta, \tag{7.103}$$

where ζ is a point of the boundary, M_1 and M_2 are known continuous functions inside the region, which are completely determined by this region. The operator on the r.h.s. of (7.103) will be denoted by $M(z, g)$. Analogously, the operators M for the regions $D_0, D_1, \ldots, D_n$ and the boundary condition (7.102) will be denoted by $M_k(z, g_k)$ ($k = 0, 1, \ldots, n$).

We now add the constants B_k to the $f_k(z)$ on the r.h.s. of (7.100). Then the system of integral equations for the functions $g_k(x)$ becomes

$$\text{on } L_m\text{:}\quad g_m(z) + \sum_{k \neq m} M_k(z; g_k) = f_m(z) \qquad (m = 0, 1, \ldots, n). \tag{7.104}$$

This system does not have a solution for arbitrary r.h.s.

Let the system (7.94) be replaced by:

$$\text{on } L_m\text{:}\quad g_m(z) + \sum_{k \neq m} \{M_k(z; g_k) - N_k(g_k)\} = f_m(z), \tag{7.105}$$

where

$$N_k(g_k) = \int_{L_k} [\alpha_k(\zeta)\, g_k(\zeta) + \beta_k(\zeta)\, \overline{g_k(\zeta)}]\, d\zeta, \tag{7.106}$$

and the $\alpha_k(\zeta)$ and $\beta_k(\zeta)$ are continuous functions of ζ satisfying the conditions

$$\int_{L_k} \alpha_k(\zeta)\, d\zeta = 1, \qquad \int_{L_k} \beta_k(\zeta)\, d\zeta = 0.$$

This system (7.105) has a solution for arbitrary r.h.s. Solving it and finding the $g_m(z)$ ($m = 0, 1, \ldots, n$), we obtain

$$B_m = \sum_{k \neq m} N_k(g_k).$$

Having constructed the functions $W_0, W_1, \ldots, W_n$, we can find the function defined by (7.101). As the $N_k(g)$ we may take

$$\begin{aligned} N_k(g) &= M_k(\infty, g), \\ N_0(g) &= M_0(a, g), \end{aligned} \qquad (k = 1, 2, \ldots, n).$$

where a is an arbitrary fixed point inside L_0.

If the contours $L_0, L_1, \ldots, L_n$ are sufficiently far from each other, then (7.105) can be solved by the method of successive approximations. In this way, we obtain the generalized Schwarz algorithm for the plane problem of the theory of elasticity.

§6 Application of the theory of symmetric integral equations

6.1 *The Sturm-Liouville problem*

Consider an ordinary linear differential operator

$$L(y) = \frac{d}{dx}\left[p(x)\frac{dy}{dx}\right] + q(x)y, \tag{7.107}$$

with $x \in [a, b]$ and $p(x) \geqslant 0$. We consider the functions $y(x)$ which, at the ends of the interval $[a, b]$, satisfy the conditions

$$\alpha y(a) + \beta y'(a) = 0, \qquad \gamma y(b) + \delta y'(b) = 0. \tag{7.108}$$

Let us assume that the only function $y(x)$ satisfying

$$L(y) = 0$$

and the boundary conditions is identical equal to zero. Now, consider the equation

$$L(y) + \lambda r(x)y = 0, \tag{7.109}$$

where $r(x)$ is a continuous positive function and λ is a numeric parameter. We require the values of λ for which there exist non-zero solutions of (7.109) satisfying the boundary conditions (7.108).

Let $G(x, s)$ be the Green's function of $L(y)$ under the conditions (7.108). We write

$$\sqrt{(r(x)\, y(x))} = \varphi(x) \quad \text{and} \quad \sqrt{(r(x)\, r(s))}\, G(x, s) = K(x, s).$$

The function $\varphi(x)$ differs from the unknown $y(x)$ by the known factor $\sqrt{r(x)}$ and satisfies the following integral equation with a symmetric kernel

$$\varphi(x) - \lambda \int_a^b K(x, s)\, \varphi(s)\, ds = 0. \tag{7.110}$$

The characteristic values of this equation are the required values of λ. They are all simple.

Example 1. Let $L(y) = y''$ and $r(x) = 1$. Let the boundary conditions (7.108) be

$$y(0) = y(l) = 0.$$

The Green's function of y'' has the form

$$G(x, s) = \begin{cases} \dfrac{(l - s)x}{l} & \text{for} \quad x \leqslant s, \\[2ex] \dfrac{(l - x)s}{l} & \text{for} \quad s \leqslant x. \end{cases} \tag{7.111}$$

In this case, the characteristic values of (7.110) are

$$\lambda_n = n^2\pi^2$$

and the corresponding eigenfunctions are

$$y_n(x) = \sqrt{2} \sin n\pi x.$$

Example 2. Consider the operator of Example 1 for boundary conditions

$$y(0) = 0 \qquad y(1) + hy'(1) = 0.$$

The characteristic values of the corresponding integral equation (7.110) are the squares of the roots of

$$\operatorname{tg} \mu + h\mu = 0.$$

6.2 *The fundamental vibrations of a string*

Consider a string of density $\rho(x)$, which lies on the interval $[0, 1]$ of the x-axis and has tension T. The equation of transverse vibrations for the string has the form

$$\rho(x) \frac{\partial^2 u}{\partial t^2} - T \frac{\partial^2 u}{\partial x^2} = F(x, t),$$

where $u(x, t)$ is the displacement, and $F(x, t)$ is the force acting on the string {see [1], Chapter II}. We set $F(x, t) = 0$, and seek the solutions of this equation, which are periodic in t, and vanish at the ends of the interval $[0, 1]$

$$u(x, t) = v(x) \sin (\nu t + \varepsilon).$$

The function $v(x)$ must satisfy the homogeneous integral equation

$$v(x) - \frac{\nu^2}{T} \int_0^l \rho(s)\, G(x, s)\, v(s)\, ds = 0,$$

where $G(x, s)$ is defined by (7.111).

The last equation can be made symmetric. Multiplying both sides of the last equality by $\sqrt{(\rho(x))}$ and writing

$$v(x)\sqrt{\rho(x)} = \varphi(x), \qquad G(x, s)\sqrt{(\rho(x)\, \rho(s))} = K(x, s),$$

we arrive at the equation

$$\varphi(x) - \lambda \int_0^l K(x, s)\, \varphi(s)\, ds = 0, \tag{7.112}$$

the kernel of which is symmetric. This has the form of (7.110). Its characteristic values are positive. After finding its characteristic values λ_k, we obtain the fundamental frequencies of the string from

$$\frac{\nu_n}{2\pi} = \frac{1}{2\pi}\sqrt{(\lambda_n T)}.$$

For a string of unit length ($l = 1$), the density of which varies according to the linear law $\rho(x) = \rho_0(1 + x)$, we set

$$\lambda = \frac{\nu^2 \rho_0}{T}, \qquad \varphi(x) = \sqrt{(1 + x)}\, v(x)$$

and

$$K(x, s) = \sqrt{\{(1 + x)(1 + s)\}}\, G(x, s).$$

Then the equation for this string takes the form (7.112). With an error less than 2%, we obtain $\lambda_1 = 6.398$ and $\lambda_2 = 49.9$.

6.3 *The stability of a rod under compression*

The equation for the bending of the elastic axis of a rod, when it is compressed by forces applied at its ends, has the form

$$E\frac{d^2}{dx^2}\left[J(x)\frac{d^2y}{dx^2}\right] + P\frac{d^2y}{dx^2} = 0, \tag{7.113}$$

where $J(x)$ is the moment of inertia of the section of the rod with abscissa x, E is Young's Modulus and P is the value of the compressive force acting at the ends of the rod. If these ends are hinged joints, then

$$y(0) = y(l) = y''(0) = y''(l) = 0$$

(we assume that the ends of the rod have abscissae $x = 0$ and $x = l$).

We write $P/E = \lambda$ and $d^2y/dx^2 = u$. Then, (7.113) takes the form

$$\frac{d}{dx}\left(J^2(x)\frac{du}{dx}\right) + J(x)\,J''(x)u + \lambda J(x)u = 0. \tag{7.114}$$

The boundary conditions for u are

$$u(0) = u(l) = 0. \tag{7.115}$$

Let $v(x) = u(x)\sqrt{J(x)}$, then $v(x)$ satisfies the symmetric integral equation

$$v(x) - \lambda \int_0^l K(x, s)\, v(s)\, ds = 0, \tag{7.112$_1$}$$

where

$$K(x, s) = \begin{cases} \dfrac{x(l-s)}{l\sqrt{J(x)\,J(s)}} & \text{for} \quad x \leqslant s, \\[2ex] \dfrac{s(l-s)}{l\sqrt{J(x)\,J(s)}} & \text{for} \quad x \geqslant s. \end{cases}$$

The first characteristic value λ_1 has the following approximate lower bound

$$\lambda_1 = \frac{1}{\sqrt{A_2}}, \quad \text{where} \quad A_2 = 2\int_0^l dx \int_0^s K^2(x, s)\, ds.$$

The corresponding approximate estimate for the critical force takes the form

$$P_1 = E\lambda_1 \approx E/\sqrt{A_2}.$$

6.4 *The fundamental vibrations of a membrane*

Consider a membrane under equilibrium conditions which occupies a finite region D with a sufficiently smooth contour in the (x, y)-plane.

The problem of the vibration of a membrane with clamped boundary reduces to the solution of the wave equation (ignoring the trivial solution; see [1], Chapter II, §5)

$$\frac{\partial^2 U}{\partial x^2} + \frac{\partial^2 U}{\partial y^2} = \frac{1}{a^2}\frac{\partial^2 U}{\partial t^2}$$

with zero boundary conditions

$$u|_L = 0$$

for given initial displacement and velocity.

The solution of this problem can be expressed in the form

$$U = \sum_{n=1}^{+\infty} (A_n \cos a\sqrt{(\lambda_n)}t + B_n \sin a\sqrt{(\lambda_n)}t)\ \Phi_n(x, y),$$

where the $\Phi_n(x, y)$ are the non-zero solutions of

$$\Delta\Phi_n + \lambda_n\Phi_n = 0 \qquad (7.116)$$

under the boundary condition

$$\Phi_n\big|_L = 0. \qquad (7.117)$$

The values λ_n must be chosen so that non-trivial solutions of the equation (7.116) and the boundary condition (7.117) exist.

Let $G(P, Q)$ be the Green's function of the Laplace operator in the region D for the Dirichlet problem {see [1], Chapter II §6}.

The functions Φ_n are the non-trivial solutions of the integral equation

$$\Phi(P) - \lambda\iint_D G(P, Q)\ \Phi(Q)\ d\xi\ d\eta = 0, \qquad (7.118)$$

where P is a point with coordinates (x, y) and Q a point with coordinates (ξ, η).

The eigenvalues and eigenfunctions of the problem of the vibration of a membrane are the eigenvalues and eigenfunctions of the symmetric integral equation (7.118). This is an equation with a weak singularity {see Chapter II}.

The equation (7.118) has a countable set of eigenvalues $\lambda_n \xrightarrow[n\to+\infty]{} +\infty$ and eigenfunctions. If D is a rectangle bounded by the straight lines $x = 0$, $x = l$, $y = 0$, and $y = m$, then the eigenvalues are

$$\lambda_{\sigma,\tau} = \pi^2(\sigma^2/l^2 + \tau^2/m^2) \qquad (\sigma, \tau = 0, 1, 2, \ldots),$$

where σ and τ do not vanish simultaneously.

6.5 *The pressure of a rigid stamp on an elastic half-space*

We consider the half-space $z < 0$. We assume that, on a portion S of its bounding plane $z = 0$, there is an absolutely rigid body (stamp) pressing on to the half-space with a force Q parallel to the z-axis. It will be assumed that there is no friction between the half-space and the stamp. The remaining portion S' of the boundary is free from the action of external forces. We require the stress and strain field in the half-space. This problem reduces to an integral equation of the first kind.

Let $z = \varphi(x, y)$ be the equation of that part of the surface of the stamp which is in contact with the half-space. We introduce on S the function

$$\Phi(x, y) = ax + by + c - \varphi(x, y),$$

where the quantity $ax + by + c$ represents the rigid body displacement of the stamp.

The boundary conditions enables one to reduce the problem to finding a harmonic function $\varphi_0(x, y)$ satisfying the following boundary conditions:

$$\text{on } S: \quad \varphi_0 = \Phi(x, y)$$

$$\text{on } S': \quad \frac{\partial \varphi_0}{\partial z} = 0.$$

The function $\varphi_0(x, y)$ is sought in the form

$$\varphi_0(x, y, z) = \iint_S \mu(\xi, \eta) \frac{d\xi\, d\eta}{r},$$

where $r^2 = (x - \xi)^2 + (y - \eta)^2 + z^2$ and $\mu(\xi, \eta)$ is an unknown function. The boundary conditions lead to the following symmetric integral equation of the first kind

$$\iint_S \frac{\mu(\xi, \eta)}{r}\, d\xi\, d\eta = \Phi(x, y) \quad \text{on} \quad S.$$

The solution of this equation is unique and the system of eigenfunctions of the kernel $1/r$ is complete.

After finding the characteristic values and eigenfunctions of the symmetric kernel $1/r$, the function $\mu(\xi, \eta)$ can be obtained in the usual way as a series which converges in the mean.

§7 Certain applications of singular integral equations

7.1 *Mixed problem of potential theory*

We require the function, harmonic in a certain plane region D, which takes, on part of its contour, the value of the function, and on the remainder the value of its normal derivative. This is a special case of the general Hilbert problem, in which it is required to find a harmonic function which satisfies on the contour a known linear combination of the function and its normal derivative.

On each of the closed curves forming the contour L, we assume that there are arcs on which the function is given, as well as arcs on which the normal derivative is given. Let $\gamma_1, \gamma_3, \ldots, \gamma_{2n-1}$ be the arcs of the contours on which the required harmonic function are given and let

$$u = f(s), \tag{7.119}$$

on these arcs, where s is the arc length of the contour. We denote by γ_2, $\gamma_4, \ldots, \gamma_{2n}$, the arcs on which the normal derivative are given, and assume that

$$\frac{\partial U}{\partial v} = f_1(s), \tag{7.120}$$

on these arcs, where v is the outward normal to the contour.

To solve the problem, it is sufficient to find the value of $U(x, y)$ on the arcs $\gamma_2, \gamma_4, \ldots, \gamma_{2n}$. Then, the problem, which is called *the Poincaré problem*, reduces to the Dirichlet problem for a multi-connected region (see §§2.2 and 5.1).

We denote by $V(x, y)$ the harmonic function adjoint to $U(x, y)$. The Poincaré problem can be formulated as following: Find an analytic function $\phi(z) = U(x, y) + iV(x, y)$ $(z = x + iy)$ where the values of its real part are given on a part of the contour, and the imaginary part is given on the remainder:

$$\left.\begin{aligned} &\text{on } \gamma_{2k-1}: \quad U = f(s), \\ &\text{on } \gamma_{2k}: \quad V = f^*(s) + C_k \end{aligned}\right\} \qquad (k = 1, \ldots, n),$$

where the C_k are arbitrary constants and

$$f^*(s) = \int_{\alpha_k}^{s} f_1(s)\, ds \qquad (\alpha_k \text{ is the beginning of the arc } \gamma_{2k}).$$

Let $T(z; \zeta)$ be the Schwarz kernel for the region D. It has the representation

$$\frac{1}{2\pi} T(z; \zeta)\, d\sigma = \frac{1}{\pi i} \frac{d\zeta}{\zeta - z} + P(z; \zeta)\, d\sigma,$$

where $P(z, \zeta)$ is a function which is continuous with respect to both variables z and ζ in the closed region $\bar{D}$ and is regular with respect to z in the region D. We write

$$\omega(z) = \frac{1}{2\pi} \sum_{k=1}^{n} \int_{\gamma_{2k-1}} f(\sigma)\, T(z; \zeta)\, d\sigma.$$

In this way, we obtain the following singular integral equation, from which we can determine $U(x, y)$ on the contours $\gamma_2, \gamma_4, \ldots, \gamma_{2n}$

$$\text{on } \gamma_{2m}: \quad -\frac{1}{\pi} \int_{\gamma_{2k}} \sum_{k=1}^{n} U(\sigma) \operatorname{Re}\left(\frac{d\zeta}{\zeta - t}\right) d\sigma + \sum_{k=1}^{n} \int_{\gamma_{2k}} U(\sigma) \operatorname{Im}\{P(t; \zeta)\}\, d\sigma$$
$$= f^*(s) - \operatorname{Im}\{\omega(t)\} + C'_m, \qquad (m = 1, \ldots, n), \quad (7.121)$$

where $C'_m = C_m - C$ with C an arbitrary constant.

Solving this equation, we determine $\varphi(z)$ as

$$\varphi(z) = \frac{1}{2\pi i} \int_L U(\sigma)\, T(z; \zeta)\, d\sigma + iC.$$

The required $U(x, y)$ is the imaginary part of $\varphi(z)$. The equation (7.121) is a singular equation which can be reduced to an equation of Fredholm type (see Chapter VI, §9).

7.2 *Mixed problems in the half-plane*

Consider the half-plane $y > 0$. On the finite sections $\gamma_2, \gamma_4, \ldots, \gamma_{2n}$ of its boundary $y = 0$, we are given the values of the normal derivative of the function which is harmonic in the upper half-plane. On the remaining sections of the boundary, we are given the value of this function.

We denote by z the complex coordinates of a point in the upper half-plane, and by $f(\xi)$ the value of the required harmonic function $U(x, y)$ on the intervals $\gamma_1, \gamma_2, \ldots, \gamma_{2n-1}$. As before, we denote the beginning of the arc

γ_{2k} as α_k, and the end as β_k ($k = 1, 2, \ldots, n$). We denote by $u(\xi)$ the required value of the function $U(x, y)$ on the arcs $\gamma_2, \gamma_4, \ldots, \gamma_{2n}$. We obtain for $U(\xi)$ the singular integral equation

$$\frac{1}{\pi} \int\limits_{\gamma_2 + \ldots + \gamma_{2n}} \frac{u(\xi)\, d\xi}{\xi - t} = B(t), \tag{7.122}$$

where t is an arbitrary point on one of arcs γ_{2k} and

$$B(t) = -f^*(t) + \operatorname{Im}\{\omega(t)\} - C'_m \qquad (t \in \gamma_{2m}).$$

The functions $f^*(t)$ and $\omega(t)$ and the constants C'_m can be determined in the same way as before. The general solution of (7.122) has the form

$$\text{on } \gamma_{2m}:\quad u(t) = \frac{1}{\pi \sqrt{\left[\prod_{k=1}^{n} (t - \alpha_k)(\beta_k - t)\right]}} \times \int\limits_{\gamma_2 + \ldots + \gamma_{2n}} B(\xi) \sqrt{\left[\prod_{k=1}^{n} (\xi - \alpha_k)(\beta_k - \xi)\right]} \frac{d\xi}{\xi - t} \tag{a}$$

$$+ \frac{Q_{n-1}(t)}{\sqrt{\left[(-1)^{n-1} \prod_{k=1}^{n} (t - \alpha_k)(\beta_k - t)\right]}}, \tag{7.123}$$

where $Q_{n-1}(t)$ is a polynomial of degree $n - 1$ with real-valued coefficients. The formula (7.123) contains $2n$ arbitrary constants {n coefficients of the polynomial $Q_{n-1}(t)$ and n arbitrary constants C'_m, contained in $B(t)$}. These constants can be chosen so that $u(t)$ is continuous.

7.3 *The problem of two elastic half-planes in contact*

Consider two elastic half-planes {the upper ($y > 0$) and the lower ($y < 0$)}, which are in frictionless contact along the segment $-a \leqslant x \leqslant a$. The segment $|x| \leqslant a$ is an infinitely narrow slit separating the two half-planes. Each half-plane is subjected to the same normal pressure with intensity h. We require the stress fields in both half-planes under the assumption that the stress at infinity is zero.

The problem reduces to the determination of an analytic function in the upper (lower) half-plane $\Phi(z)$ which satisfies the following boundary conditions on the x-axis

$$\left.\begin{aligned} \operatorname{Re}\{\Phi(z)\} &= 0 && \text{for} \quad y = 0 \quad \text{and} \quad |x| > a, \\ \operatorname{Im}\{\Phi(z)\} &= -\frac{h}{2} && \text{for} \quad y = 0 \quad \text{and} \quad |x| < a. \end{aligned}\right\} \tag{7.124}$$

Thus $\Phi(z)$ is the solution of the Poincaré problem {§7.1}. In order to find $\Phi(z)$, we must consider the singular integral equation (7.121). Solving this equation, we obtain

$$\Phi(z) = \frac{ih}{2}\left(\frac{z}{\sqrt{(z^2 - a^2)}} - 1\right). \tag{7.125}$$

For the lower half-plane, $\Phi(z)$ remains the same.

In order to find the stress and strain fields, it is sufficient to know the Goursat functions $\phi(z)$ and $\psi(z)$ (see §3.1). These functions are determined with the aid of (7.125) along with

$$\begin{aligned} \varphi'(z) &= i\Phi(z), \\ \psi(z) &= \varphi(z) - z\varphi'(z). \end{aligned}$$

This method is also applicable in the more general case, when there are several slits on the common boundary of the two elastic half-planes.

7.4 *The pressure of a rigid stamp on an elastic half-plane*

We consider a rigid stamp pressing on an elastic half-plane $y > 0$. Let the shape of the stamp pressing on the half-plane be $y = f(x)$ for $|x| \leqslant a$. Suppose also that the parts $|x| > a$ of the boundary are free from stresses and that the stamp is supported by forces normal to its boundary. The stress field in the half-plane is required.

As before, let $\varphi(z)$ and $\psi(z)$ be Goursat functions and let $\phi'(z) = p(x, y) + iq(x, y)$. The function $\psi(z)$ is defined by

$$z\varphi'(z) + \psi(z) = \varphi(z).$$

For $p(x, y)$, we obtain the following singular integral equation

$$\frac{1}{\pi}\int_{-a}^{a} \frac{p(\xi)\, d\xi}{\xi - t} = \frac{2\mu}{\kappa + 1} f'(t), \tag{7.126}$$

where μ and κ are elastic constants of the half-plane $y < 0$, and t is an arbitrary point of the interval $(-a, a)$ $\{$if $t \notin (-a, a)$, then $p = 0\}$. The solution of (7.126) is given by

$$p(t) = p_0(t) + \frac{A}{\sqrt{(a^2 - t^2)}}, \tag{7.127}$$

where

$$p_0(t) = -\frac{2\mu}{\pi(\kappa + 1)\sqrt{(a^2 - t^2)}} \int_{-a}^{a} \frac{f'(\xi)\sqrt{(a^2 - \xi^2)}}{\xi - t}\, d\xi.$$

The constant A can be determined, if we know the resultant vector P of forces acting on the stamp:

$$A = \frac{P}{2\pi} - \frac{1}{\pi} \int_{-a}^{a} p_0(t)\, dt.$$

After that $\varphi'(z)$ can be obtained as

$$\varphi'(z) = -\frac{1}{\pi i} \int_{-\infty}^{+\infty} \frac{p(\xi, 0)}{\xi - z}\, d\xi.$$

For the case of a plane stamp $f(\xi) = \text{const.}$,

$$p(t) = \frac{P}{2\pi\sqrt{(a^2 - t^2)}} \quad \text{and} \quad \varphi'(z) = -\frac{P}{2\pi\sqrt{(a^2 - z^2)}}.$$

In the case when there are several stamps on the half-plane, the problem is solved, in principle, in a similar way. However, the calculations are rather more complicated.

7.5 *The mixed problem of the theory of elasticity*

We consider a certain plane region D which is, generally speaking, multi-connected and filled with an elastic medium; and assume that on part of its contour, which we denote by M, the displacement of the medium is given, and that on the remaining part L, the forces, acting on the elastic medium, are given. We assume that the whole contour $C = L + M$ is sufficiently smooth, and suppose that L consists of several arcs $\gamma_1, \gamma_2, \ldots, \gamma_n$ which are not connected to each other. The complex coordinates of the beginning and end of the arc γ_k are denoted by α_k and β_k, respectively ($k = 1, \ldots, n$). The problem of the determination of the stresses in such a medium is called the mixed problem of the theory of elasticity. This problem reduces to finding the Goursat functions $\varphi(z)$ and $\psi(z)$.

According to (7.47) and (7.48), the following conditions must be satisfied on the contour

$$\kappa\varphi(\zeta) - \overline{\zeta\varphi'(\zeta)} - \overline{\psi(\zeta)} = 2\mu(u + iv) \quad \text{for} \quad \zeta \in M, \tag{7.128}$$

$$\varphi(\zeta) + \overline{\zeta\varphi'(\zeta)} + \overline{\psi(\zeta)} = i\int (X_\nu + iY_\nu)\, ds + b_k \quad \text{for} \quad \zeta \in \gamma_k. \tag{7.129}$$

If we introduce the functions $\delta(\zeta)$ and $f(\zeta)$ defined by

$$\delta(\zeta) = \begin{cases} 0 & \text{on} \quad M, \\ \kappa + 1 & \text{on} \quad L, \end{cases}$$

$$f(\zeta) = \begin{cases} -i\displaystyle\int (X_\nu + iY_\nu)\, ds & \text{on} \quad L, \\ 2\mu(u + iv) & \text{on} \quad M, \end{cases}$$

then the conditions (7.128) and (7.129) can be rewritten in the form

$$\kappa\varphi(\zeta) - \delta(\zeta)\,\varphi(\zeta) - \overline{\zeta\varphi'(\zeta)} - \overline{\psi(\zeta)} = f(\zeta) + c(\zeta). \tag{7.130}$$

Let $T(z; \zeta)$ be the Schwarz kernel for the region D, and

$$P(z; \zeta) = \frac{1}{4\pi} T(z; \zeta) - \frac{1}{\zeta - z}\frac{d\zeta}{d\sigma},$$

where $d\sigma$ is an element of arc length on the contour c {see §2.5}, and where $c(\zeta)$ is constant over every connected component of the contour C.

For $\varphi(\zeta)$, we obtain the following singular integral equation:

$$(\kappa - 1)\,\varphi(t) - \frac{\kappa + 1}{\pi i}\int_L \frac{\varphi(\zeta)}{\zeta - t}\, d\zeta = \Phi(t), \tag{7.131}$$

where t is a point on L and

$$\Phi(t) = \frac{\kappa + 1}{4\pi}\int_L \varphi(\zeta)\, P(t; \zeta)\, d\sigma - \frac{1}{2\pi}\int_L \overline{\varphi(\zeta)}\, K(t; \zeta)\, d\sigma + 2F(t) + 2\sum_{k=1}^{n} b_k F_k(t). \quad (7.132)$$

Here

$$F(z) = \frac{1}{4\pi}\int_C f(\zeta)\, T(z; \zeta)\, d\sigma, \qquad F_k(z) = -\frac{1}{4\pi}\int_{\gamma_k} T(z; \zeta)\, d\sigma$$

and

$$K(z, \zeta) = \frac{d}{d\sigma}\left\{[(\zeta - z)\, T(z; \zeta) - zT(a; \zeta)]\, \frac{\overline{d\sigma}}{d\zeta}\right\}, \quad (7.133)$$

where a is an arbitrary fixed point of the region D.

The r.h.s. of (7.131) contains an unknown function under the integral sign. However, the kernels of these integrals are only weakly singular. The solution of (7.131), continuous at the points α_k ($k = 1, \ldots, n$), can be expressed in the form

$$\varphi(t) = -\frac{\kappa - 1}{4\pi}\,\Phi(t) - \frac{\kappa + 1}{4\kappa i}\prod_{k=1}^{n}\left(\frac{t - \alpha_k}{t - \beta_k}\right)^m \int_L \prod_{k=1}^{n}\left(\frac{\zeta - \beta_k}{\zeta - \alpha_k}\right)^m \frac{\Phi(\zeta)\, d\zeta}{\zeta - t}\,\frac{\Phi(\zeta)\, d\zeta}{\zeta - t} \quad (7.134)$$

where $m = (1/2\pi i) \ln (-1/\kappa)$.

The constants b_k, introduced in (7.129) and (7.132), can be chosen so that the solution (7.134) is continuous at the points β_k. Using the expression (7.134) for the real and imaginary parts of $\varphi(\zeta)$, we can obtain a system of two Fredholm integral equations, which is always solvable. The function $\varphi(z)$ can be determined from its boundary values using a Cauchy integral.

If the region D can be mapped conformally onto a circle by a rational function, then the mixed problem of the theory of elasticity for this region can be solved in finite form using quadrature.

7.6 *The problem of flow past an arc of given shape*

In the complex plane, consider a smooth open arc AB past which flows fluid with velocity components at infinity defined by U and V. We denote by α and β the complex coordinates of the points A and B. Let $w(z) = \varphi(x, y) + i\psi(x, y)$ be the complex velocity potential of the flow. We require the function $w(z)$ which is analytic outside AB and satisfies the conditions

$$\lim_{z\to\infty} [\omega'(z) - (U - iV)] = 0, \tag{7.135}$$

$$\operatorname{Im} w(z) = \text{const.} \quad \text{if} \quad z \in AB. \tag{7.136}$$

It is also assumed that the velocity of the flow along the arc AB, except perhaps at the point A, is finite. We use the notation

$$\omega(z) = w'(z) - (U - iV). \tag{7.137}$$

The function $\omega(z)$ is sought in the form of a Cauchy-type integral

$$\omega(z) = \frac{1}{2\pi i}\int_{AB} \frac{T(\zeta)\, d\sigma}{\zeta - z}, \tag{7.138}$$

where $d\sigma = |d\zeta|$ with ζ a variable point on AB. The function $T(\zeta)$ is called *the vortex function.*

The vortex function satisfies the following singular integral equation

$$\frac{1}{2\pi}\int_{AB} \frac{T(\zeta)\sin(r, n)}{r}\, d\sigma = V\cos\vartheta - U\sin\vartheta, \tag{7.139}$$

where (r, n) is the angle between the direction of the vector $t - \zeta$ $\{t$ is a point of the arc $AB\}$ and the normal at t, and ϑ is the angle between the tangent to AB at t and the x-axis.

The equation (7.139) has an infinite number of solutions. The condition that the velocity at the point B must be finite leads to

$$T(\beta) = 0. \tag{7.140}$$

The solution of the equation (7.139), satisfying this condition, must be substituted into (7.138).

In the case, when AB is the interval $(-a, a)$ of the real variable axes, (7.139) takes the form

$$\frac{1}{2\pi}\int_{-a}^{a}\frac{T(\zeta)\,d\zeta}{\zeta - t} = -V.$$

Its solution, satisfying the condition

$$T(a) = 0,$$

has the form

$$T(t) = -2V\sqrt{\left(\frac{a-t}{a+t}\right)}.$$

Then, $\omega(z)$ will be

$$\omega(z) = iV - V\sqrt{\left(\frac{a-z}{a+z}\right)}$$

with

$$w'(z) = U - V\sqrt{\left(\frac{a-z}{a+z}\right)}.$$

An explicit solution of the problem can also be obtained in the case when AB is an arc of a circle.

Chapter VIII

INTEGRAL EQUATIONS WITH CONVOLUTION KERNELS

§1 General introduction

1.1 *The basic equation and special cases*

In this chapter, we describe the properties of equations which are particular cases of

$$\mu\varphi - \boldsymbol{k}\varphi \equiv \mu\varphi(x) - \sum_{j=1}^{n} b_j(x) \int_{-\infty}^{\infty} k_j(x-t)\,\varphi(t)\,dt = f(x), \tag{8.1}$$

$$-\infty < x < \infty,$$

where the $b_j(x)$ are bounded measurable functions which have finite limits as $x \to \pm\infty$ and $\mu = \text{const}$.

The most important special cases of (8.1) are:

1) Volterra equations (equations of this type have been considered in §3 of Chapter 1)

$$\mu\varphi(x) - \int_{0}^{x} k(x-t)\,\varphi(t)\,dt = f(x), \qquad x > 0$$

$$\left(j = 1, \qquad b_1(x) = \begin{cases} 1, & x > 0, \\ 0, & x < 0, \end{cases} \qquad k(x), \qquad f(x) = 0 \quad \text{for} \quad x < 0 \right). \tag{8.2}$$

2) Equations defined on an infinite interval

$$\mu\varphi(x) - \int_{-\infty}^{\infty} k(x-t)\,\varphi(t)\,dt = f(x) \qquad (j = 1, \qquad b_1(x) = 1). \tag{8.3}$$

3) The Wiener-Hopf equation

$$\mu\varphi(x) - \int_0^{\infty} k(x - t)\,\varphi(t)\,dt = f(x)$$

$$\left(j = 1, \qquad b_1(x) = \begin{cases} 1, & x > 0, \\ 0, & x < 0, \end{cases} \quad f(x) = 0 \quad \text{for} \quad x < 0\right). \tag{8.4}$$

4) Dual equations

$$\mu\varphi(x) - \int_{-\infty}^{\infty} k_1(x - t)\,\varphi(t)\,dt = f(x), \qquad x < 0,$$

$$\mu\varphi(x) - \int_{-\infty}^{\infty} k_2(x - t)\,\varphi(t)\,dt = f(x), \qquad x > 0, \tag{8.5}$$

$$\left(j = 2, \qquad b_1(-x) = b_2(x) = \begin{cases} 1, & x > 0, \\ 0, & x < 0, \end{cases}\right).$$

5) Equations defined on a finite interval

$$\mu\varphi(x) - \int_0^{T} k(x - t)\,\varphi(t)\,dt = f(x), \qquad 0 \leqslant x \leqslant T$$

$$\left(j = 1, \qquad b_1(x) = \begin{cases} 1, & x \in [0, T] \\ 0, & x \notin [0, T] \end{cases}\right). \tag{8.6}$$

The adjoint equations corresponding to (8.1), (8.2) and (8.5) are

$$\boldsymbol{k}\varphi = \mu\varphi(x) - \sum_{j=1}^{n} \int_{-\infty}^{\infty} b_j(t)\,k_j(t - x)\,\varphi(t)\,dt = f(x),$$

$$-\infty < x < \infty, \quad (8.1_1)$$

$$\mu\varphi(x) - \int_x^{\infty} k(t - x)\,\varphi(t)\,dt = f(x), \qquad x > 0, \tag{8.2$_1$}$$

$$\mu\varphi(x) - \int_{-\infty}^{0} k_1(t - x)\,\varphi(t)\,dt - \int_{0}^{\infty} k_2(t - x)\,\varphi(t)\,dt = f(x),$$

$$-\infty < x < \infty. \quad (8.5_1)$$

The equations, adjoint to (8.3), (8.4), (8.6), can be obtained by replacing $k(x - t)$ with $k(t - x)$.

Interest in equations (8.1)–(8.6) centres around the fact that these equations arise in the study of some important problems (see [16], [17], [23], [25], [26], [27], [28], [29], [46]).

1.2 *The symbol. Conditions for normal solvability*

The properties of (8.1) depend mostly on the properties of its symbol

$$\sigma(x, \lambda) = \mu - \sum_{j=1}^{n} b_j(x)\,K_j(\lambda), \tag{8.7}$$

where

$$K_j(\lambda) = \int_{-\infty}^{\infty} k_j(x) e^{i\lambda x}\,dx$$

is the Fourier transform of the kernel $k_j(x)$. The following theorem is valid.

THEOREM 1.1. (see [30]). *If the kernels $k_j(x)$ of* (8.1) *are summable on the whole axis, then:*

a) *In order that* (8.1) *should be normally solvable in the space $L_p(-\infty, \infty)$, $1 < p < \infty$, and have finite symbol, it is necessary and sufficient that the limit values of the symbol do not vanish as $x \to \pm\infty$:*

$$\sigma(\pm\infty, \lambda) = \mu - \sum_{j=1}^{n} b_j(\pm\infty)\,K_j(\lambda) \neq 0, \qquad -\infty < \lambda < \infty. \tag{8.8}$$

b) *If (a) is satisfied, then the index of* (8.1) *is defined in the spaces $L_p(-\infty, \infty)$, $1 < p < \infty$, by the formula*

$$\chi_\mu = \frac{1}{2\pi} \arg \frac{\sigma(-\infty, \lambda)}{\sigma(+\infty, \lambda)} \Bigg|_{-\infty}^{\infty} \tag{8.9}$$

c) *If μ coincides with a certain value of the functions $\sigma(+\infty, \lambda)$, $-\infty < \lambda < \infty$, then either the operator $(\mu\boldsymbol{I} - \boldsymbol{k})$ {where $\boldsymbol{I}$ is the identity operator}* *is not normally solvable, or the space of zeros of the operator* $\mu \boldsymbol{I} - \boldsymbol{k}$ *and* $(\mu \boldsymbol{I} - \boldsymbol{k})^*$ *are infinite dimensional.*

d) *If* (8.8) *is satisfied, then one of the regularizators (see Chapter 6, §5) of* (8.1) *is*

$$\boldsymbol{R}\varphi = \begin{cases} \dfrac{1}{\mu}\varphi(x) + \displaystyle\int_{-\infty}^{\infty} q_-(x-t)\,\varphi(t)\,dt, & x < 0, \\[2ex] \dfrac{1}{\mu}\varphi(x) + \displaystyle\int_{+\infty}^{\infty} q_+(x-t)\,\varphi(t)\,dt, & x > 0, \end{cases} \tag{8.10}$$

where the functions $q_\pm(x) \in L(-\infty, \infty)$, *and are defined by their Fourier transforms*

$$Q_\pm(\lambda) = \frac{\dfrac{1}{\mu}\displaystyle\sum_{j=1}^{n} b_j(\pm\infty)\,K_j(\lambda)}{\mu - \displaystyle\sum_{j=1}^{n} b_j(\pm\infty)\,K_j(\lambda)}.$$

All these properties remain valid without any essential changes if for some h

I. $e^{-hx}\,k_j(x) \in L(-\infty, \infty)$,

II. $e^{-hx}\,f(x), \qquad e^{-hx}\,\varphi(x) \in L_p(-\infty, \infty)$.

In fact, the substitutions $\varphi(x) = e^{hx}\,\varphi_1(x)$, $f(x) = e^{hx}\,f_1(x)$ transform (8.1) into

$$\mu\varphi_1(x) - \sum_{j=1}^{n} b_j(x) \int_{-\infty}^{\infty} k_{j1}(x-t)\,\varphi_1(t)\,dt = f_1(x),$$

where the kernels $k_{j1}(x) = e^{-hx}\,k_j(x) \in L(-\infty, \infty)$. A wider class of equations for which a theorem, similar to Theorem 1.1., is valid is examined in [32].

1.3 *Equations with elementary solutions*

In the case, when the functions $k_j(x)$ are rational and the coefficients $b_j(x)$ are piecewise constant, (8.1) can be solved very simply by means of the method described below (see [31]). Another method of solution is described in [36].

With piecewise constant coefficients $b_j(x)$, (8.1) can be rewritten in the form

$$\varphi(x) + \int_{-\infty}^{\infty} n_j(x - t)\,\varphi(t)\,dt = f(x),$$

$$\alpha_j < t < \alpha_{j+1}, \qquad -\infty = \alpha_0 < \alpha_1 < \ldots < \alpha_s = +\infty, \tag{8.1'}$$

where the intervals (α_j, α_{j+1}) are chosen so that all $b_j(x)$ are constants on them. The functions $n_j(x)$ are linear combinations of the functions $k_j(x)$, and hence, their Fourier transforms $N_j(\lambda)$ are also rational. Let

$$1 + N_j(\lambda) = \frac{\prod_{v=1}^{p_j} (\lambda - \alpha_{vj}^{+})^{p_{vj}} \prod_{v=1}^{q_j} (\lambda - \alpha_{vj}^{-})^{q_{vj}}}{\prod_{v=1}^{r_j} (\lambda - \beta_{vj}^{+})^{r_{vj}} \prod_{v=1}^{s_j} (\lambda - \beta_{vj}^{-})^{s_{vj}}},$$

$$\sum p_{vj} + \sum q_{vj} = \sum r_{vj} + \sum s_{vj},$$

where $\operatorname{Im} \alpha_{v,j}^{+}$ and $\operatorname{Im} \beta_{v,j}^{+} > 0$, $\operatorname{Im} \alpha_{v,j}^{-}$ and $\operatorname{Im} \beta_{v,j}^{-} < 0$. The solution of (8.1_1) can be expressed in the form

$$\varphi(x) = \boldsymbol{R}f + \sum_{v=1}^{p_j} e^{-i\alpha_{vj}^{-}x}\left(\sum_{k=0}^{p_{vj}-1} M_{vjk}x^k\right) + \sum_{v=1}^{q_j} e^{-i\alpha_{vj}^{+}x}\left(\sum_{k=1}^{q_{vj}-1} N_{vjk}x^k\right),$$

$$\alpha_j < x < \alpha_{j+1}, \qquad 0 < j < s - 1. \tag{8.11}$$

For the intervals $(-\infty, \alpha_1)$ and $(\alpha_{s-1}, +\infty)$, corresponding to $j = 0$ and $j = s - 1$, respectively, the first and the second sum vanish. $R\varphi$ is an operator of the same form as (8.1) but with kernels $a_j(x - t)$, the Fourier transforms of which are defined by

$$1 + A_j(\lambda) = [1 + N_j(\lambda)]^{-1}.$$

After the substitution of (8.11) into (8.1), we obtain a system of linear algebraic equations, which is equivalent to (8.1). The coefficients M_{vjk}, N_{vjk} can be determined from this algebraic system.

As an example of the application of this method, we consider the equation

$$\varphi(t) + 4\left(\int_{-\infty}^{0} e^{-|t-\tau|}\,\varphi(\tau)\,d\tau + \int_{1}^{\infty} e^{-|t-\tau|}\,\varphi(\tau)\,d\tau\right)$$

$$= \frac{2}{3}e^{5t} + \frac{e^t}{3}(3 + 2e^{-6}) \equiv f_1, \qquad t < 0, \quad \text{(a)}$$

$$\varphi(t) + 3\left(\int_{-\infty}^{0} e^{-2|t-\tau|}\,\varphi(\tau)\,d\tau + \int_{0}^{\infty} e^{-2|t-\tau|}\,\varphi(\tau)\,d\tau\right)$$

$$= \frac{3}{7}e^{-5t} + \frac{e^{-2t}}{7}(3 + 7e^{-3}) \equiv f_2, \qquad t > 0.$$

We continue $\varphi(t)$, letting it be zero on the interval $[0, 1]$, and then rewrite the given equation in the form (8.1_1)

$$\varphi(t) + 4\int_{-\infty}^{\infty} e^{-|t-\tau|}\,\varphi(\tau)\,d\tau = f_1(t), \qquad t < 0,$$

$$\varphi(t) = 0, \qquad 0 < t < 1,$$

$$\varphi(t) + 3\int_{-\infty}^{\infty} e^{-2|t-\tau|}\,\varphi(\tau)\,d\tau = f_2(t), \qquad t > 1.$$

After simple calculations we obtain $1 + N_0(\lambda) = (\lambda^2 + 9)/(\lambda^2 + 1)$, $1 + N_2(\lambda) = (\lambda^2 + 16)/(\lambda^2 + 4)$, $N_1(\lambda) = 0$. Hence

$$1 + A_0(\lambda) = [1 + N_0(\lambda)]^{-1} = 1 - \frac{8}{\lambda^2 + 9},$$

$$1 + A_2(\lambda) = [1 + N_2(\lambda)]^{-1} = 1 - \frac{12}{\lambda^2 + 4}, \qquad A_1(\lambda) = 0,$$

and

$$
R\varphi = \begin{cases} \varphi - \dfrac{4}{3}\displaystyle\int_{-\infty}^{\infty} e^{-3|t-\tau|}\,\varphi(\tau)\,d\tau, & t < 0, \\ \varphi, & 0 < t < 1, \\ \varphi - \dfrac{3}{2}\displaystyle\int_{-\infty}^{\infty} e^{-2|t-\tau|}\,\varphi(\tau)\,d\tau, & t > 1. \end{cases}
$$

The zeros of the functions $1 + N_0(\lambda)$ and $1 + N_2(\lambda)$ are:

$$
\alpha_{10}^{+} = 3i, \qquad \alpha_{10}^{-} = -3i, \qquad \alpha_{12}^{+} = 4i, \qquad \alpha_{12}^{-} = -4i.
$$

Further, if $t < 0$, then

$$
Rf = f_1 - \frac{4}{3}\left[\int_{-\infty}^{0} e^{-3|t-\tau|}\,f_1(\tau)\,d\tau + \int_{1}^{\infty} e^{-3|t-\tau|}\,f_2(\tau)\,d\tau\right] = e^{5t} + ae^{3t},
$$

If $t > 1$, then

$$
Rf = f_2 - \frac{3}{2}\left[\int_{-\infty}^{0} e^{-4|t-\tau|}\,f_1(\tau)\,d\tau + \int_{1}^{\infty} e^{-4|t-\tau|}\,f_2(\tau)\,(d\tau)\right]
$$
$$
= e^{-5t} + be^{-4t}.
$$

Here a and b denote numbers, the explicit values of which are not important, According to (8.11)

$$
\varphi(t) = e^{5t} + Ae^{3t}, \qquad t < 0,
$$
$$
\varphi(t) = e^{-5t} + Be^{-4t}, \qquad t > 1.
$$

Substituting this into the given equation, we obtain the relationships: $2A + 5e^{-2}B = 0, 5A + 2e^{-5}B = 0$, from which it follows that $A = B = 0$. Hence, the solution of the equation is the function

$$
\varphi(t) = \begin{cases} e^{5t}, & t < 0, \\ e^{-5t}, & t > 1. \end{cases}
$$

It can be shown that the homogeneous equation has no non-trivial solutions.

1.4 *Equations which reduce to the form* (8.1)

We list several types of equations which can be reduced to the form (8.1) by a simple substitution (see [42]).

a) Using the substitution $x = e^{-t}$, $y = e^{-\tau}$, the equation

$$\varphi(x) - \int_0^A \frac{1}{y} k\left(\frac{x}{y}\right) \varphi(y)\, dy = f(x), \qquad 0 < x < A,$$

can be reduced to the form

$$\varphi_1(t) - \int_{-\ln A}^{\infty} k_1(t - \tau)\, \varphi_1(\tau)(\tau)\, d\tau = f_1(t), \tag{8.12}$$

in which $\varphi_1(t) = \varphi(e^{-t})$, $f_1(t) = f(e^{-t})$, $k_1(t) = k(e^{-t})$. If $A = \infty$, then we obtain an equation defined on the whole axis; if $A = 1$, then the equation is defined on the half-axis. We can reduce the equation with the kernel $(1/x)\, k(x/y)$ to the form (8.12) by the same substitution when

$$\varphi_1(t) = e^{-t}\, \varphi(e^{-t}), \qquad f_1(t) = e^{-t}\, \varphi(e^{-t}), \qquad k_1(t) = k(e^{-t}).$$

b) Dual equations of the form

$$\int_0^{\infty} k_1(xy)\, \varphi(y)\, dy = f(x), \qquad 0 \leqslant x < 1,$$

$$\int_0^{\infty} y^{\alpha} k_2(xy)\, \varphi(y)\, dy = g(x), \qquad 1 \leqslant x < \infty,$$

often arise in the theory of electricity. By means of the substitution $x = e^{-t}$, $y = e^{\tau}$, it can be reduced to the form

$$\int_{-\infty}^{\infty} q_2(t - \tau)\, \varphi_1(\tau)\, d\tau = g_1(t), \qquad -\infty < t < \infty,$$

$$\int_{-\infty}^{\infty} q_1(t - \tau)\, \varphi_1(\tau)\, d\tau = f_1(t), \qquad 0 < t < \infty,$$

where $\varphi_1(\tau) = \varphi(e^\tau)$, $q_1(t) = e^{-t} k_1(e^{-t})$, $q_2(t) = e^{-(\alpha+1)t} k_2(e^{-t})$, $g_1(t) = e^{-(\alpha+1)t} g(e^{-t})$, $f_1(t) = e^{-t} f(e^{-t})$.

c) The equation

$$\varphi(x) = f(x) + \int_0^\infty y^{-1-\alpha}(x+y)^\alpha \varphi(y)\, dy, \qquad 0 < x < \infty,$$

can be reduced to the form

$$\varphi_1(t) = f_1(t) + \int_{-\infty}^{\infty} \left[2 \operatorname{ch} \frac{t-\tau}{2}\right]^\alpha \varphi_1(t)\, dt$$

by the substitution $x = e^t$, $y = e^\tau$, $\varphi_1(t) = \varphi(e^t)e^{-(\alpha t/2)}$, $f_1(t) = f(e^t)e^{-(\alpha t/2)}$.

d) If $k(x)$ is a homogeneous function of degree α, then the equation

$$\varphi(x) = f(x) + \int_0^\infty y^{-(\alpha+1)} k(x+y)\, \varphi(y)\, dy$$

can be reduced to the form

$$\varphi_1(t) = f_1(t) + \int_{-\infty}^{\infty} k\left[2 \operatorname{ch} \frac{t-\tau}{2}\right]^\alpha \varphi_1(t)\, dt$$

by the substitution used in (c). This substitution also works for the equation with the kernel $x^{-\alpha}y^{-1} k(x+y)$, when $k(x)$ is a homogeneous function of degree α.

1.5 *Function spaces*

Henceforth, E^+ will denote any one of the following spaces:

$L_p(0, +\infty)$—the space of functions which are summable on $(0, \infty)$ with exponent p;

$M(0, +\infty)$—the space of functions which are (essentially) bounded on $(0, \infty)$;

$C(0, +\infty)$—the space of functions which are bounded and continuous on $(0, +\infty)$;

$C_u(0, +\infty)$—the space of functions which are uniformly continuous on $(0, \infty)$;

$C_0(0, +\infty)$—the space of continuous functions which tend to zero as $x \to +\infty$.

By E (and $E(a, b)$), we shall denote any one of the above spaces defined on the whole axis $(-\infty, +\infty)$ (and the finite interval $[a, b]$). By E', we denote the space of functions $f(x)$ such that, for $x > 0$, $f(x) \in E^+$ and, for $x < 0$, $f(-x) \in E^+$. By $e^{hx}E$, $e^{hx}E^+$ and $e^{hx}E'$, we denote the space of functions $f(x)$ such that $e^{-hx} f(x) \in E, E^+$ and E', respectively. For the spaces E, E^+ and E', except L_p when $p > 1$, the integrals

$$\int_{-\infty}^{\infty} k(x - t)\, \varphi(t)\, dt \left(\int_{0}^{\infty} k(x - t)\, \varphi(t)\, dt \right) \tag{8.13}$$

are interpreted in the usual Lebesgue sense when the kernels $k(x)$ are summable. For the spaces L_p with $p > 1$, the sense is

$$\operatorname*{Lim}_{A \to +\infty} \int_{-A}^{A} \left(\operatorname*{Lim}_{A \to +\infty} \int_{0}^{A} \right),$$

where Lim denotes the limit in the sense of the mean with exponent p. The integral operator (8.13) is bounded in all spaces E, E^+ and E' and its norm does not exceed $\|\boldsymbol{k}\|_L$. This operator is bounded in the spaces $L_2(-\infty, \infty)$ $\{$and $L_2(0, +\infty)\}$, if $k(x) \in L_2(-\infty, +\infty)$ $\{$and $L_2(0, +\infty)\}$ and has a bounded Fourier transform $K(\lambda)$. The norm of this operator is equal to $\sup |K(\lambda)|$. (For other conditions on the boundedness of the operator (8.13) see [24], [45]).

In §6, when describing the properties of equations with kernels of different exponential growth as $x \to +\infty$ and $x \to -\infty$, we shall refer to the spaces $\{\alpha, \beta\}$ (see [7], [42], [48]). By definition, $f(x) \in \{\alpha, \beta\}$ if $e^{-\alpha x} f(x) \in L_2(0, +\infty)$ and $e^{-\beta x} f(x) \in L_2(-\infty, 0)$.

The simplest properties of these spaces are:

1) If $\alpha \leqslant \alpha_1$, $\beta_1 \leqslant \beta$, then $\{\alpha, \beta\} \in \{\alpha_1, \beta_1\}$.
2) If $f_k(x) \in \{\alpha_k, \beta_k\}$ then $\sum f_k(x) \in \{\max \alpha_k, \min \beta_k\}$.
3) $\{\alpha, \alpha\} \cap \{\beta, \beta\} = \{\min(\alpha, \beta), \max(\alpha, \beta)\}$.
4) If $k(x) \in \{\alpha, \beta\}$ and $\varphi(x) \in \{\alpha_1, \beta_1\}$, then (8.13) has sense only if $\alpha_1 \leqslant \beta$ and $\alpha \leqslant \beta_1$. The integral itself belongs to the space $\{\max(\alpha_1, \alpha), \min(\beta_1, \beta)\}$.

5) If $f(x) \in \{\alpha, \beta\}, \alpha \leqslant y \leqslant \beta$, then the Fourier transform of $f(x)$ is defined:

$$F(\zeta) = F_y f = \int_{-\infty}^{\infty} f(x) e^{i\zeta x}\, dx, \qquad \operatorname{Im} \zeta = y.$$

If $\alpha < \beta$, then $F(\zeta)$ is holomorphic on the strip $\alpha < \operatorname{Im} \zeta < \beta$ and the integral

$$\int_{-\infty}^{\infty} |F(x + iy)|^2\, dx, \qquad \alpha \leqslant y \leqslant \beta$$

is uniformly bounded with respect to y.

Conversely, if $F(\zeta)$ satisfies these conditions, then it is the Fourier transform of the function

$$f(x) = \frac{1}{2\pi} \int_{-\infty + iy}^{\infty + iy} e^{-i\zeta x} F(\zeta)\, d\zeta \in \{\alpha, \beta\}.$$

The class of such functions $F(\zeta)$ is denoted by $\{\{\alpha, \beta\}\}$.

Examples of the spaces $\{\alpha, \beta\}$:

1) $\{0, 0\} = L_2(-\infty, +\infty)$.
2) $\{\alpha, \alpha\} = e^{\alpha x} L_2(-\infty, \infty)$; if $f(x) \in \{\alpha, \alpha\}$, then $F(\lambda + i\alpha) \in L_2(-\infty, \infty)$.
3) $\{-\alpha, \alpha\} = e^{-\alpha|x|} L_2(-\infty, \infty)$.
4) $\{\alpha, +\infty\}$ (or $\{-\infty, \alpha\}$) consists of functions, which vanishes for $x < 0$ (or, for $x > 0$) and belong to $e^{\alpha x} L(0, +\infty)$ for $x > 0$ (or, to $e^{\alpha x} L_2(-\infty, 0)$ for $x < 0$). The Fourier transforms of functions from $\{\alpha, +\infty\}$ (or from $\{-\infty, \alpha\}$) are holomorphic in the half-plane $\operatorname{Im} \zeta > \alpha$ (or, in $\operatorname{Im} \zeta < \alpha$).

§2 Examples

2.1 *The basic problem of the theory of radiation* (see [41], [46])

This problem consists in determining radiation intensity. In the case of an isotropic parallel plane dispersive medium, the problem reduces to the solution of the integral equation

$$B(\tau) = \frac{\lambda}{2} \int_0^{\tau_0} B(\tau') Ei(|\tau - \tau'|)\, d\tau' + g(\tau), \qquad 0 \leqslant \tau \leqslant \tau_0,$$

where $Ei(x) = \int_1^{\infty} (e^{-xy}\, dy/y)$, λ is a parameter characterizing the dispersive power of the medium, $g(\tau)$ is a known function defined by the nature of the radiation source, τ_0 is the optical intensity of the medium, and $B(\tau)$ is the function to be determined in terms of which the radiation intensity can be easily evaluated.

If the optical intensity is infinite (this assumption is made in the study of dispersion in a stellar atmosphere) then we obtain the Wiener–Hopf equation. If the source of radiation is absent, then $g(\tau) \equiv 0$, and the problem reduces to the homogeneous form. The problem of neutron scattering has an analogous formulation (see [26]).

2.2 *The problem of linear smoothing and forecasting* (see [17], [23], [62])

Let the input signal to a linear stationary system be the sum of a signal $s(t)$ and a noise $n(t)$, both of which are assumed to be stationary random functions. The system transforms the input signal into an output $X_{\text{output}}(t)$, which can be expressed in the form

$$X_{\text{output}}(t) = \int_0^\infty [s(t - \tau) + n(t - \tau)]\, w(\tau)\, d\tau,$$

where $w(\tau)$ is an impulse transforming function, which depends on the structure of the system. It is necessary to determine the transforming function $w(\tau)$ so that the expectation of $\varepsilon(t) = s(t + T) - X_{\text{output}}(t)$, is a minimum $\{\varepsilon(t)$ defines pure smoothing when $T = 0$, and pure forecast when $n(t) \equiv 0\}$. This problem reduces to an integral equation of the first kind with respect to the unknown function $w(\tau)$. If the statistical characteristics of the signal and the noise are known for past time (from $-\infty$ up to the present), then we obtain the Wiener-Hopf equation

$$\psi(t + T) = \int_0^\infty \varphi(t - \tau)\, w(\tau)\, d\tau, \qquad t > 0.$$

If these statistical characteristics are only known over a finite interval of time $[t - T_0, t]$, then we obtain the equation

$$\psi(t + T) = \int_0^{T_0} \varphi(t - \tau)\, w(\tau)\, d\tau, \qquad 0 \leqslant t \leqslant T_0.$$

In both these equations $\psi(t)$ and $\varphi(t)$ are known functions which are defined in terms of the correlation and auto-correlation functions of the signal and noise.

2.3 *The electromagnetic coastal effect* (see [16])

In this problem, the effect on dry land of electromagnetic waves from the sea is examined. Let the coast be a straight infinite line, choose the x-axis perpendicular to the coast and the y-axis along the coast, and assume that the oncoming waves have the form $F^0(x)e^{isy}$. It is necessary to determine on the surface of the dry land the field in the form $F(x)e^{isy}$. We obtain the following equation for the unknown function $F(x)$

$$F(x) = 2P^0(x) + \pi\alpha i \int_0^\infty H_0^{(1)}(m\,|x - t|)\, F(t)\, dt, \qquad x > 0,$$

where $H_0^{(1)}$ is the Hankel function of the first kind, and α and m are constants.

2.4 *A problem in the theory of hereditary elasticity* (see [29], [56])

The determination of a deformation $\varepsilon(t)$ of a linearly stressed body for a given tension $\sigma(t)$ can be reduced, in the theory of hereditary elasticity, to the equation

$$\varepsilon(t) = \frac{\sigma(t)}{E} + \chi \int_0^t k(t - \tau)\, \varepsilon(\tau)\, d\tau, \qquad t > 0,$$

where E and χ are constants, characterizing the elastic properties of the medium, $k(t - \tau)$ is the kernel of relaxation, characterizing the hereditary properties of the medium, and t is the time of deformation.

2.5 *The potential of a conducting disk*

The problem of determining the potential $v(\rho, z)$ of a charged conducting disc reduces to the equations

$$\int_0^\infty f(u)\, J_0(\rho u)\, du = g(\rho), \qquad 0 < \rho < 1,$$

$$\int_0^\infty u f(u)\, J_0(\rho u)\, du = 0, \qquad \rho > 1$$

$\left(\text{here } v(\rho, z) = \int_0^\infty f(u)\, J_0(\rho u) e^{-uz}\, du\right).$

The substitution $u = e^{-\tau}$ and $\rho = e^t$ transforms these equations to the form

$$\int_{-\infty}^{\infty} f_1(\tau)\, J_0(e^{t-\tau})e^{t-\tau}\, d\tau = g_1(t), \qquad -\infty < t < 0,$$

$$\int_{-\infty}^{\infty} f_1(\tau)\, J_0(e^{t-\tau})e^{2(t-\tau)}\, d\tau = 0, \qquad 0 < t < \infty,$$

$f_1(t) = f(e^{-t}), g_1(t) = e^t\, g(e^t)$ (see [42], page 423).

§3 Equations defined on a semi-infinite interval with summable kernels {see [22]}

3.1 *The solvability conditions*

When studying the properties of the equation

$$\varphi(x) = f(x) + \int_0^{\infty} k(x - t)\, \varphi(t)\, dt, \qquad 0 < x < \infty, \tag{8.14}$$

we assume that the following conditions are satisfied:

A) $k(x) \in L(-\infty, +\infty)$

B) $1 - K(\lambda) \neq 0, \qquad -\infty < \lambda < \infty$

Under these conditions, all the results of §1 concerning (8.1) are also valid for (8.14). However, these properties can be supplemented.

THEOREM 3.1. *In any of the spaces E^+ (see* §1), (8.14) *is normally solvable, and has finite index*

$$\chi = -\frac{1}{2\pi} \arg\,[1 - K(\lambda)]\Big|_{-\infty}^{\infty}.$$

Conversely, if in some space E^+, (8.14) *is normally solvable, has finite index and condition A is satisfied, then condition B is also satisfied.*

Note. Theorem 3.1 is valid in the space $L_2(0, \infty)$ under the following suppositions: $k(x) \in L_2(-\infty, \infty)$, $K(\lambda)$ is continuous and $K(\pm\infty) = 0$ (see [9]).

THEOREM 3.2. *The number of linearly independent solutions of the homogeneous equation*

$$\varphi(x) = \int_0^\infty k(x - t)\, \varphi(t)\, dt \tag{8.14$_1$}$$

is equal to χ, *if* $\chi > 0$, *and to zero, if* $\chi \leqslant 0$. *These solutions are one and the same in any of the spaces* E^+, *and they are*

a) *summable, absolutely continuous, and vanish as* $x \to +\infty$, *and*

b) *can be chosen so that*

$$\varphi_{k+1} = \frac{d\varphi_k}{dx}, \qquad \varphi_k(0) = 0, \qquad k = 0, 1, \ldots, \chi - 2,$$

$$\frac{d\varphi_{\chi-1}}{dx} = \gamma(x), \qquad \varphi_{\chi-1}(0) \neq 0,$$

where $\gamma(x)$ *is defined below by* (8.17), (8.18), (8.20).

Functions satisfying the conditions a) and b) are said to form a D-chain of length x.

Corollaries which follow from Theorems 3.1 and 3.2 are:

If $\chi = 0$, then (8.14) is uniquely solvable in E^+ for arbitrary r.h.s. from E^+. If $\chi > 0$, then, as before, (8.14) is solvable for any $f \in E^+$, but the solution will not be unique. If $\chi < 0$, then the homogeneous equation adjoint to (8.14$_1$) has $|\chi|$ linearly independent solutions. For the solvability of the non-homogeneous equation (8.14) in E^+, it is necessary and sufficient that its r.h.s. $f(x)$ should satisfy the conditions

$$\int_0^\infty f(x)\, \psi_j(x)\, dx = 0 \qquad (j = 0, 1, \ldots, |\chi| - 1), \tag{8.15}$$

where $\psi_j(x)$ is a basis for the solutions of the adjoint homogeneous equation.

3.2 *Factorization*

All known methods of solution for (8.14) are based on the so called factorization of the function $[1 - K(\lambda)]^{-1}$, i.e. the representation of this function as the product of functions which are holomorphic in certain half-planes. In the case of summable kernels, it takes the form:

THEOREM 3.3. *Let* $K(x) \in L(-\infty, +\infty)$ *and* $1 - K(\lambda) \neq 0$ *for* $-\infty < \lambda < \infty$. *If the index* $\chi \geqslant 0$, *then the only valid factorization is*

$$[1 - K(\lambda)]^{-1} = \sigma_-(\lambda)\left(\frac{\lambda - i}{\lambda + i}\right)^{\chi} \sigma_+^0(\lambda), \tag{8.16}$$

where $\sigma_-(\lambda)$, $\sigma_+^0(\lambda)$ *are holomorphic in the upper and lower half-plane, respectively, continuous up to the boundaries and have no zeros there.*

For simplicity of presentation, we write

$$\sigma_+(\lambda) = \left(\frac{\lambda - i}{\lambda + i}\right)^{\chi} \sigma_+^0(\lambda).$$

The following representations hold for $\sigma_+(\lambda)$:

$$\sigma_+(\lambda) = 1 + \int_0^\infty \gamma(x)e^{i\lambda x}\, dx, \qquad \sigma_-(\lambda) = 1 + \int_0^\infty \gamma_\tau(x)e^{i\lambda x}\, dx,$$

$$\gamma(x), \qquad \gamma_\tau(x) \in L(0, +\infty). \tag{8.17}$$

Under the suppositions of Theorem 3.3, $\sigma_\pm(\lambda)$ can be expressed in terms of $K(\lambda)$ by means of formulas

$$\ln \sigma_+(\lambda) = -\frac{1}{2\pi i}\int_{-\infty}^{\infty} \frac{\ln [1 - K(\mu)]}{\mu - \lambda}\, d\mu, \qquad \operatorname{Im} \lambda > 0, \tag{8.18}$$

$$\ln \sigma_-(\lambda) = \frac{1}{2\pi i}\int_{-\infty}^{\infty} \frac{\ln [1 - K(\mu)]}{\mu - \lambda}\, d\mu, \qquad \operatorname{Im} \lambda < 0. \tag{8.18$_1$}$$

In both formulas, we chose logorithmic branches which vanish as $\lambda \to \infty$. The integrals must be interpreted in terms of their Cauchy principal values. Under the same assumptions, ln $[1 - K(\lambda)]$ can be expressed as the Fourier integral of a summable function

$$\ln[1 - K(\lambda)] = \int_{-\infty}^{\infty} e^{i\lambda x} l(x)\, dx, \qquad l \in L(-\infty, \infty),$$

and $\sigma_{\pm}$ can be expressed in terms of $l(x)$ by

$$\sigma_+(\lambda) = e^{\int_0^{\infty} l(x) e^{i\lambda x} dx}, \qquad \sigma_-(\lambda) = e^{\int_{-\infty}^{0} l(x) e^{i\lambda x} dx}, \tag{8.19}$$

where $\gamma(x)$ and $\gamma_\tau(x)$ are defined by

$$\gamma(x) = \frac{1}{2\pi} \int_{-\infty + hi}^{\infty + hi} e^{-i\lambda x}[\sigma_+(\lambda) - 1]\, d\lambda,$$
$$\gamma_\tau(x) = \frac{1}{2\pi} \int_{-\infty - hi}^{\infty - hi} e^{-i\lambda x}[\sigma_-(\lambda) - 1]\, d\lambda. \tag{8.20}$$

This is valid when, for some $h > 0$, the Cauchy principal values of the integrals on the r.h.s. exist (this is the case if, for example, $\sigma_{\pm}(\lambda) = 1 + c/\lambda + 0(1/|\lambda|^{\alpha})$, $\alpha > 1$).

In the special, but important case, when $K(\lambda)$ is rational, the factorization is elementary.

Let, for instance,

$$k(x) = \sum_{j=1}^{r} \sum_{k=0}^{r_j - 1} a_{jk} x^k e^{-i\lambda_j x}\, \theta(x) + \sum_{j=1}^{q} \sum_{k=0}^{q_j - 1} b_{jk} x^k e^{-i\mu_j x}\, \theta(-x),$$

Then

$$\operatorname{Im} \lambda_j < 0, \qquad \operatorname{Im} \mu_j > 0, \qquad \theta(x) = \begin{cases} 0, & x < 0, \\ 1, & x > 0. \end{cases}$$

$$[1 - K(\lambda)]^{-1} = \frac{\prod_{j=1}^{r} (\lambda - \lambda_j)^{r_j} \prod_{j=1}^{q} (\lambda - \mu_j)^{q_j}}{\prod_{j=1}^{p} (\lambda - \alpha_j)^{p_j} \prod_{j=1}^{s} (\lambda - \beta_j)^{s_j}}, \quad \operatorname{Im} \alpha_j < 0, \quad \operatorname{Im} \beta_j > 0,$$

$$\sum s_j + \sum p_j = \sum r_j + \sum q_j.$$

If $m_+ = \sum_{j=1}^{q} q_j$, $n_+ = \sum_{j=1}^{s} s_j$, then

$$\left.\begin{aligned} \sigma_+(\lambda) &= \frac{\prod_{j=1}^{q}(\lambda - \lambda_j)^{q_j}(\lambda - i)^{m_+ - n_+}}{\prod_{j=1}^{p}(\lambda - \alpha_j)^{p_j}}; \\ \sigma_-(\lambda) &= \frac{\prod_{j=1}^{r}(\lambda - \mu_j)^{r_j}(\lambda - i)^{n_+ - m_+}}{\prod_{j=1}^{s}(\lambda - \beta_j)^{s_j}}. \end{aligned}\right\} \qquad (8.17_1)$$

The functions $\gamma(x)$ and $\gamma_\tau(x)$ can be obtained by factorizing $\sigma_+(\lambda)$ and $\sigma_-(\lambda)$ into partial fractions. If

$$\sigma_+(\lambda) = 1 + \sum_{j=1}^{p} \sum_{k=1}^{p_j} \frac{A_{jk}}{(\lambda - \alpha_j)^k}, \qquad \sigma_-(\lambda) = 1 + \sum_{j=1}^{s} \sum_{k=1}^{s_j} \frac{B_{jk}}{(\lambda - \beta_j)^k},$$

then

$$\left.\begin{aligned} \gamma(x) &= \sum_{j=1}^{p} \sum_{k=1}^{p_j} \frac{i^k}{(k-1)!} A_{jk} x^{k-1} e^{-i\alpha_j x}, \\ \gamma_\tau(x) &= \sum_{j=1}^{p} \sum_{k=1}^{s_j} \frac{(-i)^k}{(k-1)!} B_{jk} x^{k-1} e^{i\beta_j x}. \end{aligned}\right\} \qquad (8.20_1)$$

Other examples of simple factorizations are given in [22], [25], [27].

3.3 *Solution of the non-homogeneous equation*

THEOREM 3.4. *Let the index χ of* (8.14) *be non-negative and let $\gamma(x)$ and $\gamma_\tau(x)$ be functions defining the factorization $[1 - K(\lambda)]^{-1}$ (see* (8.17)–(8.20)). *One of the solutions of* (8.14) (*which is unique for $\chi = 0$*) *in any of the spaces E^+ is given by*

$$\varphi(x) = f(x) + \int_0^\infty \gamma(x, t) f(t)\, dt, \qquad (8.21)$$

in which the resolvent kernel $\gamma(x, t)$ *is defined by*

$$\gamma(x, t) = \gamma(x - t) + \gamma_\tau(t - x) + \int_0^{\min(x,t)} \gamma(x - r)\,\gamma_\tau(t - r)\,dr$$

$$(\gamma(t), \gamma_\tau(t) \equiv 0 \quad \text{for} \quad t < 0). \quad (8.22)$$

If the kernel $k(x)$ is even, then $\chi = 0$, $\gamma(x) = \gamma_\tau(x)$ and

$$\gamma(x, t) = \gamma[|x - t|] + \int_0^{\min(x,t)} \gamma(x - r)\,\gamma(t - r)\,dr. \quad (8.22_1)$$

If the kernel is Hermitian, i.e., $k(-t) = \overline{k(t)}$, then $\chi = 0, \gamma_\tau(x) = \overline{\gamma(x)}$ and

$$\gamma(x, t) = \gamma(x - t) + \overline{\gamma(t - x)} + \int_0^{\min(x,t)} \gamma(x - r)\,\overline{\gamma(t - r)}\,dr. \quad (8.22_2)$$

THEOREM 3.5. *Let the index* χ *of* (8.14) *be negative and the solvability conditions* (8.15) *be satisfied. Then the only solution of* (8.14) *is given by*

$$\varphi(x) = f(x) + \int_0^\infty \gamma_\tau(t, x)\,f(t)\,dt, \quad (8.21_1)$$

in which $\gamma_\tau(t, x)$ *is the resolvent of the adjoint equation.*

We see from the preceding formulas that the resolvent $\gamma(x, t)$ (or $\gamma_\tau(x, t)$) is completely defined by the functions $\gamma(x)$ and $\gamma_\tau(x)$, which are determined by (8.17)–(8.20).

Note. The Fourier transform of the solution of the non-homogeneous equation (8.14) in the space $L(-\infty, \infty)$ can be expressed by

$$\Phi(\lambda) = \sigma_+ p_+[\sigma_- F(\lambda)] \quad (8.21_2)$$

where $F(\lambda)$ is the Fourier transform of $f(x)$, and p_+ is an operator which transforms

$$\Omega(\lambda) = \int_{-\infty}^{\infty} e^{i\lambda x}\,\omega(x)\,dx, \qquad \omega \in L,$$

into

$$p_+\Omega = \int_0^\infty e^{i\lambda x}\,\omega(x)\,dx.$$

In other words: if $\Omega = \Omega^+ - \Omega^-$, and Ω^+ (or Ω^-) is a holomorphic function in the upper (or lower) half-plane, then $p_+\Omega = \Omega^+$ (see the examples in §5).

3.4 *Solution of the homogeneous equation*

THEOREM 3.6. *Let the index of* (8.14) *be positive. Then the complete system of linearly independent solutions of the homogeneous equation* (8.14_1) *in any of the spaces* E^+ *is the system of functions* $g_1(x), g_2(x), \ldots, g_\chi(x)$, *the Fourier transforms of which are*

$$G_1(\lambda) = i\,\frac{\sigma_+(\lambda)}{\lambda - i},\; G_2(\lambda) = i^2\,\frac{\sigma_+(\lambda)}{(\lambda - i)^2}, \ldots, G_\chi(x) = i^\chi\,\frac{\sigma_+(\lambda)}{(\lambda - i)^\chi}$$

The basis for the solutions, viz. $\varphi_k(x)$, $k = 0, 1, \ldots, \chi - 1$, mentioned in Theorem 3.3, can be expressed in terms of the functions $g_k(x)$ by

$$\varphi_s(x) = \sum_{k=0}^{s} C_s^k g_{\chi-k}, \qquad s = 0, 1, \ldots, \chi - 1.$$

The functions g_k and φ_k can also be determined from the system

$$\frac{dg_1}{dx} = g_1(x) + \gamma(x), \qquad \frac{dg_k}{dx} = g_k(x) + g_{k-1}(x),$$

$$g_k(0) = 0, \qquad k = 2, 3, \ldots, \chi, \qquad g_1(0) = 1,$$

$$\frac{d\varphi_k}{dx} = \varphi_{k+1}(x), \qquad \varphi_k(0) = 0, \qquad k = 0, 1, \ldots, \chi - 2,$$

$$\frac{d\varphi_{\chi-1}}{dx} = \gamma(x), \qquad \varphi_{\chi-1}(0) = 1.$$

Methods for the approximate solution of (8.14) are examined in [13], [14], [18], [21].

Equations of the form (8.1) with kernels, which are summable with respect to a given weight, are examined in [1], [8], [38], [39], [44], [47], [48], [50]. (see also §§6, 7).

§4 Dual equations with summable kernels and their adjoints {see [11]}

4.1 *Reduction of the dual equation to an equivalent equation defined on a semi-infinite interval*

The dual equation

$$\left.\begin{aligned} \varphi(x) - \int_{-\infty}^{\infty} k_1(x-t)\,\varphi(t)\,dt &= f(x), \qquad x < 0, \\ \varphi(x) - \int_{-\infty}^{\infty} k_2(x-t)\,\varphi(t)\,dt &= f(x), \qquad x > 0, \end{aligned}\right\} \tag{8.24}$$

is examined under the assumptions:

A_1. $k_1(x), k_2(x) \in L(-\infty, \infty)$.

B_1. $1 - K_j(x) \neq 0, j = 1, 2, -\infty < \lambda < \infty$.

C_1. The r.h.s. and the solution belong to one and the same space of type E' (see §1.5).

Under these conditions, (8.24) can be reduced to an equivalent equation defined on a semi-infinite interval. If we put

$$g(x) = \varphi(x) - \int_{-\infty}^{\infty} k_1(x-t)\,\varphi(t)\,dt - f(x), \qquad -\infty < x < \infty, \tag{8.25}$$

then:

a) $\varphi(x)$ is uniquely defined in terms of $g(x)$ by

$$\varphi(x) = f(x) + g(x) + \int_{-\infty}^{\infty} q(x-t)[g(t) + f(t)]\,dt, \tag{8.26}$$

where $q(x) \in L(-\infty, \infty)$ and has the Fourier transform

$$Q(\lambda) = \frac{K_1(\lambda)}{1 - K_1(\lambda)}. \tag{8.27}$$

b) $g(x)$ satisfies

$$g(x) - \int_{0}^{\infty} k_0(x - t)\, g(t)\, dt = \int_{-\infty}^{\infty} k_0(x - t)\, f(t)\, dt, \tag{8.28}$$

$$x > 0, \qquad g(x) = 0, \qquad x < 0,$$

in which

$$k_0(x) \in L(-\infty, \infty) \quad \text{and} \quad K_0(\lambda) = \frac{K_2(\lambda) - K_1(\lambda)}{1 - K_1(\lambda)}. \tag{8.29}$$

Equations (8.24) and (8.28) are equivalent in the sense that (8.25) and (8.26) define a one-one correspondence between their solutions. If $\varphi, f \in E'$, then $g \in E^+$ and, conversely, if $g \in E^+$ and $f \in E'$ then $\varphi \in E'$.

4.2 *The formula for the index and properties of the basis for the solutions of the homogeneous equation*

Theorems 3.1–3.2 remain valid for (8.24), if $K(\lambda)$ is replaced by

$$\frac{K_2(\lambda) - K_1(\lambda)}{1 - K_1(\lambda)}$$

and the space E^+ by E'.

The formula for the index takes the form

$$\chi = \frac{1}{2\pi} \arg \frac{1 - K_1(\lambda)}{1 - K_2(\lambda)}\bigg|_{-\infty}^{\infty}. \tag{8.30}$$

There are some changes in the properties of the basis for the solutions of the homogeneous equation: a non-trivial solution exists, if $\chi > 0$.

The homogeneous equation has one and the same solution in the spaces E'. We can pick out a basis for the solutions, viz. $\varphi_0, \varphi_1, \ldots, \varphi_{\chi-1}$, such that:
a) the φ_k $(k = 0, 1, \ldots, \chi - 2)$ are summable, absolutely continuous and vanishes as $x \to \pm\infty$;
b) $\varphi_{\chi-1}$ is summable, vanishes as $x \to \pm\infty$ and becomes absolutely continuous after the subtraction of

$$\theta(x) = \begin{Bmatrix} 1, & x > 0 \\ 0, & x < 0 \end{Bmatrix}$$

from it;

c)

$$\varphi_{k+1} = \frac{d\varphi_k}{dx} \qquad (k = 0, 1, \ldots, \chi - 1),$$

$$\frac{d}{dx}[\varphi_{\chi-1}(x) - \theta(x)] \in L(-\infty, \infty).$$

With the aid of the transformation (8.26), this basis can be obtained from the basis for the solutions of the homogeneous equation (8.28).

4.3 *The equation adjoint to the dual*

The equation

$$\varphi(x) - \int_{-\infty}^{0} k_1(t - x)\, \varphi(t)\, dt - \int_{0}^{\infty} k_2(t - x)\, \varphi(t)\, dt = f(x), \tag{8.31}$$

$$-\infty < x < \infty,$$

under the assumptions A_1, B_1, C_1 (see the beginning of this section), is equivalent to

$$\varphi(x) = f(x) + \int_{-\infty}^{\infty} q(t - x)\, f(t)\, dt + \int_{0}^{\infty} k_0(t - x)\, \varphi(t)\, dt, \tag{8.32}$$

$$-\infty < x < \infty,$$

in which $q(x)$ and $k_0(x)$ have the abovementioned sense.

The r.h.s. of (8.32) contains only the values of $\varphi(x)$ for $x > 0$. Solving this equation for $x > 0$, we obtain the values of $\varphi(x)$ for $x < 0$. Thus, the solution of the adjoint equation reduces to the solution of the equation defined on a semi-infinite interval.

THEOREM 4.1. *If conditions A_1 and B_1 are satisfied, then* (8.31) *is normally solvable in any of the spaces E and has finite index*

$$\chi_\tau = \frac{1}{2\pi} \arg \frac{1 - K_2(\lambda)}{1 - K_1(\lambda)}\bigg|_{-\infty}^{\infty}. \tag{8.33}$$

If, in one of the spaces E, (8.31) *is normally solvable, has finite index, and condition* A_1 *is satisfied, then condition* B_1 *is satisfied.*

THEOREM 4.2. *The homogeneous equation* (8.31) *has the same solutions in all the spaces E. The number of linearly independent solutions is equal to* χ_τ, *if* $\chi_\tau > 0$ *and to zero if* $\chi_\tau \leqslant 0$. *These solutions are summable, absolutely continuous and vanish at infinity. They can be chosen so that*

$$\psi_j(x) = \int_{-\infty}^{x} \psi_{j+1}(t)\, dt \qquad (j = 1, 2, \ldots, \chi_\tau - 1). \tag{8.34}$$

§5 Examples

In this section, we consider examples illustrating the general ideas of the two preceding sections.

Example 5.1. It is necessary to solve the equation

$$\varphi(x) + 4\int_0^\infty e^{-|x-t|}\, \varphi(t)\, dt = e^{-x}, \qquad 0 < x < \infty, \tag{8.35}$$

where $K(\lambda) = 8/(\lambda^2 + 1)$ and thus $1 + K(\lambda) = (\lambda^2 + 9)/(\lambda^2 + 1)$.

Since the symbol $1 + K(\lambda)$ is rational, it is natural to use the simpler formulas (8.17_1) and (8.20_1):

$$\sigma_+(\lambda) = \frac{\lambda + i}{\lambda + 3i}, \qquad \sigma_-(\lambda) = \frac{\lambda - i}{\lambda - 3i}$$

and $\gamma(x) = -2e^{-3x}$, $\gamma_\tau(x) = 2e^{-3x}$ (the kernel is even). On the basis of (8.22_1)

$$\gamma(x, t) = -2e^{-3|x-t|} + 4\int_0^{\min(x,t)} e^{-3(x-r)}e^{-3(t-r)}\, dr$$

$$= -\tfrac{2}{3}e^{-3(x+t)} - \tfrac{4}{3}e^{-3|x-t|}.$$

It follows from (8.21) that

$$\varphi(x) = e^{-x} - \frac{2}{3}\int_0^\infty [2e^{-3|x-t|} + e^{-3(x+t)}]e^{-t}\,dt = \frac{e^{-3x}}{2},$$

The index χ of an equation is equal to the difference between the number of poles and the number of zeros of $1 + K(\lambda)$ in the upper half-plane. In this case $\chi = 0$. Hence, the homogeneous equation has no solutions and the non-homogeneous is always solvable.

This example has a simpler solution, if we make use of (8.21_1):

$$\Phi(\lambda) = \sigma_+(\lambda)\, p_+[\sigma_-(\lambda)\, F(\lambda)],$$

where $F(\lambda) = i/(\lambda + i)$, $\sigma_-(\lambda)\, F(\lambda) = [i/2(\lambda + i)] + [i/2(\lambda - 3i)]$. Hence, $p_+[\sigma_- F] = [i/2(\lambda + i)]$. Therefore, $\Phi(\lambda) = [i/2(\lambda + 3i)]$ and $\varphi(x) = \frac{1}{2}e^{-3x}$.

Example 5.2. Consider (8.14) with its symbol defined by

$$1 - K(\lambda) = e^{-1/(\lambda^2+\alpha^2)}, \qquad \alpha > 0. \tag{8.36}$$

We determine $\sigma_\pm(\lambda)$ using (8.18), (8.18_1)

$$\ln \sigma_+(\lambda) = \frac{1}{2\pi i}\int_{-\infty}^{\infty} \frac{d\mu}{(\mu^2 + \alpha^2)(\mu - \lambda)} = \frac{-1}{2\alpha i(\lambda + \alpha i)},$$

$$\ln \sigma_-(\lambda) = -\frac{1}{2\pi i}\int_{-\infty}^{\infty} \frac{d\mu}{(\mu^2 + \alpha^2)(\mu - \lambda)} = \frac{1}{2\alpha i(\lambda - \alpha i)}.$$

A simpler way to determine $\sigma_\pm(\lambda)$ involves the use of the decomposition

$$[1 - K(\lambda)]^{-1} = \exp\left[-\frac{1}{2\alpha i(\lambda + \alpha i)}\right] \exp\left[\frac{1}{2\alpha i(\lambda - \alpha i)}\right].$$

It now follows that

$$\sigma_+(\lambda) = 1 + [e^{[i/2\alpha(\lambda+\alpha i)]} - 1] = 1 + \sum_{k=1}^{\infty} \frac{i^k}{(2\alpha)^k(\lambda + \alpha i)^k}.$$

This series is the Fourier transform of

$$\gamma(x) = \sum_{k=1}^{\infty} \frac{x^{k-1}e^{-\alpha x}}{(2\alpha)^k\, k!\,(k-1)!} = -i\sqrt{\left(\frac{1}{2\alpha x}\right)}e^{-\alpha x}\, J_1\left(i\sqrt{\frac{2x}{\alpha}}\right),$$

where J_1 is a Bessel function.

The solution of this equation in any of the spaces E^+ is given by (8.21) and (8.22_1).

We now assume that the r.h.s. $f(x) \in L(0, \infty)$ and that its Fourier transform $F(\lambda)$ has only a finite number of simple poles $\beta_1, \beta_2, \ldots, \beta_s$ with residuals $B_1, B_2, \ldots, B_s$. Thus the function

$$\sigma_-(\lambda)\, F(\lambda) - \sum_{k=1}^{s} \frac{B_k \sigma_-(\beta_k)}{\lambda - \beta_k}$$

is holomorphic in the lower half-plane so that

$$p_+[\sigma_-(\lambda)\, F(\lambda)] = \sum_{k=1}^{s} \frac{B_k \sigma_-(\beta_k)}{\lambda - \beta_k}.$$

It follows from (8.21_2) that

$$\Phi(\lambda) = \sigma_+(\lambda) \sum_{k=1}^{s} \frac{B_k \sigma_-(\beta_k)}{\lambda - \beta_k}$$

and

$$\varphi(x) = \frac{1}{2\pi} \sum_{k=1}^{s} B_k \sigma_-(\beta_k) \int_{-\infty}^{\infty} e^{-ix\lambda} \frac{\sigma_+(\lambda)}{\lambda - \beta_k}\, d\lambda$$

$$= \frac{1}{2\pi} \sum_{k=1}^{s} B_k e^{[1/2\alpha i(\beta_k - \alpha i)]} \int_{-\infty}^{\infty} e^{-ix\lambda} \frac{e^{-[1/2\alpha i(\lambda + \alpha i)]}}{\lambda - \beta_k}\, d\lambda$$

$$= \sum_{k=1}^{s} B_k e^{[1/2\alpha i(\beta_k - \alpha i)] - i\beta_k x} \left[1 - i\int_0^x \sqrt{\left(\frac{1}{2\alpha t}\right)} J_1\left(i\sqrt{\frac{2t}{\alpha}}\right) e^{-(\alpha - i\beta_k)t}\, dt \right]$$

(8.37)

Example 5.3. Consider the dual equation

$$\varphi(x) + \frac{3}{2}\int_{-\infty}^{\infty} e^{-|x-t|}\,\varphi(t)\,dt = e^{x}, \qquad x < 0,$$

$$\varphi(x) - \frac{5}{6}\int_{-\infty}^{\infty} e^{-3|x-t|}\,\varphi(t)\,dt = 2e^{-x}, \qquad x > 0. \tag{8.38}$$

According to the results of the preceding section, this equation is equivalent to the Wiener-Hopf equation (8.28) with

$$K_1(\lambda) = -\frac{3}{\lambda^2+1}, \qquad K_2(\lambda) = \frac{5}{\lambda^2+9},$$

$$K_0(\lambda) = \frac{8}{\lambda^2+9}, \qquad 1 - K_0(\lambda) = \frac{\lambda^2+1}{\lambda^2+9}.$$

The Fourier transform of the r.h.s. of the equation, which is equivalent to (8.28), is

$$F_1(\lambda) = F(\lambda)\,K_0(\lambda) = \frac{8}{\lambda^2+9}\left[\frac{2i}{\lambda+i} - \frac{i}{\lambda-i}\right].$$

The factorization of $[1 - K_0(\lambda)]^{-1}$ yields: $\sigma_+(\lambda) = (\lambda + 3i)/(\lambda + i)$, $\sigma_-(\lambda) = (\lambda - 3i)/(\lambda - i)$ (see Example 5.1). On the basis of (8.21_2), the Fourier transform of the solution of (8.28) equals

$$G(\lambda) = \sigma_+ p_+[\sigma_- F_1].$$

Since $\sigma_- F_1 = -\frac{3}{2}i/(\lambda + 3i) + 4i/(\lambda + i) - \frac{5}{2}i/(\lambda - i) - [2/(\lambda - i)^2]$, we obtain

$$p_+[\sigma_- F_1] = -\frac{3}{2}\frac{i}{\lambda+3i} + \frac{4i}{\lambda+i} \quad \text{and} \quad G(\lambda) = \frac{5}{2}\frac{i}{\lambda+i} - \frac{8}{(\lambda+i)^2}.$$

Then, from (8.26), it follows that the Fourier transform of the solution of (8.38) can be expressed $\{$if $f \in L(-\infty, \infty)\}$ in the form

$$\Phi(\lambda) = \frac{F(\lambda) + G(\lambda)}{1 - K_1(\lambda)} = -\frac{69}{4}\frac{i}{\lambda+2i} + \frac{25}{4}\frac{i}{\lambda-2i} + \frac{16i}{3(\lambda+i)}.$$

Therefore,

$$\varphi(x) = \begin{cases} -\frac{69}{4}e^{-2x} + \frac{16}{3}e^{-x}, & x > 0, \\ -\frac{25}{4}e^{2x}, & x < 0. \end{cases}$$

The index of (8.38) is equal to zero since its kernels are even. Thus, the above solution is unique.

Example 5.4. We seek the solution of the homogeneous equation

$$\varphi(x) - \int_0^\infty k(x - t)\, \varphi(t)\, dt = 0,$$

where

$$k(x) = \begin{cases} \frac{48}{5}e^{-2x} - \frac{50}{3}e^{-3x}, & x > 0, \\ \frac{1}{15}e^{3x}, & x < 0. \end{cases}$$

The symbol of this equations equals

$$1 - K(\lambda) = \frac{(\lambda + i)(\lambda + 2i)^2}{(\lambda - 2i)(\lambda^2 + 9)}.$$

Hence, the index of this equation equals 2 (the difference between the number of poles and zeros of the symbol in the upper half-plane). Factorization (see (8.17_1)) yields

$$\sigma_+(\lambda) = \frac{(\lambda - i)^2(\lambda + 3i)}{(\lambda + 2i)^2(\lambda + i)}, \qquad \sigma_-(\lambda) = \frac{(\lambda - 2i)(\lambda - 3i)}{(\lambda - i)^2}.$$

On the basis of (8.23), the Fourier transforms of the solutions of this equation are

$$G_1(\lambda) = i\,\frac{\sigma_+(\lambda)}{\lambda - i} = i\,\frac{(\lambda - i)(\lambda + 3i)}{(\lambda + 2i)^2(\lambda + i)},$$

$$G_2(\lambda) = i^2\,\frac{\sigma_+(\lambda)}{(\lambda - i)^2} = -\frac{\lambda + 3i}{(\lambda + 2i)^2(\lambda + i)}.$$

Taking inverse Fourier transforms, we obtain

$$g_1(x) = -4e^{-x} + 5e^{-2x} + 3xe^{-2x},$$

$$g_2(x) = 2e^{-x} - 2e^{-2x} - xe^{-2x}.$$

The solutions, forming the D-chain, are

$$\varphi_0(x) = g_2(x),$$
$$\varphi_1(x) = g_2(x) + g_1(x) = -2e^{-x} + 3e^{-2x} + 2xe^{-2x}.$$

§6 Dual equations with kernels of exponential type {see [8], [48]}

6.1 *Reduction to a boundary value problem*

We assume that the kernels of

$$\begin{aligned} \varphi(x) - \int_{-\infty}^{\infty} k_1(x-t)\,\varphi(t)\,dt &= f(x), \qquad x < 0, \\ \varphi(x) - \int_{-\infty}^{\infty} k_2(x-t)\,\varphi(t)\,dt &= f(x), \qquad x > 0, \end{aligned} \tag{8.39}$$

satisfy the conditions:

A_2. $k_1(x) \in \{\alpha_1, \beta_1\}$, $k_2(x) \in \{\alpha_2, \beta_2\}$ (the spaces $\{\alpha, \beta\}$ and $\{\{\alpha, \beta\}\}$ are discussed in §1.5).

B_2. Their Fourier transforms $F_{\alpha_i}k_i$, $F_{\beta_i}k_i$ are bounded, satisfy the Hölder condition, and vanish at infinity.

A solution $\varphi(x)$ is sought in the class $\{\min(\beta_1, \beta_2), \max(\alpha_1, \alpha_2)\}$, which denotes the widest class in which the integrals of (8.39) have sense.

Of necessity, the r.h.s. belongs to the class

$$\{\max(\alpha_2, \min(\beta_1, \beta_2)), \min(\beta_1, \max(\alpha_1, \alpha_2))\}.$$

For instance, if $\alpha_i = \beta_i = 0$, then $f(x)$, $\varphi(x) \in \{0, 0\} = L_2(-\infty, \infty)$; if $\alpha_i = \beta_i = \alpha$, then $f(x)$, $\varphi(x) \in \{\alpha, \alpha\} = e^{\alpha x}L_2$; if $-\alpha_i = \beta_i = h$, then $f(x), \varphi(x) \in e^{h|x|}L_2$.

The Wiener-Hopf equation (8.14) can be obtained as a particular case of (8.39) when $f(x) = 0$ for $x < 0$ and $k_1(x) \equiv 0$. In this case, $k_1(x) \in \{-\infty, \infty\}$. The solution $\varphi(x)$ is sought in the space $\{\beta_2, +\infty\}$, and the r.h.s. $f(x) \in \{\max(\alpha_2, \beta_2), \infty\}$.

The most commonly used method of solution for (8.39), under the above assumptions, consists in the reduction of (8.39) to the equivalent Riemann boundary value problem of the theory of analytic functions (see [8], [48]).

Starting with different relationships between the α_i, β_i, we obtain different boundary value problems {on one or several straight lines}. In some cases, these problems (and (8.39) along with them) can be solved in closed form. We illustrate two cases in order to demonstrate the techniques of the reduction to a boundary value problem. This technique consists of two stages. The first is the preparation of the equation for the application of the Fourier transform.

With this aim in view, we transform (8.39) so that both its equations hold on the infinite interval. We set

$$\psi_+(x) = \begin{cases} \varphi(x) - \int\limits_{-\infty}^{\infty} k_1(x-t)\,\varphi(t)\,dt + f_-(x), & \text{if} \quad x > 0, \\ 0, & \text{if} \quad x < 0; \end{cases}$$

$$\psi_-(x) = \begin{cases} \varphi(x) - \int\limits_{-\infty}^{\infty} k_2(x-t)\,\varphi(t)\,dt - f_+(x), & \text{if} \quad x < 0, \\ 0, & \text{if} \quad x > 0; \end{cases}$$

$$f_+(x) = \begin{cases} f(x), & x > 0, \\ 0, & x < 0, \end{cases} \qquad f_-(x) = \begin{cases} -f(x), & x < 0, \\ 0, & x > 0. \end{cases}$$

Using the functions $\psi_\pm(x)$, we can rewrite equation (8.39) in the form

$$\begin{aligned} &\varphi(x) - \int\limits_{-\infty}^{\infty} k_1(x-t)\,\varphi(t)\,dt + f_-(x) = \psi_+(x), \\ &\varphi(x) - \int\limits_{-\infty}^{\infty} k_2(x-t)\,\varphi(t)\,dt - f_+(x) = \psi_-(x), \quad (-\infty < x < \infty). \end{aligned} \tag{8.40}$$

The functions $\psi_\pm(x)$ are unknown (for $x > 0$ and $x < 0$, respectively), but we know that

$$\psi_-(x) \in \{-\infty, \min(\beta_2, \max(\alpha_1, \alpha_2))\},$$

$$\psi_+(x) \in \{\max(\alpha_1, \min(\beta_1, \beta_2)), \infty\}.$$

(For the Wiener-Hopf equation, we have $\psi_+(x) \equiv 0$, $\psi_-(x) \in \{-\infty, \min(\beta_2, \alpha_2)\}$).

The next step depends on the relationship between the numbers α_i, β_i.

6.2 *Case 1*

If $\max(\alpha_1, \alpha_2) = \min(\beta_1, \beta_2) = \alpha$, then all members of (8.40) belong to one and the same space $\{\alpha, \alpha\}$. More precisely, $\varphi(x) \in \{\alpha, \alpha\}$; $f_-, \psi_- \in \{-\infty, \alpha\} \subset \{\alpha, \alpha\}$; $f_+, \psi_+ \in \{\alpha, +\infty\} \subset \{\alpha, \alpha\}$.

Now, we move to the second stage—the application of the Fourier transform to (8.40). As a consequence of its application, we obtain

$$\begin{aligned} [1 - K_1(\zeta)]\, \Phi(\zeta) + F_-(\zeta) &= \Psi_+(\zeta), \\ [1 - K_2(\zeta)]\, \Phi(\zeta) - F_+(\zeta) &= \Psi_-(\zeta). \end{aligned} \tag{8.41}$$

If $k_i \in L_2(-\infty, \infty)$, then ζ is real and we obtain a boundary value problem on the real axis. This case has been studied in [34], where it was shown that (8.40) reduces to a boundary value problem on the real axis. For the Wiener-Hopf equation, we obtain

$$[1 - K(\zeta)]\, \Phi_+(\zeta) - F_+(\zeta) = \Psi_-(\zeta). \tag{8.41$_1$}$$

Eliminating $\Phi(\zeta)$ from (8.41), we obtain an equation for the two unknown $\Psi_\pm(\zeta)$

$$\Psi_+(\zeta) = \frac{1 - K_1(\zeta)}{1 - K_2(\zeta)}\, \Psi_-(\zeta) + \frac{[1 - K_1(\zeta)]\, F_+(\zeta) - [1 - K_2(\zeta)]\, F_-(\zeta)}{1 - K_2(\zeta)}. \tag{8.42}$$

This leads to the following problem: find $\Psi_+(\zeta) \in \{\{\alpha, \infty\}\}$, and $\Psi_-(\zeta) \in \{\{-\infty, \alpha\}\}$ such that (8.42) holds and $[\Psi_+(\zeta) - F_-(\zeta)]/[1 - K_1(\zeta)] = \Phi(\zeta) \in \{\{\alpha, \alpha\}\}$. We note that this last condition holds automatically, if $1 - K_1(\zeta) \neq 0$, $\zeta = \lambda + i\alpha$.

A more detailed formulation of this problem is the following: find a piecewise holomorphic function $\Psi(\zeta)$ with the line of jumps $\zeta = \lambda + i\alpha$, $-\infty < \lambda < \infty$, such that (8.42) holds and such that:

1) $\displaystyle\int_{-\infty}^{\infty} |\Psi(x + iy)|^2\, dx \leqslant C = \text{const.}$ for all y.

2) $\displaystyle\left|\frac{\Psi_+(\zeta) - F_-(\zeta)}{1 - K_1(\zeta)}\right|^2$ is summable along the line $\zeta = \lambda + i\alpha$, $-\infty < \lambda < \infty$.

The solution of this problem can be obtained in closed form. If $\Psi_+(x)$ is its solution, then the solution $\varphi(x)$ of (8.39) is determined by

$$\varphi(x) = \frac{1}{2\pi}\int_{-\infty}^{\infty} e^{-i\zeta x}\frac{\Psi_+(\zeta) - F_-(\zeta)}{1 - K_1(\zeta)}\,d\zeta. \tag{8.43}$$

For the Wiener-Hopf equation, we obtain

$$\varphi(x) = \frac{1}{2\pi}\int_{-\infty}^{\infty} e^{-i\zeta x}\frac{\Psi_-(\zeta) + F_+(\zeta)}{1 - K(\zeta)}\,d\zeta. \tag{8.43$_1$}$$

The final result is given by

THEOREM 6.1. *If conditions* A_2 *and* B_2 *are satisfied (see the beginning of the present section) and, in addition, the functions* $1 - K_i(\zeta)$ $(i = 1, 2)$ *do not vanish on the straight line* $\zeta = \lambda + i\alpha$, $-\infty < \lambda < \infty$, *then, under the condition*

$$\chi = \frac{1}{2\pi}\arg\frac{1 - K_1(\zeta)}{1 - K_2(\zeta)}\bigg|_{-\infty}^{\infty} \geqslant 0, \tag{8.44}$$

all solutions of (8.39) *are given by* (see [6], [7]):

$$\Psi_+(\zeta) = X^+(\zeta)[\Omega^+(\zeta)] + \frac{P_{\chi-1}(\zeta)}{[\zeta - i(\alpha - 1)]^\chi}, \tag{8.45}$$

$$\Gamma(z) = \frac{1}{2\pi i}\int_{-\infty+i\alpha}^{\infty+i\alpha} \ln\left[\left(\frac{\tau - i(\alpha+1)}{\tau - i(\alpha - 1)}\right)^{-\chi} G(\tau)\right]\frac{d\tau}{\tau - z}, \tag{8.46}$$

$$\Omega(z) = \frac{1}{2\pi i}\int_{-\infty+i\alpha}^{\infty+i\alpha} \frac{g(\tau)\,d\tau}{X^+(\tau)(\tau - z)}; \qquad X^+(\zeta) = e^{\Gamma^+(\zeta)}, \tag{8.47}$$

where Γ^+, Ω^+ *are the limit values of functions* $F^+(z)$ *and* $\Omega^+(z)$ *as* $z \to \zeta + i0$, $\zeta = \lambda + i\alpha$; $G(\tau)$ *and* $g(\tau)$ *are the coefficient and the r.h.s. of* (8.42); $P_{\chi-1}(\zeta)$ *is an arbitrary polynomial of degree not exceeding* $\chi - 1$ *for* $\chi > 0$ *and zero for* $\chi = 0$. *If* $\chi < 0$ *then, for the solvability of* (8.39), *it is necessary and sufficient that the conditions should be*

$$\int_{-\infty+i\alpha}^{\infty+i\alpha} \frac{g(\tau)\,d\tau}{X^{+}(\tau)[\tau - i(\alpha - 1)]^{k}} = 0 \qquad (k = 0, 1, \dots, |\chi| - 1). \tag{8.48}$$

If these conditions hold then the solution is unique and is given by the above formulas with $P_{\chi-1} \equiv 0$.

6.3 *Case 2*

Let $\alpha = \max(\alpha_1, \alpha_2) < \min(\beta_1, \beta_2) = \beta$. {The homogeneous Wiener-Hopf equations in the spaces $e^{hx}E^{+}$ have been studied in detail in [22] under similar assumptions on the kernels.} This case is much more difficult than the preceding one. We now have $\varphi \in \{\beta, \alpha\}$; f_{+}, $\psi_{+} \in \{\beta, \infty\}$; f_{-}, $\psi_{-} \in \{-\infty, \alpha\}$.

Expressing φ in the form $\varphi = \varphi_{+} - \varphi_{-}$ (see §6.1), we rewrite (8.40) in the form

$$\left.\begin{aligned}
&\left[\varphi_{+}(x) - \int_{-\infty}^{\infty} k_1(x - t)\,\varphi_{+}(t)\,dt - \psi_{+}(x)\right] \\
&\qquad - \left[\varphi_{-}(x) - \int_{-\infty}^{\infty} k_1(x - t)\,\varphi_{-}(t)\,dt - f_{-}(x)\right] = 0, \\
&\left[\varphi_{+}(x) - \int_{-\infty}^{\infty} k_2(x - t)\,\varphi_{+}(t)\,dt - f_{+}(x)\right] \\
&\qquad - \left[\varphi_{-}(x) - \int_{-\infty}^{\infty} k_2(x - t)\,\varphi_{-}(t)\,dt - \psi_{-}(x)\right] = 0.
\end{aligned}\right\} \tag{8.49}$$

Here, the square brackets contain, when the subscript "+" has been used, a function belonging to the space $\{\beta, \beta\}$; and a function belonging to the space $\{\alpha, \alpha\}$, when the subscript "−" has been used.

For the first equation, we denote these functions by $\omega_{+}(x)$ and $\omega_{-}(x)$. Then $\omega_{+}(x) - \omega_{-}(x) = 0$, $-\infty < x < \infty$. From this relation it follows (see §1.5) that $\omega^{+}(x), \omega^{-}(x) \in \{\alpha, \beta\}$.

Let $F_y\omega_+ = \Omega_1(\lambda + iy)$, $\alpha \leqslant y \leqslant \beta$, $-\infty < \lambda < \infty$ be the Fourier transform of the function ω_+. Then $\Omega_1(\zeta) \in \{\{\alpha, \beta\}\}$ (see §1).

We note that $F_\beta\omega_+ = \Omega_1(\lambda + i\beta)$ and $F_\alpha\omega_- = \Omega_1(\lambda + i\alpha)$. More explicitly,

$$\left.\begin{aligned} [1 - K_1(\zeta)]\,\Phi_+(\zeta) - \Psi_+(\zeta) &= \Omega_1(\zeta), \qquad \zeta = \lambda + i\beta, \\ [1 - K_1(\zeta)]\,\Phi_-(\zeta) - F_-(\zeta) &= \Omega_1(\zeta), \qquad \zeta = \lambda + i\alpha. \end{aligned}\right\} \tag{8.50}$$

The same reasonings applied to the second equation in (8.49) yields

$$\left.\begin{aligned} [1 - K_2(\zeta)]\,\Phi_+(\zeta) - F_+(\zeta) &= \Omega_2(\zeta), \qquad \zeta = \lambda + i\beta, \\ [1 - K_2(\zeta)]\,\Phi_-(\zeta) - \Psi_-(\zeta) &= \Omega_2(\zeta), \qquad \zeta = \lambda + i\alpha. \end{aligned}\right\} \tag{8.51}$$

Thus, (8.40) reduces to the problem: find the functions

$$\Phi_+, \Psi_+ \in \{\{\beta, \infty\}\}, \qquad \Phi_-, \Psi_- \in \{\{-\infty, \alpha\}\}, \qquad \Omega_1, \Omega_2 \in \{\{\alpha, \beta\}\},$$

which satisfy the relations (8.50)–(8.50_1). This problem reduces to two Riemann problems which can be solved successively and the solution can be obtained in a closed form (see [48]). The final result is given by

THEOREM 6.2. *If conditions* A_2, B_2 *are satisfied and the functions*

$$1 - K_1(\lambda + i\alpha), \qquad 1 - K_2(\lambda + i\beta) \qquad (-\infty < \lambda < \infty),$$

do not vanish everywhere, then the homogeneous equation has exactly max $(0, \nu, \chi)$ *linearly independent solutions in the space* $\{\beta, \alpha\}$, *where*

$$\chi = \frac{1}{2\pi} \arg \frac{1 - K_1(\lambda + i\alpha)}{1 - K_2(\lambda + i\beta)}\bigg|_{-\infty}^{\infty}$$

and ν *is the number of common zeros of the functions* $1 - K_i(\zeta)$ $(i = 1, 2)$ *in the strip* $\alpha < \operatorname{Im}\zeta < \beta$. *If* $\chi - \nu \geqslant 0$, *then the non-homogeneous equation is solvable for any r.h.s.* $f(x) \in \{\beta, \alpha\}$; *if* $\chi - \nu < 0$, *then the r.h.s. must satisfy* $|\chi - \nu|$ *necessary and sufficient conditions in order to ensure solvability for the non-homogeneous equation. The solution* $\varphi(x)$ *can be expressed in terms of the functions* $\Phi_\pm(\zeta)$ *by*

$$\varphi = \frac{1}{2\pi}\left[\int_{-\infty + i\beta}^{\infty + i\beta} \Phi_+(\zeta)e^{-i\zeta x}\,d\zeta - \int_{-\infty + i\alpha}^{\infty + i\alpha} \Phi_-(\zeta)e^{-i\zeta x}\,d\zeta\right].$$

If max $(\alpha_1, \alpha_2) > \min(\beta_1, \beta_2)$, then (8.39) reduces to a boundary value problem which has remained unsolved for the general case (the exceptional case is $\alpha_2 \leqslant \beta_2 < \alpha_1 \leqslant \beta_1$) (see [8], [48]).

The equation

$$\varphi(x) - \int_{-\infty}^{0} k_1(x-t)\varphi(t)\,dt - \int_{0}^{\infty} k_2(x-t)\,\varphi(t)\,dt = f(t),$$

$$-\infty < x < \infty,$$

can be reduced to a boundary value problem in a similar way. If $k_1(x)$, $k_2(x)$ satisfy the previous conditions, then the solution $\varphi(x)$ must be sought in the class $\{\beta_2, \alpha_1\}$ and the r.h.s. value must be taken from the class $\{\max(\beta_2, \alpha_1, \alpha_2), \min(\alpha_2, \beta_1, \beta_2)\}$.

For example, let $\alpha_2 \leqslant \beta_2 = \alpha_1 \leqslant \beta_1$. Then

$$\varphi(x) \in \{\beta_2, \beta_2\}, \qquad f(x) \in \{\beta_2, \alpha_2\}.$$

We rewrite this equation in the form

$$\left[\varphi_+(x) - \int_{-\infty}^{\infty} k_2(x-t)\,\varphi_+(t)\,dt\right] - \left[\varphi_-(x) - \int_{-\infty}^{\infty} k_1(x-t)\,\varphi_-(t)\,dt\right] = f(x).$$

All its components belong to the class $\{\beta_2, \beta_2\}$ (see §1).

Applying the Fourier transform F_{β_2}, we obtain the problem: find the functions $\Phi_+(\zeta) \in \{\{\beta_2, +\infty\}\}$, $\Phi_-(\zeta) \in \{\{-\infty, \beta_2\}\}$, where

$$[1 - K_2(\zeta)]\,\Phi_+(\zeta) - [1 - K_1(\zeta)]\,\Phi_-(\zeta) = F(\zeta),$$

$$\zeta = \lambda + i\beta_2, \qquad -\infty < \lambda < \infty.$$

This problem was considered above (see case I for the dual equation). The solution of this equation is given by (8.49) with α replaced by β_2. For other possible cases see [8], [46].

§7 The Wiener-Hopf method[1])

7.1 *A description of the method*

The Wiener-Hopf method was historically the first method for the solution of (8.14). The new method presented above for the solution of (8.14) possesses some advantage over the Wiener-Hopf method: It enables one to examine (8.14) for a wider class of functions, and leads to simpler problems in the theory of analytic functions. Nevertheless, it is worthwhile describing the Wiener-Hopf method. Firstly, because it can be applied to the solution of certain partial differential equations (see [27], [49]), and, secondly, because this method is widely used in the study of physical problems (see [25], [26]).

The Wiener-Hopf method consists in replacing the integral equation by the following problem: find functions $\Phi_+(z)$ and $\Phi_-(z)$ which are holomorphic in the half-planes $\operatorname{Im} z > \alpha$ and $\operatorname{Im} z < \beta$, $\alpha < \beta$, respectively, where they satisfy the relation

$$A(z)\,\Phi_+(z) + B(z)\,\Phi_-(z) + C(z) = 0 \tag{8.51}$$

on the strip $\alpha < \operatorname{Im} z < \beta$, and their behaviour at infinity is known. Here $A(z)$, $B(z)$, $C(z)$ are known functions, holomorphic in the strip $\alpha < \operatorname{Im} z < \beta$. We seek the solution of this problem in the following way: if $A(z)/B(z)$ has a finite number of zeros in the strip $\alpha < \operatorname{Im} z < \beta$ then, under some supplementary conditions, it can be expressed in the form

$$\frac{A(z)}{B(z)} = \frac{X_+(z)}{X_-(z)} \prod_{j=1}^{k} (z - \alpha_j), \tag{8.52}$$

where the α_j are zeros of this function in the strip $\alpha < \operatorname{Im} z < \beta$; and X_+ and X_- are functions which are holomorphic in the half-planes $\operatorname{Im} z > \alpha$, $\operatorname{Im} z < \beta$, and have no zeros there. The substitution of (8.52) in (8.51) yields

$$X_+(z) \prod_{j=1}^{k} (z - \alpha_j)\,\Phi_+(z) + X_-(z)\,\Phi_-(z) + \frac{C(z)\,X_-(z)}{B(z)} = 0. \tag{8.53}$$

Under certain conditions (see [27], [44]), the last component in (8.53) can be expressed as the difference $\Omega_+(z) - \Omega_-(z)$ of functions which are holomorphic in the previous half-planes. Then

$$X_+(z) \prod_{j=1}^{k} (z - \alpha_j)\,\Phi_+(z) + \Omega_+(z) = -X_-(z)\,\Phi_-(z) + \Omega_-(z).$$

[1]) This method is outlined in [25], [27], [40], [42] as well as in [3] and [44].

The l.h.s. and r.h.s. of this equality are holomorphic in $\operatorname{Im} z > \alpha$ and $\operatorname{Im} z < \beta$, respectively. If, from the conditions connected with this problem, it follows that both sides have a rate of growth at infinity, for example, with orders n and m, then, according to a theorem of Liouville, they are polynomials of degree not more than $\chi = [\min(n, m)]$. The solution of this problem is given by

$$\Phi_+(z) = X_+^{-1}(z) \frac{P(z) - \Omega_+(z)}{\prod_{j=1}^{k} (z - \alpha_j)},$$

$$\Phi_-(z) = X_-^{-1}(z)[-P(z) + \Omega_-(z)],$$

where $P(z)$ is an arbitrary polynomial of degree χ.

Now, we present conditions under which the representation (8.52) holds along with the representation for the last component of (8.53) as the difference of functions which are holomorphic in the mentioned planes.

a) The last component of (8.53), viz.

$$F(z) = \frac{C(z)\, X_-(z)}{B(z)}$$

can be expressed as the difference of two functions, holomorphic in the half-planes $\operatorname{Im} z > \alpha$ and $\operatorname{Im} z < \beta$, respectively, if $F(z)$ is continuous in the strip $\alpha \leqslant \operatorname{Im} z \leqslant \beta$, holomorphic inside it, and $F(z) = O(|z|^{-\sigma})$, $\sigma > 0$, as $\operatorname{Re} z \to 0$.

b) The representation (8.52) holds if

$$F(z) = \ln \frac{A(z)}{B(z) \prod_{j=1}^{k} (z - \alpha_j)}$$

satisfies the conditions of a).

Other conditions are considered in [3], [22], [42]. Examples of representations are given in [22], [25], [27], [42]. In [27], an extensive bibliography on the Wiener-Hopf method is presented.

7.2 *The reduction of* (8.14) *to a boundary value problem*

Let the kernel $k(x)$ and the r.h.s. $f(x)$ of this equation satisfy:

a) $k(x) \in e^{-h|x|} L_2(-\infty, \infty)$ (or $k(x) \in e^{-h|x|} L(-\infty, \infty)$);

b) $f(x) \in e^{\mu x} L_2(0, \infty)$, where $\mu < h$.

If, under these conditions, we seek the solution $\varphi(x) \in e^{\mu x} L_2(0, \infty)$, then (8.14) is equivalent to a problem of the type considered in §7.1

$$[1 - K(z)]\, \Phi_+(z) - \Psi_-(z) = F_+(z), \qquad \mu < \operatorname{Im} z < h, \tag{8.54}$$

under the supplementary constraint that $|\Phi_+(z)| \to 0$ as $|z| \to \infty$. As usual, $K(z)$, $\Phi_+(z)$, $F_+(z)$ are the Fourier transforms of the kernel, the solution and the r.h.s. of (8.14).

7.3 *An example*

Consider the equation

$$\varphi(x) - \frac{1}{2}\int_0^\infty e^{-|x-t|}\, \varphi(t)\, dt = Ae^{\alpha x}, \qquad 0 < \alpha < 1.$$

Since $e^{\alpha x} \in e^{\mu x} L_2(0, \infty)$ for any $\mu > \alpha$ and $e^{-|x|} \in e^{-h|x|} L_2(-\infty, \infty)$ for any $h < 1$, where h is arbitrary as long as it satisfies the inequality $\alpha < h < 1$, the condition (8.54) takes the form

$$\frac{z^2}{1 + z^2}\, \Phi_+(z) - \Psi_-(z) = \frac{iA}{z - i\alpha}, \qquad 0 < \alpha < \operatorname{Im} z < h < 1.$$

In the given problem, we have

$$X_+(z) = \frac{z^2}{z + i}, \qquad X_-(z) = z - i.$$

with $C(z)\, X_-(z) = [A(z - i)]/[z - i\alpha] = Ai[1 + \{i(\alpha - 1)/(z - i\alpha)\}]$. Hence, $(z^2/z + i)\, \Phi_+(z) + [A(\alpha - i)]/[z - i\alpha] = (z - i)\, \Psi_-(z) + iA$. Since $\Phi_+(z) \to 0$ as $|z| \to \infty$, the l.h.s. increases at infinity more slowly than z. Thus, according to Liouville's Theorem, the l.h.s. is constant for all z. Therefore,

$$\Phi_+(z) = \frac{A(\alpha - 1)(z + i)}{z^2(z - i\alpha)} + C\frac{z + i}{z^2}, \qquad C = \text{const.}$$

Applying the inverse Fourier transform, we obtain

$$\varphi(x) = C(1 + x) + A(\alpha - 1)\left[\frac{1}{\alpha^2}(\alpha + 1)e^{\alpha x} - \frac{1}{\alpha^2}\right].$$

§8 **Equations with degenerate symbol** {see [7]}

8.1 *The Riemann problem*

In the preceding sections (§§1, 3, 4, 6), it was assumed that conditions of type-B are satisfied, which guarantee the normal solvability of given equations and the finiteness of their indices. Conditions of type-B require that the symbol of the equation does not vanish anywhere. Below, we present some results which hold when this condition does not apply. Up till now, the problems relating to this case have not been completely resolved. When this condition is broken, only the simplest cases have been studied. The basic tool for the study of this problem is the theory of the Riemann boundary value problem for the 'exceptional' case of Gahov when its coefficient has zeros and poles.

Below, we present some basic results from this theory. A formulation for the problem is:

Problem R. Find functions $\Phi^{\pm}(\zeta)$ which satisfy:
a) $\Phi^{+}(\zeta) \in \{\{0, \infty\}\}$, $\Phi^{-}(\zeta) \in \{\{-\infty, 0\}\}$ (see §1), $\Phi^{\pm}(\zeta) \to 0$ as $\zeta \to \infty$;
b) the limits of $\Phi^{\pm}(\zeta)$ on the real axis satisfy

$$\Phi^{+}(\lambda) = G(\lambda)\,\Phi^{-}(\lambda) + g(\lambda), \qquad -\infty < \lambda < \infty. \tag{8.55}$$

We assume that the coefficient $G(\lambda)$ and the free term have the following form:

$$\text{I.}\quad G(\lambda) = \frac{\prod_{i=1}^{r} (\lambda - a_i)^{\alpha_i}\, p_{+}(\lambda)\, p_{-}(\lambda)}{\prod_{i=1}^{s} (\lambda - b_i)^{\beta_i}\, q_{+}(\lambda)\, q_{-}(\lambda)}\, G_1(\lambda), \tag{8.56}$$

where α_i, β_i are non-negative integers; a_i, b_i are real numbers; $p_{+}(\lambda)$, $q_{+}(\lambda)$, $p_{-}(\lambda)$, $q_{-}(\lambda)$ are polynomials of degree $m_{+}, n_{+}, m_{-}, n_{-}$ respectively, the zeros of which lie in the upper half-plane, when the subscript is '+', and lower half-plane, when the subscript is '−'; $G_1(\lambda)$ satisfies the Hölder condition, $G_1(\infty) = 1$ and $\arg G_1(\lambda)|_{-\infty}^{\infty} = 0$.

II. We write $\chi = m_{+} - n_{+}$, $h = n_{-} - m_{-}$, $n = \sum_{i=1}^{s} \beta_i$, $m = \sum_{i=1}^{r} \alpha_i$ and $\nu = n - \chi + h - m$, and assume that

$$g(\lambda) = \frac{g_1(\lambda)}{\prod_{i=1}^{s} (\lambda - b_i)^{\beta_i}},$$

with $g_1(\lambda)$ such that

$$\frac{g_1(\lambda)\, q_-(\lambda)}{p_-(\lambda) e^{\Gamma^+(\lambda)}} = Q(\lambda) + \psi(\lambda),$$

where $Q(\lambda)$ is a polynomial of degree not exceeding $n + h - 1$, if $\nu \geqslant 0$, and of degree not exceeding $m + \chi - 1$, if $\nu < 0$; $\psi(\lambda) \in L_2(-\infty, \infty)$; and the definition of $\Gamma^+(\lambda)$ is given below.

The solution of (8.55) is given by the following formulas in which $P_k(z)$ is an arbitrary polynomial of degree not exceeding k, if $k \geqslant 0$, and $P_k(z) \equiv 0$, if $k < 0$:

The homogeneous problem:

a) If $\nu \geqslant 0$ ($G(\lambda)$ has a zero of order ν at infinity), then

$$\begin{aligned} \Phi_0^+(z) &= \frac{\prod_{j=1}^{r} (z - a_j)^{\alpha_j}\, p_-(z) e^{\Gamma^+(z)}}{q_-(z)} P_{\chi-n-1}(z), \\ \Phi_0^-(z) &= \frac{\prod_{j=1}^{s} (z - b_j)^{\beta_j}\, q_+(z) e^{\Gamma^-(z)}}{p_+(z)} P_{\chi-n-1}(z), \end{aligned} \tag{8.57}$$

where $\Gamma^{\pm}(z)$ are the limit values of the function

$$\Gamma(z) = \frac{1}{2\pi i} \int_{-\infty}^{\infty} \frac{\ln G_1(\lambda)\, d\lambda}{\lambda - z} \quad \text{for} \quad \operatorname{Im} z \to \pm 0.$$

The problem has exactly max $(0, \chi - n)$ linearly independent solutions.

b) If $\nu < 0$, ($G(\lambda)$ is equivalent to $\lambda^{|\nu|}$ at infinity), then

$$\begin{aligned} \Phi_0^+(z) &= \frac{\prod_{j=1}^{r} (z - a_j)^{\alpha_j}\, p_-(z)}{q_-(z)} e^{\Gamma^+(z)} P_{\chi-n+\nu-1}(z), \\ \Phi_0^-(z) &= \frac{\prod_{j=1}^{s} (z - b_j)^{\beta_j}\, q_+(z)}{p_+(z)} e^{\Gamma^-(z)} P_{\chi-n+\nu-1}(z). \end{aligned} \tag{8.58}$$

The problem has exactly max $(0, \chi - n + \nu)$ linearly independent solutions.

The non-homogeneous problem:

a) If $\nu \geqslant 0$, then

$$\Phi^+(z) = Y^+(z) + \Phi_0^+(z), \qquad \Phi^-(z) = Y^-(z) + \Phi_0^-(z), \tag{8.59}$$

where the $\Phi_0^\pm(z)$ are defined by (8.58) and

$$\left.\begin{aligned} Y^+(z) &= \frac{p_-(z)e^{\Gamma^+(z)}}{\prod_{j=1}^{s}(z-b_j)^{\beta_j}\, q_-(z)}\,[\Omega^+(z) + Q(z) - Q_\rho(z)],\\ Y^-(z) &= \frac{q_+(z)e^{\Gamma^-(z)}}{\prod_{j=1}^{r}(z-a_j)^{\alpha_j}\, p_+(z)}\,[\Omega^-(z) - Q_\rho(z)]. \end{aligned}\right\} \tag{8.60}$$

Here $\Omega^\pm(z)$ are the values of

$$\Omega(z) = \frac{1}{2\pi i}\int_{-\infty}^{\infty}\frac{\psi(\lambda)\,d\lambda}{\lambda - z}$$

for $\operatorname{Im} z > 0$ and $\operatorname{Im} z < 0$, $Q_\rho(z)$ is the Hermite interpolatory polynomial of degree $\rho = m + n - 1$ which is constructed in accordance with the conditions (if the derivatives exist)

$$\frac{d^k}{dz^k}[\Omega^+(z) + Q(z) - Q_\rho(z)] = 0, \tag{8.61}$$

$$z = b_j, \qquad j = 1, 2, \ldots, s, \qquad k = 0, 1, \ldots, \beta_j - 1,$$

$$\frac{d^k}{dz^k}[\Omega^-(z) - Q_\rho(z)] = 0,$$

$$z = a_j, \qquad j = 1, 2, \ldots, r, \qquad k = 0, 1, \ldots, \alpha_j - 1.$$

(In other words, the polynomial $Q_\rho(z)$ is chosen so that $Y^\pm(\lambda) \in L_2(-\infty, \infty)$). In order that the polynomial $Q_\rho(z)$ with the prescribed properties (for $\chi - n \geqslant 0$) should exist, it is sufficient that the functions $g_1(\lambda)$ and $G_1(\lambda)$ should have derivatives of order α_j and β_j at the points a_j and b_j, respectively (see [6]).

If $\chi - n < 0$, then the functions $Y^\pm$ are, generally speaking, not bounded at infinity. In this case, the solution exists for the r.h.s., $g(\lambda)$, satisfying

$|\chi - n|$ supplementary conditions, which guarantee that $Y^{\pm}(z)$ vanishes at infinity (see [6], p. 121).

b) If $\nu < 0$, then

$$\Phi^+(z) = Y_1^+(z) + \Phi_0^+(z), \qquad \Phi^-(z) = Y_1^-(z) + \Phi_0^-(z), \tag{8.62}$$

where $\Phi_0^{\pm}(z)$ is the solution of the corresponding homogeneous problem (for $\nu < 0$); and

$$\left.\begin{aligned} Y_1^+(z) &= \frac{p_-(z)e^{\Gamma^+(z)}}{\prod_{i=1}^{s}(z - b_i)^{\beta_i}\, q_-(z)}[\Omega^+(z) - Q_\rho(z)], \\ Y_1^-(z) &= \frac{q_+(z)e^{\Gamma^-(z)}}{\prod_{i=1}^{r}(z - a_i)^{\alpha_i}\, p_+(z)}[\Omega^-(z) - Q(z) - Q_\rho(z)]. \end{aligned}\right\} \tag{8.63}$$

The polynomial Q_ρ has been defined above.

If $\chi - n + \nu < 0$, then the solution exists if and only if the r.h.s. $g(\lambda)$ satisfies $|\chi - n + \nu|$ supplementary conditions.

Conclusion: If χ is regarded as the index of the problem, then in comparison with the case of normal solvability (see [6]), *the number of linearly independent solutions is decreased by the number of zeros of $G(\lambda)$ (including zeros at $z = \infty$). The zeros of this coefficient do not influence the actual number of solutions. The number of solvability conditions is increased by the number of poles of $G(\lambda)$.*

Henceforth, we shall assume that the coefficients and r.h.s. values of the problems, to which the integral equations under consideration reduce, satisfy all the conditions cited in the formulation of the problem.

We seek the solutions of the equations considered below in L_2. Such solutions of the corresponding boundary value problems will vanish at infinity.

8.2 *The Wiener-Hopf equation of the second kind*

Consider the equation

$$\varphi(x) - \int_0^\infty k(x - t)\,\varphi(t)\,dt = f(x), \qquad x > 0. \tag{8.64}$$

We assume that its kernel $k(x) \in L_2(-\infty, \infty)$ and that the Fourier transform $K(\lambda)$ of this kernel is bounded and vanishes at infinity. We assume that the symbol $1 - K(\lambda)$ of this equation vanishes at a finite number of points on the real axis and that the assumptions and notation of Problem R are valid:

$$[1 - K(\lambda)]^{-1} = \frac{p_+(\lambda)\, p_-(\lambda)}{\prod_{j=1}^{s} (\lambda - b_j)^{\beta_j}\, q_+(\lambda)\, q_-(\lambda)} G_1(\lambda). \tag{8.65}$$

On the basis of the method described in §6, (8.64) reduces to the equivalent boundary value problem

$$\Phi^+(\lambda) = \frac{\Phi^-(\lambda)}{1 - K(\lambda)} + \frac{F(\lambda)}{1 - K(\lambda)}. \tag{8.66}$$

If $\Phi^+(\lambda)$, $\Phi^-(\lambda)$ are solutions of this problem then

$$\varphi(x) = \frac{1}{2\pi} \int_{-\infty}^{\infty} \Phi^+(\lambda) e^{-i\lambda x}\, d\lambda. \tag{8.67}$$

THEOREM 8.1. *If the coefficient and the free term* $F(\lambda)/(1 - K(\lambda))$ *of* (8.66) *satisfy all conditions of Problem R, then* $\nu = 0$ *and:*
a) *For* $\chi - n \geqslant 0$, *the homogeneous equation has exactly* $\chi - n$ *linearly independent solutions in* $L_2(0, \infty)$, *the Fourier transforms of which are given by* (8.67) *with* $\alpha_j = 0$. *If* $\chi - n < 0$, *then the homogeneous equation has only the trivial solution.*
b) *The non-homogeneous equation is solvable, if* $\chi - n \geqslant 0$. *If* $\chi - n < 0$, *then the non-homogeneous equation is solvable if and only if the r.h.s. satisfies* $|\chi - n|$ *supplementary conditions. If these conditions are satisfied, then the solution is unique. When the equation is solvable, the Fourier transforms of the solutions are given by* (8.59), (8.60), (8.61).

8.3 *The Wiener-Hopf equation of the first kind*

The equation

$$\int_0^{\infty} k(x - t)\, \varphi(t)\, dt = f(x), \qquad 0 < x < \infty, \tag{8.68}$$

reduces to the equivalent problem

$$\Phi^+(\lambda) = \frac{\Phi^-(\lambda)}{K(\lambda)} + \frac{F(\lambda)}{K(\lambda)}. \tag{8.69}$$

Suppose that $K(\lambda)$ is such that, under the notations and assumptions of Problem R,

$$[K(\lambda)]^{-1} = \frac{p_+(\lambda)\, p_-(\lambda)}{\prod_{j=1}^{s} (\lambda - b_j)^{\beta_j}\, q_+(\lambda)\, q_-(\lambda)}\, G_1(\lambda).$$

THEOREM 8.1_1. *Let $K(\lambda) \sim (1/\lambda^k)$ as $\lambda \to \infty$, where $k > 0$ is an integer. If all the mentioned conditions are satisfied, then $\nu = -k < 0$ and:*
a) *The homogeneous equation has exactly* $\max(0, \chi - n - k)$ *linearly independent solutions. The Fourier transforms of these solutions are given by* (8.58).
b) *The non-homogeneous equation is always solvable for $\chi - n - k \geqslant 0$. If $\chi - n - k < 0$, then the r.h.s. must satisfy $|\chi - n - k|$ necessary and sufficient solvability conditions. If these conditions are satisfied, then the equation has a unique solution. Under the solvability conditions the Fourier transforms $\Phi^+(\lambda)$ of the solutions are given by* (8.62)–(8.63) *(with $\nu = -k < 0$).*

Here, the interpolating polynomial Q_ρ, $\rho = n - 1$, is constructed according to the conditions

$$\frac{d^k}{d\lambda^k}[\Omega^+(\lambda) - \Omega_\rho(\lambda)]_{\lambda = b_j} = 0$$

$$(j = 1, 2, \ldots, s;\ k = 0, 1, \ldots, \beta_j - 1).$$

The solution of the equation is given by (8.67).

8.4 *The dual equation of the second kind*

This equation

$$\left.\begin{aligned} \varphi(x) - \int_{-\infty}^{\infty} k_1(x - t)\, \varphi(t)\, dt &= f(x), \qquad x < 0, \\ \varphi(x) - \int_{-\infty}^{\infty} k_2(x - t)\, \varphi(t)\, dt &= f(x), \qquad x > 0, \end{aligned}\right\} \tag{8.70}$$

reduces to the equivalent boundary value problem of type R

$$\omega^+(\lambda) = \frac{1 - K_1(\lambda)}{1 - K_2(\lambda)}\omega^-(\lambda) + \frac{K_2(\lambda) - K_1(\lambda)}{1 - K_2(\lambda)} F(\lambda) \tag{8.71}$$

under the supplementary condition

$$\Phi(\lambda) = \frac{F(\lambda) + \omega^+(\lambda)}{1 - K_1(\lambda)} \in L_2(-\infty, \infty). \tag{8.72}$$

If this condition is satisfied, then

$$\varphi(x) = \frac{1}{2\pi}\int_{-\infty}^{\infty} e^{-i\lambda x}\frac{\omega^+(\lambda) + F(\lambda)}{1 - K_1(\lambda)}\,d\lambda. \tag{8.73}$$

We assume now that the functions $1 - K_i(\lambda)$ $(i = 1, 2)$ have zeros of integer order on the real axis. Let

$$\left.\begin{aligned} 1 - K_1(\lambda) &= \prod_{j=1}^{r}(\lambda - a_j)^{\alpha_j}\prod_{j=1}^{p}(\lambda - c_j)^{\gamma_j} K_{11}(\lambda), \\ 1 - K_2(\lambda) &= \prod_{j=1}^{s}(\lambda - b_j)^{\beta_j}\prod_{j=1}^{p}(\lambda - c_j)^{\gamma_j} K_{12}(\lambda), \end{aligned}\right\} \tag{8.74}$$

where α_j, γ_j, β_j are positive integers.

By c_j, we denote the common zeros of the functions $1 - K_i(\lambda)$, $(i = 1, 2)$ and by γ_j, the smallest multiplicity of the zero at the point c_j.

We assume that the coefficient of (8.71) can be expressed in the form

$$G(\lambda) = \frac{\prod_{j=1}^{r}(\lambda - a_j)^{\alpha_j}\, p_+(\lambda)\, p_-(\lambda)}{\prod_{j=1}^{s}(\lambda - b_j)^{\beta_j}\, q_+(\lambda)\, q_-(\lambda)} G_1(\lambda), \tag{8.75}$$

where $G_1(\lambda)$ satisfies the conditions mentioned in the formulation of Problem R (in the present case, $G(\infty) = 1$ and $\nu = 0$).

To the conditions imposed on the coefficients of the problem, we add condition (8.72). In order that (8.72) be satisfied, it is sufficient that $l = \left\{\sum_{j=1}^{p}(\gamma_j - 1)\right\}$ supplementary conditions

$$\frac{d^k}{d\lambda^k}[\omega^+(\lambda) + F(\lambda)]_{\lambda=c_j} = 0 \tag{8.76}$$

$$(j = 1, 2, \ldots, p;\ k = 1, 2, \ldots, \gamma_j - 1)$$

be satisfied (assuming all functions considered are differentiable).

THEOREM 8.2. *If all the mentioned conditions are satisfied, then* $\nu = 0$ *and:*

a) *The homogeneous equation* (8.70) *has* $\max(0, \chi - n - l)$ *linearly independent solutions in* $L_2(-\infty, \infty)$. *The functions* $\omega^{\pm}(\lambda)$, *in terms of which the solutions can be expressed (see* (8.73)), *have the form*

$$\left.\begin{aligned} \omega_0^+(z) &= \prod_{j=1}^{r}(z - a_j)^{\alpha_j}\prod_{j=1}^{p}(z - c_j)^{\gamma_j}\, p_-(z)\, q_-^{-1}(z)e^{\Gamma^+(z)}\, P_{\chi-n-l-1}(z),\\ \omega_0^-(z) &= \prod_{j=1}^{s}(z - b_j)^{\beta_j}\prod_{j=1}^{p}(z - c_j)^{\gamma_j}\, q_+(z)\, p_+^{-1}(z)e^{\Gamma^-(z)}\, P_{\chi-n-l-1}(z). \end{aligned}\right\} \tag{8.77}$$

b) *The non-homogeneous equation is always solvable if* $\chi - n - l \geqslant 0$. *If* $\chi - n - l < 0$, *then it is solvable if and only if its r.h.s. satisfies supplementary conditions guaranteeing that* $Y^{\pm}$ *vanish at infinity. When this equation is solvable, the solution is given by* (8.73), *where*

$$\omega^+(z) = \omega_0^+(z) + Y^+(z). \tag{8.78}$$

The same methods can be applied to the dual equation of the first kind and to the adjoint one {see [7]}. Equations with degenerate symbols and with kernels of exponential type are studied in [38], [39]. Equations with degenerate symbols in the spaces of generalized functions are considered in [19], [35].

§9 Examples

Example 1. Consider the Wiener-Hopf equation of the second kind

$$\varphi(x) - \frac{5}{4}\int_0^\infty k(x - t)\,\varphi(t)\,dt = f(x), \qquad x > 0, \tag{8.79}$$

where

$$k(x) = \begin{cases} e^{-2x}, & x > 0, \\ e^{2x}, & x < 0, \end{cases} \qquad K(\lambda) = \frac{4}{\lambda^2 + 4},$$

$$f(x) = \tfrac{2}{3}e^{-x} - \tfrac{5}{4}e^{-2x}, \qquad x > 0, \qquad F(\lambda) = \frac{2i}{3(\lambda + i)} - \frac{5i}{4(\lambda + 2i)}.$$

The equivalent problem has the form

$$\Phi^+(\lambda) = \frac{\lambda^2 + 4}{\lambda^2 - 1}\Omega^- + \frac{\lambda^2 + 4}{\lambda^2 - 1}F(\lambda). \tag{8.80}$$

In the present case $G_1(\lambda) = 1$, hence $e^{\Gamma^\pm(z)} = 1$, $p_+(\lambda) = \lambda - 2i$, $p_-(\lambda) = \lambda + 2i$, $q_+(\lambda) = q_-(\lambda) = 1$, $b_1 = 1$, $b_2 = -1$, $\beta_1 = \beta_2 = 1$, $n_+ = n_- = 0$, $m_+ = m_- = 1$, $\chi = m_+ - n_+ = 1$, $n = 2$, $\chi - n = -1 < 0$, $\rho = m + n - 1 = 1$. Since $\chi - n < 0$, the homogeneous equation has only the trivial solution. In order to construct the function Y^+, we consider the function

$$\frac{g_1(\lambda)\, q_-(\lambda)}{p_-(\lambda)e^{\Gamma^+(\lambda)}} = i\left[-\frac{7}{12} - \frac{2i}{\lambda + i} + \frac{5i}{\lambda + 2i}\right].$$

Hence, $\Omega^+(\lambda) = 2/(\lambda + i) - 5/(\lambda + 2i)$. According to (8.60), we have

$$Y^+(\lambda) = \frac{\lambda + 2i}{\lambda^2 - 1}\left[\frac{2}{\lambda + i} - \frac{5}{\lambda + 2i} - Q_1(\lambda)\right].$$

Since the interpolation polynomial $Q_1(\lambda) = A\lambda + B$ is defined by the conditions $Q_1(\pm 1) = \psi^+(\pm 1)$, we obtain the following system for the coefficients A and B

$$\frac{2}{1 + i} - \frac{5}{1 + 2i} - A - B = 0,$$

$$\frac{2}{-1 + i} - \frac{5}{-1 + 2i} + A - B = 0,$$

which yields $A = 0$, $B = i$. Hence $Y^+(\lambda) = -i/(\lambda + i)$. Since the homogeneous problem has only the trivial solution, $\Phi^+(\lambda) = Y^+(\lambda) = -i/(\lambda + i)$. Thus $\varphi(x) = e^{-x}$, $x > 0$.

The non-homogeneous problem is solvable for a given $f(x)$. In the general case, the solvability condition takes the form $Y^+(\lambda) \in L_2(-\infty, \infty)$.

Example 2. Find the solution of the Wiener-Hopf equation of the first kind

$$\int_0^\infty k(x - t)\,\varphi(t)\,dt = f(x), \qquad 0 < x < \infty, \tag{8.81}$$

where $k(x) = e^x + 4xe^x - 2x^2e^x$ for $x < 0$ and $k(x) = 0$, for $x > 0$, $K(\lambda) = i\{1/(\lambda - i) + [4i/(\lambda - i)^2] - [4/(\lambda - i)^3]\}$; $f(x) = e^{-x}$ for $x > 0$, $f(x) = 0$ for $x < 0$. The equivalent problem has the form

$$\Phi^+(\lambda) = -i\frac{(\lambda - i)^3}{(\lambda + i)^2}\Phi^-(\lambda) - i\frac{(\lambda - i)^3}{(\lambda + i)^2}F(\lambda), \qquad F(\lambda) = \frac{i}{\lambda + i}. \tag{8.82}$$

In this problem $p_+ = (\lambda - i)^3, p_- = 1, q_+ = 1, q_- = i(\lambda + i)^2, G_1(\lambda) = 1$, $e^{\Gamma^\pm(\lambda)} = 1, n_+ = 3, n_- = 0, m_+ = 0, m_- = 2, m = n = 0, \chi = 3 - 0 = 3$, $K(\lambda) \sim (1/\lambda)$ and $k = 1, \nu = -k = -1 < 0, \chi - n - k = 2$.

Hence, the homogeneous equation has two solutions. Here we have

$$\frac{g_1(\lambda)\,q_-(x)}{p_-(\lambda)\,e^{\Gamma^+(\lambda)}} = (\lambda - i)^3\,F(\lambda) = i(\lambda + i)^2 + 6(\lambda + i) + 12i - \frac{8}{\lambda + i},$$

hence

$$\Omega^+(\lambda) = \frac{-8}{\lambda + 1}.$$

It follows from §8.1 that the leading term of this expression can have the exponent $m + \chi - 1 = 2 (\nu < 0)$ at infinity. This in fact is the case. The solution of the homogeneous problem is given by (8.58):

$$\Phi_0^+(\lambda) = \frac{1}{(\lambda + i)^2}(A\lambda + B), \qquad \Phi_0^- = \frac{1}{(\lambda - i)^3}(A\lambda + B).$$

Solutions of the non-homogeneous problem can be determined from (8.62), (8.63) ($Q_\rho \equiv 0$ since $m + n - 1 < 0$)

$$\Phi^+(\lambda) = \Phi_0^+(\lambda) + \frac{8i}{(\lambda + i)^2(\lambda + i)}.$$

Thus, we obtain finally

$$\varphi(x) = C_1 e^{-x} + C_2 x e^{-x} - 4x^2 e^{-x}.$$

Example 3. Find the conditions under which the following equation is solvable

$$\left.\begin{aligned} &\varphi(x) - \int_{-\infty}^{\infty} e^{-|x-\tau|}\,\varphi(x)\,d\tau = f(x), \qquad x < 0,\\ &\varphi(x) - (1-i)\int_{-\infty}^{\infty} e^{-(x-\tau)}\,\theta(x-\tau)\,\varphi(\tau)\,d\tau = f(x), \qquad x > 0,\\ &\theta(x) = \begin{cases} 1, & x > 0,\\ 0, & x < 0;\end{cases} \end{aligned}\right\} \tag{8.83}$$

$$1 - K_1(\lambda) = \frac{\lambda^2 - 1}{\lambda^2 + 1}, \qquad 1 - K_2(\lambda) = \frac{\lambda - 1}{\lambda + i}.$$

The corresponding problem is

$$\omega^+ = \frac{\lambda + 1}{\lambda - i}\,\omega^- + (i + 1)\,F(\lambda), \tag{8.84}$$

where $p_+(\lambda) = p_-(\lambda) = 1$, $q_+(\lambda) = \lambda - i$, $q_-(\lambda) = 1$, $G_1 = 1$, $\Gamma^\pm = 1$, $m_+ = m_- = n_- = 0$, $n_+ = 1$, $\chi = -1$, $h = 0$, $l = 1$, $\chi - n - l = -2$, $m = 1$, $n = 0$, $\nu = 0$. In addition, we have

$$\frac{(i + 1)\,F(\lambda)\,q_-(\lambda)}{p_-(\lambda)e^{\Gamma^+}} \equiv (i + 1)\,F(\lambda) = Q(\lambda) + \psi(\lambda),$$

where $Q(\lambda)$ is the polynomial of degree not more than $n + h - 1 = -1$, i.e. $Q \equiv 0$.

Let $(i + 1)\,F(\lambda) = \Omega_+(\lambda) - \Omega_-(\lambda)$. Since $\chi - n - l = -2 < 0$, the homogeneous problem has only a trivial solution. The solution of the non-homogeneous problem has the form (on the basis of (8.60))

$$\omega^+ = \Omega_+ - Q_\rho, \qquad \omega^- = \frac{\lambda - i}{\lambda + 1}\,[\Omega^- - Q_\rho],$$

where Q_ρ is a polynomial of degree $m + n - 1 = 0$. In addition, we must have $\omega^{\pm}(\lambda) \in L_2$, and hence, $Q_\rho = 0$. Therefore, $F(\lambda)$ must be such that

1) $\omega^+ + F(\lambda)|_{\lambda=1} = 0,$ (see 8.76)

2) $\dfrac{\omega^+ + F(\lambda)}{1 - K_1(\lambda)} \in L_2.$ (see 8.72)

Both these conditions will be satisfied if $\omega^+(\lambda) + F(\lambda)$ is divisible by $\lambda^2 - 1$. Since $\omega^+ = \Omega_+$, this condition reduces to the divisibility of

$$\Omega_+ + \frac{1}{1+i}[\Omega_+ - \Omega_-] = \frac{\Omega_+(2+i) - \Omega_-}{i+1} \tag{8.85}$$

by $\lambda^2 - 1$.

For example, this condition is satisfied if

$$F(\lambda) = \frac{1}{i+1}\left[\frac{\lambda+1}{(2+i)(\lambda+i-2)} - \frac{\lambda+1}{(\lambda-i)^2}\right].$$

In this case

$$\frac{\omega^+ + F(\lambda)}{1 - K_1} = -\frac{4(\lambda+i)}{(\lambda+i-2)^2(\lambda-i)}.$$

Taking the inverse Fourier transform, we obtain

$$\varphi(x) = 2(1-i)e^x\theta(-x) + (1-3i)e^{-x}(\cos 2x + i\sin 2x)\,\theta(x)$$
$$+ 2(1+i)\,xe^{-x}(\cos 2x + i\sin 2x)\,\theta(x).$$

§10 Systems of equations on a semi-infinite interval {see [10]}

10.1 *The basic assumptions*

In this section, we present basic results of the theory for systems of integral equations of the form

$$\varphi(x) - \int_0^\infty k(x-t)\,\varphi(t)\,dt = f(x), \qquad x > 0. \tag{8.86}$$

Here $\varphi(x)$ and $f(x)$ are n-dimensional vectors

$$\begin{bmatrix} \varphi_1(x) \\ \varphi_2(x) \\ \cdot \\ \cdot \\ \cdot \\ \varphi_n(x) \end{bmatrix} \quad \text{and} \quad \begin{bmatrix} f_1(x) \\ f_2(x) \\ \cdot \\ \cdot \\ \cdot \\ f_n(x) \end{bmatrix},$$

$k(x)$ is the $(n \times n)$ matrix

$$\begin{bmatrix} k_{11}(x) & k_{12}(x) & \dots & k_{1n}(x) \\ k_{21}(x) & k_{22}(x) & \dots & k_{2n}(x) \\ \cdot & \cdot & \cdot & \cdot \\ k_{n1}(x) & k_{n2}(x) & \dots & k_{nn}(x) \end{bmatrix}.$$

The most developed theory for such systems is valid under the following assumptions:

A) Elements of the matrix $k(x)$ are summable on the whole axis.

B) Det $[I - K(\lambda)] \neq 0, -\infty < \lambda < \infty$, where I is the unit matrix and

$$K(\lambda) = \int_{-\infty}^{\infty} e^{i\lambda x} k(x)\, dx$$

is the Fourier transform of $k(x)$.

C) The components of the r.h.s. $f(x)$ and of the solution $\varphi(x)$ belong to one of the spaces E^+ (see §1.5). The space of n-tupled vector-functions with components from E^+ will be denoted by E_n^+.

If elements of the matrix $k(x)$ belong to the space $e^{\alpha x} L(-\infty, \infty)$ (i.e. $k_{ij}(x)e^{-\alpha x} \in L(-\infty, \infty)$), then the equation is reduced to an equation with a summable matrix after multiplication by $e^{-\alpha x}$ (see §1.2) and, furthermore, we must suppose that $f(x), \varphi(x) \in e^{\alpha x} E_n^+$.

The theory for the system (8.36) is similar to the theory for the scalar equation (8.14), especially with regard to the existence of solutions. The technique of solution for the system is much more complicated than in the case of the scalar equation. An elementary solution for the system can be obtained only in some rare cases.

10.2 *Factorization of matrix-functions*

The theory for the systems (8.86) is based on the *factorization of matrix-functions* (compare §3).

By *the standard left factorization* of a continuous non-singular matrix-function $\mathfrak{M}(\lambda)$, $-\infty < \lambda < \infty$, we denote

$$\mathfrak{M}(\lambda) = \mathfrak{N}_+(\lambda)\,\mathfrak{N}(\lambda)\,\mathfrak{N}_-(\lambda), \tag{8.87}$$

where $\mathfrak{N}(\lambda)$ is the diagonal matrix-function

$$\mathfrak{N}(\lambda) = \left\| \left(\frac{\lambda - i}{\lambda + i}\right)^{\chi_j} \delta_{jk} \right\|,$$

with δ_{jk} the Kronecker delta symbol; $\chi_1 \geqslant \chi_2 \geqslant \ldots \geqslant \chi_n$ integer exponents; and $\mathfrak{N}_+(\lambda)$ and $\mathfrak{N}_-(\lambda)$ matrices extended so that they are holomorphic inside the upper and lower half-planes respectively, continuous up to the boundary and such that the determinants of their extensions differ from zero in their respective half-planes. If we interchange the factors $\mathfrak{N}_\pm(\lambda)$ in (8.87), then we obtain *the standard right factorization.*

Different left (or right) factorizations of a given matrix differ from one another by multipliers $\mathfrak{N}_\pm(\lambda)$, which have the same exponents χ_j. These exponents are referred to as *the left* (*or the right*) *indices* of the matrix-function.

THEOREM 10.1. *If the matrix* $\mathfrak{M}(\lambda)$ *satisfies the conditions A and B of* §10.1, *then the matrix* $I - \mathfrak{M}(\lambda)$ *admits a standard factorization, where*

$$\mathfrak{N}_\pm(\lambda) = I + \int_0^\infty n_\pm(x) e^{\pm i\lambda x}\, dx, \tag{8.88}$$

and the $n_\pm(x)$ *are matrices with elements from* $L(0, \infty)$.

From a theorem of Wiener (see [3], [10]), it follows that if the matrix $I - \mathfrak{M}(\lambda)$ satisfies the conditions of Theorem 10.1, then the matrix $[I - \mathfrak{M}(\lambda)]^{-1}$ satisfies the same conditions.

The process of factorization of matrices is much more complicated than that of functions (see §3). Only a few cases, where it can be implemented using simple tools, are known. For example, this is the case for triangular matrix-functions (see [10], §11) and for matrices with rational elements (see [2], §5-a, [5], [63]).

The problems concerning factorization are studied in a highly general form and in detail in [12]. This paper contains an extensive bibliography.

10.3 *The solvability conditions*

THEOREM 10.2. *Let conditions A and B be satisfied* (see §10.1). *Then:*
a) *If the matrix* $[I - K(\lambda)]^{-1}$ *has no negative left indices, then the system is solvable in any space* E_n^+ *for any r.h.s. from the same space. If all left indices are equal to zero, then the solution is unique.*
b) *If the matrix* $[I - K(\lambda)]^{-1}$ *has negative left indices, then the system* (8.86) *is solvable in any space* E_n^+ *if and only if*

$$\int_{-\infty}^{\infty} f(x)\, \psi_j(x)\, dx = 0, \tag{8.89}$$

where $\psi_j(x)$ *is an arbitrary solution of the adjoint homogeneous system*

$$\psi(x) - \int_0^{\infty} k'(t - x)\, \psi(t)\, dt = 0,$$

where $k'(x)$ *is adjoint to the matrix* $k(x)$.
c) *If the matrix* $[I - K(\lambda)]^{-1}$ *has positive left indices and* χ_p *is the smallest of them, then the homogeneous system has in all spaces* E_n^+ *one and the same solutions. The number* α *of linearly independent solutions is equal to the sum of positive left indices of the matrix* $[I - K(\lambda)]^{-1}$, *i.e.* $\alpha = \sum_{i=1}^{p} \chi_i$. *These solutions can be chosen so that they form p D-chains* (see §3) *of lengths* $\chi_1, \chi_2, \ldots, \chi_p$.
d) *The same results hold for the adjoint equation*

$$\psi(x) - \int_0^{\infty} k'(t - x)\, \psi(t)\, dt = f(x) \tag{8.90}$$

with the difference that its left indices are the numbers $-\chi_n \geqslant -\chi_{n-1} \geqslant \ldots \geqslant -\chi_1$. *The number of solutions of the adjoint equation is equal to* $\beta = -\sum_{=q}^{n} \chi_s$, *where* χ_q *is the first negative index of the matrix* $[I - K(\lambda)]^{-1}$.

These facts with some additions can be reformulated in the terminology of functional analysis as following:

THEOREM 10.3. *If conditions A and B are satisfied, then the operator, defined by the l.h.s. of* (8.86), *is normally solvable and has the finite index in any of the spaces* E_n^+. *In all these spaces, it has one and the same subspace of zeros, the dimension of which is equal to the sum of positive left indices of the matrix* $[I - K(\lambda)]^{-1}$. *The dimension of the subspace of zeros of the adjoint operator is equal to the absolute value of the sum of the negative left indices.*

The index of the operator (8.86) *in any of the spaces* E_n^+ *is equal to*

$$\chi = -\frac{1}{2\pi} \arg \det [I - K(\lambda)] \Big|_{-\infty}^{\infty} = \sum_{i=1}^{n} \chi_i. \tag{8.91}$$

Given condition A, a necessary and sufficient condition that (8.86) *be solvable and have finite index is the validity of condition B.*

The solution of the system is given by (8.21) and (8.22) of §3 with the functions replaced by the corresponding matrices or vectors. In more detail:

a) If all the left indices equal zero (in this case the system is uniquely solvable), then factorization takes the form

$$[I - K(\lambda)]^{-1} = \sigma_+(\lambda)\, \sigma_-(\lambda), \tag{8.92}$$

where the matrices $\sigma_\pm(\lambda)$ have the form

$$\sigma_\pm(\lambda) = I + \int_0^\infty \gamma_\pm(x)\, e^{\pm i\lambda x}\, dx, \tag{8.93}$$

where $\gamma_\pm(x)$ are summable matrices. The solution of the system in any of the spaces E_n^+ is given by (8.21) and (8.22) in which $\gamma(x, t)$ is an $(n \times n)$-matrix-function and $f(x)$ is a vector.

b) Let the matrix $[I - K(\lambda)]^{-1}$ have non-zero left indices with the solvability conditions satisfied. The standard left factorization has the form (8.87).

Let us set

$$\tilde{\sigma}_+(\lambda) = \mathfrak{N}_+(\lambda) \left\| \left(\frac{\lambda - i}{\lambda + i}\right)^{\chi_j^+} \delta_{jk} \right\|,$$

$$\tilde{\sigma}_-(\lambda) = \left\| \left(\frac{\lambda - i}{\lambda + i}\right)^{\chi_j^-} \delta_{jk} \right\| \mathfrak{N}_-(\lambda),$$

where

$$\chi_j^+ = \begin{cases} \chi_j, & \text{if} \quad \chi_j > 0, \\ 0, & \text{if} \quad \chi_j \leqslant 0, \end{cases} \qquad \chi_j^- = \begin{cases} 0, & \text{if} \quad \chi_j \geqslant 0, \\ \chi_j, & \text{if} \quad \chi_j < 0. \end{cases}$$

These matrices also satisfy the relationship (8.92) and the matrices

$$\sigma_+(\lambda) = \tilde{\sigma}_+(\lambda)\,\tilde{\sigma}_+^{-1}(\infty), \qquad \sigma_-(\lambda) = \tilde{\sigma}_-(\lambda)\,\tilde{\sigma}_-^{-1}(\infty)$$

have a representation of the form (8.93). As above, one of the solutions of (8.86) is given by (8.21) and (8.22).

c) The solution of the homogeneous system (8.87) can be constructed in the following way: let χ_p be the smallest positive left index of the matrix $[I - K(\lambda)]^{-1}$ and $\Phi_1(\lambda), \Phi_2(\lambda), \ldots, \Phi_p(\lambda)$ be the first p column vectors of the matrix

$$\tilde{\sigma}_+(\lambda) = \mathfrak{N}_+(\lambda) \left\| \left(\frac{\lambda - i}{\lambda + i}\right)^{\chi_j^+} \delta_{jk} \right\|.$$

Then the Fourier transform of the general solution of the homogeneous system is the vector

$$\Phi(\lambda) = \sum_{j=1}^{p} \left[\sum_{k=1}^{\chi_j} a_k^{(j)} (\lambda - i)^{-k} \right] \Phi_j(\lambda), \tag{8.94}$$

where the $a_k^{(j)}$ are arbitrary constants (we recall that any solution of the homogeneous system is summable). The functions $\varphi_{j0}, \varphi_{j1}, \ldots, \varphi_{j,\chi_j - 1}$, $j = 1, 2, \ldots, p$, which have the Fourier transforms

$$i(\lambda - i)^{-1}\,\Phi_j(\lambda), \ldots, i^{\chi_j}(\lambda - i)^{-\chi_j}\,\Phi_j(\lambda),$$

form the p D-chains, which were mentioned in Theorem 10.2.

10.4 *Dual equations*

Let $k_j(x)$, $j = 1, 2$, be $(n \times n)$-matrices and $\varphi(x)$, $f(x)$ be n-tuple vectors. We assume that the following conditions are satisfied:

A$_1$. Elements of the matrices $k_1(x)$ and $k_2(x)$ are summable on the whole axis.

B$_1$. Det $[I - K_j(\lambda)] \neq 0$, $-\infty < \lambda < \infty$ $(j = 1, 2)$, where $K_j(\lambda)$ is the Fourier transform of the matrix $k_j(x)$.

C_1. The r.h.s. $f(x)$ and the unknown vector-function belong to one and the same E_n.

Then the system of dual equations

$$\left.\begin{aligned} \varphi(x) - \int_{-\infty}^{\infty} k_1(x-t)\,\varphi(t)\,dt &= f(x), \qquad x < 0, \\ \varphi(x) - \int_{-\infty}^{\infty} k_2(x-t)\,\varphi(t)\,dt &= f(x), \qquad x > 0, \end{aligned}\right\} \tag{8.95}$$

reduces, by means of the one-one inverse transform in the space E'_n, viz.

$$g(x) = \varphi(x) - \int_{-\infty}^{\infty} k_1(x-t)\,\varphi(t)\,dt - f(x), \qquad -\infty < x < \infty,$$

to a system of equations on a semi-infinite interval

$$g(x) - \int_{0}^{\infty} k_0(x-t)\,g(t)\,dt = \int_{-\infty}^{\infty} k_0(x-t)\,f(x)\,dt, \qquad 0 < x < \infty, \tag{8.96}$$

in which the matrix kernel $k_0(x)$ is summable and has the matrix Fourier transform

$$K_0(\lambda) = [K_2(\lambda) - K_1(\lambda)][I - K_1(\lambda)]^{-1}. \tag{8.97}$$

If $g(x)$ is the solution of the system (8.96), then the solution of the system (8.95) is given by

$$\varphi(x) = f(x) + g(x) + \int_{-\infty}^{\infty} q(x-t)[f(t) + g(t)]\,dt, \qquad -\infty < x < \infty, \tag{8.98}$$

where the Fourier transform of the summable matrix kernel $q(x)$ is given by the function $Q(\lambda) = K_1(\lambda)[I - K_1(\lambda)]^{-1}$.

Thus, the system (8.95) reduces to the system (8.86). All results, established for the system (8.86), can be carried over to the system (8.95) without essential changes (see §4). When formulating them it is only necessary to replace the matrix $K(\lambda)$ by the matrix $K_0(\lambda)$. We only give the formula for the index of the system (8.95)

$$\chi = \frac{1}{2\pi} \arg \frac{\det [I - K_1(\lambda)]}{\det [I - K_2(\lambda)]} \Bigg|_{-\infty}^{\infty}. \tag{8.99}$$

Under the mentioned assumptions, the system, adjoint to (8.95),

$$\varphi(x) - \int_{-\infty}^{0} k_1(t - x)\, \varphi(t)\, dt - \int_{0}^{\infty} k_2(t - x)\, \varphi(t)\, dt = f(x),$$

$$-\infty < x < \infty, \quad (8.100)$$

is equivalent to

$$\varphi(x) = f(x) + \int_{-\infty}^{\infty} q(t - x)\, f(t)\, dt + \int_{0}^{\infty} k^0(t - x)\, \varphi(t)\, dt,$$

$$-\infty < x < \infty, \quad (8.101)$$

where the matrix $q(x)$ has the same sense as in (8.98), and the matrix $k^0(x)$ has the Fourier transform $[I - K_1(\lambda)]^{-1}[K_2(\lambda) - K_1(\lambda)]$. The system (8.101) is, practically speaking, a system of equations on a semi-infinite interval, since its r.h.s. contains the values of $\varphi(x)$ only for $x > 0$. Solving this system for $x > 0$, we also obtain the values of $\varphi(x)$ for $x < 0$.

For systems of equations with kernels with different exponential behaviour at infinity see [20]. For solutions of systems the kernels of which have rational Fourier transforms, see [37]. Systems of equations of the form (8.1) are considered in [33].

10.5 *Kernels with an exponential decay at infinity*

THEOREM 10.4. (see [43]). *If the elements of the matrix $k(x)$ belong to the space $e^{-h|x|}\, L(-\infty, \infty)$, i.e. if*

$$\int_{-\infty}^{\infty} e^{h|x|}\, k(x)\, dx < \infty,$$

and if

$$\det [I - K(\lambda \pm ih)] \neq 0, \qquad -\infty < \lambda < \infty,$$

then any solution of the homogeneous system (8.86) *with components from the space* $e^{hx}E_n^+$ *admits the asymptotic representation*

$$\varphi(x) = \left\{ \sum_{j=1}^{m} e^{i\alpha_j x} P_{jk}(x) \right\}_{k=\overline{1,n}} + 0(e^{-hx}), \qquad x \to +\infty, \tag{8.102}$$

where the α_j, $j = 1, 2, \ldots, m$, *are different zeros of the function* $\det [I - K(\lambda)]$ *in the strip* $-h < \operatorname{Im} \lambda < h$; *the* p_j *are their multiplicities; the* $P_{jk}(x)$ *are polynomials of degree* $\leqslant p_j - 1$ *with arbitrary coefficients.*

Analogously, if the matrices $k_1(x), k_2(x) \in e^{h|x|} L(-\infty, \infty)$ and

$$\det [I - K_j(\lambda \pm ih)] \neq 0,$$

then any solution $\varphi(x) = \{\varphi_1(x), \varphi_2(x), \ldots, \varphi_n(x)\}$ of either the homogeneous system (8.95) or (8.100) admits the asymptotic representation

$$\varphi(x) = \left\{ \sum_{j=1}^{m_2} e^{-i\alpha_j^{(2)} x} P_{jk}^{(2)}(x) \right\}_{k=\overline{1,n}} + o(e^{-hx}), \qquad x \to +\infty,$$

$$\varphi(x) = \left\{ \sum_{j=1}^{m_1} e^{-i\alpha_j^{(1)} x} P_{jk}^{(1)}(x) \right\}_{k=\overline{1,n}} + o(e^{hx}), \qquad x \to -\infty,$$

where the $\alpha_j^{(r)}$, $r = 1, 2, j = 1, 2, \ldots, m_r$, are different zeros of the function $\det [I - K_r(\lambda)]$ in the strip $-h < \operatorname{Im} \lambda < h$; the $p_j^{(r)}$ are their multiplicities; the $P_j^{(r)}(x)$ are polynomials of degree not exceeding $p_j^{(r)} - 1$.

The asymptotic character of the solutions of such an equation has been studied briefly in [3], and in detail in [22]. Both appeared before [43].

§11 Equations defined on a finite interval

In this section, we shall consider equations of the form

$$\varphi(x) = f(x) + \int_0^T k(x - y)\, \varphi(y)\, dy, \qquad 0 < x < T < \infty. \tag{8.103}$$

If the kernel is summable on the interval $(-T, T)$, then the integral operator is compact in any of the spaces $L_p(0, T)$, $1 \leqslant p \leqslant \infty$, $C(0, T)$, and the Fredholm theory is applicable to this equation (see Chapter V).

Since $k(x)$ is only defined for $(-T, T)$, the kernel can be extended arbitrarily outside this interval. Henceforth, we shall assume that the extension is such that the kernel is summable on the whole axis.

11.1 *Reduction to a boundary value problem*

In the space $L_2(0, T)$, this equation reduces to the Riemann boundary value problem for two pairs of unknown functions (see [4]). Find the functions $\Omega^{\pm}(z)$, and $\Phi^{\pm}(z)$ satisfying the conditions:

1) The functions Ω^{+}, $\Phi^{+}(\Omega^{-}, \Phi^{-})$ are analytic in the upper (lower) half-plane, and moreover, $\Omega^{\pm}(\lambda)$ vanish at infinity.

2) The limits of these functions satisfy the following relationships

$$\left.\begin{aligned} \Omega^{+}(\lambda) + e^{-i\lambda T}\,\Omega^{-}(\lambda) + [1 - K(\lambda)]\,\Phi^{-}(\lambda) &= F(\lambda) \\ \Omega^{+}(\lambda) + \Omega^{-}(\lambda) + [1 - K(\lambda)]\,\Phi^{+}(\lambda) &= e^{i\lambda T}\,F(\lambda) \end{aligned}\right\} \tag{8.104}$$

along the real axis.

If this problem is solvable, then the solution of (8.103) is given by

$$\varphi(x) = \frac{1}{2\pi}\int_{-\infty}^{\infty} \Phi^{-}(\lambda) e^{-i\lambda x}\, d\lambda, \qquad x > 0.$$

Sometimes it is possible to construct the solution of (8.104) in a closed form when particular assumptions concerning the coefficients (see [5], [2]) are made.

11.2 *Kernels with rational Fourier transforms*

In the case when the Fourier transform of the kernel is rational, (8.103) can be solved simply in closed form (see [4]). Under the assumption that $1 - K(\lambda) \neq 0$, $-\infty < \lambda < \infty$, the Fourier transform of the solution of (8.103) is given by

$$\Phi(\lambda) = \frac{1}{1 - K(\lambda)}\{F(\lambda) - \Omega^{+}(\lambda) - e^{-i\lambda T}\,\Omega^{-}(\lambda)\}, \tag{8.105}$$

in which

$$\Omega^{\pm}(\lambda) = \sum_j \sum_{k=1}^{p_j^{\pm}} \frac{M_{jk}^{\pm}}{(\lambda - b_j^{\mp})^k},$$

where $b_j^+(b_j^-)$ are the poles of the function $1 - K(\lambda)$, lying in the upper (lower) half-plane; $p_j^{\pm}$ are their multiplicities. The constants $M_{jk}^{\pm}$ are determined by the conditions

$$\frac{d^s}{d\lambda^s}[\Omega^+(\lambda) + e^{-i\lambda T}\,\Omega^-(\lambda) - F(\lambda)]_{\lambda=a_j^+} = 0$$

$$(s = 0, 1, \ldots, q_j^+ - 1),$$

$$\frac{d^s}{d\lambda^s}[\Omega^+(\lambda) + e^{-i\lambda T} - F(\lambda)]_{\lambda=a_j^-} = 0$$

$$(s = 0, 1, \ldots, q_j^- - 1),$$

where $a_j^+(a_j^-)$ are the zeros of the function $1 - K(\lambda)$, lying in the upper (lower) half-plane; $q_j^{\pm}$ are their multiplicities. They can also be determined by the substitution of the solution into the given equation.

This method can be used for the solution of equations of the first kind. All the formulas remain valid with $1 - K(\lambda)$ replaced by $K(\lambda)$ (in this case the solutions will be generalized functions: they will contain functions $\delta^{(s)}(x)$ and $\delta^{(s)}(T - x)$).

11.3 *Reduction to a differential equation with constant coefficients*

Another method of solution, which has been studied in detail in [1], consists of replacing (8.103) by a differential equation with constant coefficients: if $1 - K(\lambda) = M(\lambda)/N(\lambda)$, where $M(\lambda)$ and $N(\lambda)$ are polynomials of degree m and n respectively, then the solution of the equation (8.103) (if it exists) satisfies the differential equation

$$M\left(i\frac{d}{dx}\right)\varphi(x) = N\left(i\frac{d}{dx}\right)f(x), \qquad 0 < x < T. \tag{8.106}$$

The solution of this equation contains m arbitrary constants which can be determined after the substitution of the solution into the equation. These constants satisfy a system of linear algebraic equations, which is equivalent to the given equation. This method is also applicable to equations of the first kind, but in this case we must seek its solution in the form

$$\varphi(x) = Y(x, C_1, \ldots, C_m) + \sum_{k=0}^{n-m-1} [A_k\, \delta^{(k)}(x) + B_k\, \delta^{(k)}(x - T)],$$

where Y is a general solution of (8.106). The arbitrary constants A_k and B_k can be determined after the substitution of the solution into the given equation. When seeking the generalized solutions, the r.h.s. $f(x)$ must be expressed in the form

$$f(x) = f(x)[\theta(x) - \theta(x - 1)], \qquad \theta(x) = \begin{cases} 0, & x < 0, \\ 1, & x > 0. \end{cases}$$

When differentiating $f(x)$ we must take into account that $\theta^{(k)}(x) = \delta^{(k-1)}(x)$. For other methods of solution see [23], [31], [36] and §1 of the present chapter.

11.4 *Eigenvalues*

In this section, we shall present some results regarding the eigenvalues of (8.108). These results mainly deal with the case when $K(\lambda) \geqslant 0$, i.e. when the integral operator of the equation is positive. It is assumed that the eigenvalues are arranged in decreasing order.

THEOREM 11.1. *Suppose that:*

a) *$K(\lambda)$ is non-negative, even and decreasing for $\lambda > 0$,*

b) *if $\lambda, \mu \to \infty$ and $\lambda \sim \mu$, then $K(\lambda) \sim K(\mu)$. If $\lambda = o(\mu)$ then $K(\lambda) = o[K(\mu)]$. Then the eigenvalues of*

$$\lambda\varphi(x) = \int_S k(x - y)\, \varphi(y)\, dy, \qquad x \in S, \tag{8.107}$$

in which S is the union of a finite number of intervals, have the asymptotic distribution

$$\lambda_n \sim K\left(\frac{\pi n}{\text{mes } S}\right). \tag{8.108}$$

This statement holds for arbitrary kernels, the Fourier transforms of which are asymptotically equal to a function satisfying the conditions of the theorem.

The more general case when S is a Jordan measurable set is considered in [59].

In particular, the following result follows from Theorem 11.1:

If S is a bounded measurable set, $k(x) = |x|^\alpha$, where α is a non-integer number > -1, and the λ_n are the eigenvalues of (8.107), then

$$\lambda_n \sim \frac{2^{1+\alpha}}{\sqrt{\pi}} \frac{\Gamma\left(\frac{\alpha+1}{2}\right)}{\Gamma\left(-\frac{\alpha}{2}\right)} \left\{\frac{\text{mes } S}{\pi n}\right\}^{1+\alpha}. \tag{8.109}$$

This asymptotic result holds for positive definite kernels the Fourier transforms of which are asymptotically (when $\lambda \to \infty$) equal to the Fourier transform of the kernel $|x|^\alpha$ (see [52], [54], [59]). For instance (see [54]), if $k(x) = \cos\left[(\pi/2)\alpha\right] c\,|x|^{\alpha-1} + b(x)$, where $c > 0$, $\alpha > \frac{1}{2}$ is not an integer and $b(x)$ satisfies the Hölder condition with exponent $\beta > \alpha$, then the eigenvalues of the equation

$$\lambda\varphi(x) = \int_{-T}^{T} k(x-t)\,\varphi(t)\,dt, \qquad -T < x < T, \tag{8.110}$$

have the asymptotic distribution

$$\lambda_n \sim 2\Gamma(\alpha)\, c\left(\frac{2T}{n\pi}\right)^{\alpha}, \qquad n \to \infty.$$

THEOREM 11.2. (see [61]). *Let $K(\lambda)$ be a bounded, non-negative, even function and $\Gamma(\lambda) = -\ln K(\lambda)$. Then:*

a) *If $\Gamma(\lambda)$ is continuous, increases for large values of λ, and $\lambda^{-1}\,\Gamma(\lambda) \to 0$ as $\lambda \to \infty$, then the eigenvalues of the equation $\lambda\varphi(x) = \int_{-1}^{1} k(x-y)\,\varphi(y)\,dy$ have the asymptotic distribution*

$$\lambda_n \sim K\left[\frac{n\pi}{2} + o(n)\right].$$

b) *If $\Gamma(\lambda) \sim \alpha\lambda$, $0 < \alpha < \infty$, then*

$$\ln \lambda_n \sim -n\pi \frac{K\left[\operatorname{sch} \frac{\pi}{\alpha}\right]}{K\left[\operatorname{th} \frac{\pi}{\alpha}\right]}, \tag{8.111}$$

where

$$K(v) = \int_0^{\pi/2} (1 - v^2 \sin^2 \theta)^{-1/2}\, d\theta.$$

c) *If $\Gamma(\lambda)$ is convex for large values of λ and $\lambda^{-1}\Gamma(\lambda) \to 0$ as $\lambda \to +\infty$, then* $\ln \lambda_n \sim \Gamma(\sigma_n)$, *where σ_n is the only root of the equation* $\Gamma(\sigma) = 2n \ln n/\sigma$, $0 < \sigma < n$.

d) *If* $\ln \Gamma(\lambda) \sim \alpha \ln \lambda$, $\alpha > 1$, *then* $\ln \lambda_n \sim -(2 - 2/\alpha)n \ln n$.

e) *If* $\ln \Gamma(\lambda)/\ln \lambda \to \infty$ *as* $\lambda \to \infty$, *then* $\ln \lambda_n \sim 2n \ln n$ (*see* [59]).

We shall denote by $k_\alpha(x, y)$ the kernel which is defined in the following way: If α is an even positive integer, then $k_\alpha(x, y)$ is the Green function of the differential operator

$$L(y) = (-1)^{\alpha/2} y^{(\alpha)}, \quad y^{(s)}(-1) = y^{(s)}(1) = 0, \qquad 0 \leqslant s \leqslant \alpha/2 - 1.$$

If $0 < \operatorname{Re} \alpha < 2$, $x, y \in (-1, 1)$, then

$$k_\alpha(x, y) = \frac{\cos \frac{\alpha\pi}{2}}{\Gamma(\alpha)} |x - y|^{\alpha - 1} \max(0, x - y)$$

$$- \frac{\sin \frac{\alpha\pi}{2}}{\pi\Gamma(\alpha)} (1 - y^2)^{\alpha/2} \int_{-1}^{x} \frac{(x - t)^{\alpha - 1}}{(1 - t^2)^{\alpha/2}(t - y)}\, dt$$

(if $y < x$, then the integral must be interpreted as its Cauchy principle value). Thus, we have $k_\alpha(x, y) = k_\alpha(y, x)$. For $\alpha = 2k + \beta$, $k = 0, 1, \ldots$, $0 \leqslant \operatorname{Re} \beta < 2$ and $x < y$, we define the kernel $k_\alpha(x, y)$ as the analytical continuation of $k_\alpha(x, y)$ with respect to α. For $x > y$, we define this kernel as the analytical continuation of $k_\alpha(y, x)$.

For $\alpha = 1$

$$k_1(x, y) = \frac{1}{2\pi} \ln \frac{1 - xy + [(1 + x^2)(1 - y^2)]^{1/2}}{1 - xy - [(1 - x^2)(1 - y^2)]^{1/2}}.$$

For $\alpha = 2$

$$k_2(x, y) = \tfrac{1}{2}[1 + \min(x, y)][1 - \max(x, y)].$$

THEOREM 11.3. (see [58]). *Let the Fourier transform of the even real valued functions $k(x) \in L_p(-\infty, \infty)$, $1 \leqslant p \leqslant 2$, satisfy the conditions*

1) $\max\limits_{|\lambda| > \delta} K(\lambda) < M$ *for arbitrary* $\delta > 0$.
2) $\lim\limits_{\lambda \to 0} |\lambda|^{-\alpha} [M - K(\lambda)] = c, \qquad 0 < c < \infty, \qquad 0 < \alpha < \infty.$

Then the positive eigenvalues $\lambda_{n,A}$ of

$$\lambda\varphi(x) = A \int_{-1}^{1} k[A(x - y)]\, \varphi(y)\, dy, \qquad -1 < x < 1, \tag{8.112}$$

have the asymptotic value

$$\lambda_{n,A} = M - \frac{c}{\lambda_n A^\alpha} + o(A^{-\alpha}), \tag{8.113}$$

where the λ_n are the eigenvalues of the positive definite kernel $k_\alpha(x, y)$ described above.

For $\alpha = 2$, this theorem gives

$$\lambda_{n,A} = M - \frac{c^2\pi^2 n^2}{8A^2} + o(A^{-2}).$$

We also state a particular but important result, at least for some applications (see [51]): all the eigenvalues of

$$\lambda\varphi(x) = \frac{1}{\pi} \int_{-T}^{T} \varphi(y) \frac{\sin(x - y)}{x - y}\, dy, \qquad -T \leqslant x \leqslant T, \tag{8.114}$$

are simple and as $T \to \infty$

$$\lambda_n \sim 1 - \frac{4\sqrt{(\pi)}8^n}{n!\, e^{2T}} T^{n+1/2}.$$

(In [55], there are Tables for the first eight eigenvalues for $T = 0.5$; 1.0; 2.9; 8.0.)

The eigenfunctions corresponding to even (odd) n are even (odd) functions. The eigenfunctions $\varphi_n(x)$ can be expressed in terms of the eigenfunctions of the problem

$$(1 - x^2)\frac{d^2u}{dx^2} - 2x\frac{du}{dx} + (\chi - x^2)u = 0,$$

where u is continuous on $[-1, 1]$ in the following way: $\varphi_n(x) = u_n(x/T)$ (here the u_n correspond to the eigenvalues arranged in ascending order) (see [55], [61]).

The Fourier transform of the kernel of (8.114) coincides with the characteristic function for the interval $(-T, T)$. An equation of the form (8.107), under the assumption that the Fourier transform of its kernel coincides with the characteristic function for some bounded open set, is considered in [53].

We conclude this chapter with two statements of a general character.

THEOREM 11.4. (see [60]). *Let $k(x) = e^{\gamma(x)}$, $x \geqslant 0$, where $\gamma(x)$ is a twice continuously differentiable real valued function such that*

a) $\gamma''(x) > 0$,
b) $\lim_{x \to \infty} x\gamma'(x) = \infty$,
c) $\gamma''(x) = o(\gamma'^2)$.

Then (8.110) *with the kernel $k(|x - y|)$ has eigenvalues which are asymptotically equal to $\pm[k(2T)]^2/k'(2T)$ as $T \to \infty$. These eigenvalues are simple and the corresponding eigenfunctions tend to $\sqrt{(k2x)}$ and sign $x\sqrt{(k2x)}$, respectively. The remaining eigenvalues behave like $o([k(2T)]^2/k'(2T))$ as $T \to \infty$.*

THEOREM 11.5. (see [15]). *Assume that the kernel of* (8.110) *can be expressed in the form*

$$k(x) = \frac{1}{2\pi}\int_{-\infty}^{\infty} e^{i\lambda x} f(\lambda)\, d\lambda, \qquad f(\lambda) \in L(-\infty, \infty).$$

Let $N(a, b)$ be the number of eigenvalues lying in the interval (a, b) with $ab > 0$. If the Lebesgue measure of the sets $[f(x) = a]$, $[f(x) = b]$ is equal to zero, then

$$\lim_{T\to\infty} \frac{N(a, b)}{2T} = \frac{1}{2\pi} \operatorname{mes} [a < f(x) < b].$$

Chapter IX

MULTIDIMENSIONAL SINGULAR EQUATIONS

Theory for singular integral equations, in which an integral is taken over some manifold of dimension m (where $m > 1$) (for brevity, we call such equations 'multidimensional'), has developed comparatively slowly over a rather long period. Interest in them has increased sharply during recent years. The monograph [18], which contains results mainly by its author, as well as the results of Calderon and Zygmund, appeared in 1962. About the same time, a series of papers by different authors, in which important problems for multidimensional singular equations were formulated or solved, was published. Thus the problem concerning the index was completely solved for equations with a smooth closed domain of integration and with a non-degenerate symbol; the case of a domain of integration with a boundary was investigated; singular integro-differential equations and their applications were studied; and, equations with degenerate symbols continue to be studied. There are many results concerning multidimensional singular equations in the theory of generalized functions.

We will not consider applications of multidimensional singular equations in this present chapter. Some applications are given in [18]. Multidimensional singular equations were applied to the Cauchy problem for equations of hyperbolic type in [28] and [43]. Kupradze (see [15]) uses multidimensional singular integral equations widely in the solution of three dimensional elasticity problems. Finally, we mention the review paper of Agranovich [1], in which multidimensional singular integro-differential equations are used in the study of boundary value problems for elliptic systems of a rather general form.

As a rule, we will not cite here the bibliography given in [18].

§1 Basic concepts and theorems

1.1 *Notations*

We introduce the following notation: E_m is an m-dimensional Euclidean space with $x, y, z, \ldots$ as its points. We denote the Cartesian coordinates of a general point $x \in E_m$ by $x_1, x_2, \ldots, x_m$. Let $r = |x - y|$ and $\theta = (y - x)/r$ where θ ranges over the unit sphere with centre at the origin. We denote this sphere by S.

On occasions it will be necessary to consider the unit sphere in $(m + 1)$-dimensional Euclidean space E_{m+1}, and in that case we denote it by Σ and call it the Riemannian sphere for the space E_m. The space E_m can be mapped onto Σ by means of the stereographic transform

$$\xi_k = \frac{2x_k}{x^2 + 1}, \quad k = 1, 2, \ldots, m; \quad \xi_{m+1} = \frac{x^2 - 1}{x^2 + 1};$$
$$x^2 = |x^2| = \sum_{k=1}^{m} x_k^2, \tag{9.1}$$

where $\xi_1, \xi_2, \ldots, \xi_m, \xi_{m+1}$ are Cartesian coordinates of a point on the sphere Σ. If ξ and η are points on the Riemannian sphere, corresponding to the points x and y in the space E_m {as defined by (9.1)}, and ρ is the distance between ξ and η; viz.

$$\rho^2 = |\xi - \eta|^2 = \sum_{k=1}^{m} (\xi_k - \eta_k)^2,$$

then

$$r^2 = \tfrac{1}{4}\rho^2(x^2 + 1)(y^2 + 1). \tag{9.2}$$

1.2 *The singular integral*

Let D be a region of the space E_m. This region may be finite or infinite and it may coincide with E_m which represents the most interesting case. We consider singular integrals of the form

$$\int_D u(y) \frac{f(x, \theta)}{r^m} dy, \tag{9.3}$$

which are interpreted using

$$\int_D u(y) \frac{f(x, \theta)}{r^m} dy = \lim_{\varepsilon \to 0} \int_{D \setminus (r < \varepsilon)} u(y) \frac{f(x, \theta)}{r^m} dy, \tag{9.4}$$

where dy denotes an element of volume in E_m.

Certainly, (9.4) has sense only when the limit on the r.h.s. exists. The point x is called the *pole* of the singular integral (9.3), the function $f(x, \theta)$ its *characteristics*, and $u(y)$ its *density*. The formula (9.4) corresponds to the definition of an integral in the sense of the Cauchy principle value: the integral is evaluated over the region D except for a ball with centre at the singular point x and with arbitrarily small radius ε. If the resulting integral has a limit as $\varepsilon \to 0$, then this limit is taken to define (9.3).

Henceforth, we shall assume that the characteristic satisfies the condition

$$\int_S f(x, \theta)\, dS = 0. \tag{9.5}$$

As a rule it is easy to show that the singular integral (9.4) does not exist, if the condition (9.5) is not satisfied.

1.3 *Existence conditions for singular integrals*

We shall formulate some conditions under which the integral (9.3) exists and has certain properties. Below, we denote the open region by D, and the closed region obtained by the addition to D of its boundary points by $\bar{D}$.

THEOREM 1.1. *In the closed region $\bar{D}$, let the density $u(y)$ satisfy a Lipschitz condition with exponent α, $0 < \alpha < 1$, so that*

$$|u(y) - u(x)| \leqslant Cr^{\alpha}, \qquad C = \text{const},$$

and the condition $u(y) = 0(|y|^{-k})$, $k = const > 0$, at infinity (if D has infinite measure). If the characteristic is bounded and continuous with respect to θ and x, (9.3) exists for any $x \in D$ and is a continuous function with respect to x in D.

THEOREM 1.2. *Let the density satisfy the conditions of Theorem 1.1 and the characteristic be bounded and continuously differentiable with respect to the Cartesian coordinates of x and θ. Then, (9.3) satisfies a Lipschitz condition with one and the same exponent α in any finite internal subregion $D' \subset D$.*

A stronger theorem will be stated in §3.

Before we formulate the next theorem, we require some definitions. We say that $u \in A_{\alpha,k}$, if $u(x)$ is defined everywhere on E_m and satisfies the conditions: $u(x) = 0(|x|^{-k})$ for sufficiently large values of x and

$$|u(y) - u(x)| \leqslant C(x^2 + 1)^{-k/2} r^{\alpha}, \qquad r \leqslant 1. \tag{9.6}$$

Further, we say, that $u \in A'_{\alpha,k}$, if $u(x)$ satisfies the inequality (9.6) and the condition

$$u(x) = 0(|x|^{-k} \ln |x|)$$

at infinity.

In both definitions C, α and k are positive constants with $\alpha < 1$.

THEOREM 1.3. *Let the characteristic satisfy the conditions of Theorem 1.2 with $k < m$. Then the singular integral operator*

$$v(x) = \int_{E_m} \frac{f(x, \theta)}{r^m} u(y)\, dy \tag{9.7}$$

maps a function $u \in A_{\alpha,k}$ into a function $v \in A'_{\alpha,k}$.

THEOREM 1.4. (see [29]). *Let for some p, $1 < p < \infty$, the characteristic satisfies the inequality*

$$\int_S |f(x, \theta)|^{p'}\, dS \leqslant C = \text{const}, \qquad 1/p + 1/p' = 1. \tag{9.8}$$

If $u \in L_p(E_m)$, then (9.7) exists for almost all $x \in E_m$. In addition, $v \in L_p(E_m)$ and the singular operator (9.7) is bounded in $L_p(E_m)$: there exists a constant A_p (which may depend on p) such that

$$\|v\|_p = \left\{ \int_{E_m} |v(x)|^p\, dx \right\}^{1/p} \leqslant A_p \left\{ \int_{E_m} |u(x)|^p\, dx \right\}^{1/p} = A_p \|u\|_p. \tag{9.9}$$

Let the density and the characteristic satisfy the conditions of Theorem 1.1. We remove from D an arbitrary region σ_ε containing x. Let $r = \alpha(\varepsilon, x, \theta)$ be the equation of the boundary of this region, assume that the following limit exists

$$\lim_{\varepsilon \to 0} \frac{\alpha(\varepsilon, x, \theta)}{\varepsilon} = \beta(x, \theta) > 0,$$

and that the convergence be uniform with respect to θ. Then

$$\lim_{\varepsilon \to 0} \int_{D \setminus \sigma_\varepsilon} \frac{f(x, \theta)}{r^m} u(y)\, dy = \int_D \frac{f(x, \theta)}{r^m} u(y)\, dy - u(x) \int_S f(x, \theta) \ln \beta(x, \theta)\, dS. \tag{9.10}$$

In this formula, the first term on the r.h.s. is the singular integral defined by (9.4).

Let D be a finite region in E_m, x a point of D. Consider the integral

$$\omega(x) = \int_D \frac{\varphi(x, \theta)}{r^{m-1}} u(y)\, dy. \tag{9.11}$$

THEOREM 1.5. *Let the function $\varphi(x, \theta)$ have continuous first derivatives with respect to the Cartesian coordinates of the points x and θ in $\bar{D}$ and $u \in L_p(D)$, $1 < p < \infty$. Then* (9.11) *has generalized derivatives $\partial\omega/\partial x_k \in L_p(D)$, which are given by (where r denotes the vector directed from the point x to the point y)*

$$\frac{\partial \omega}{\partial x_k} = \int_D \frac{\partial}{\partial x_k}\left[\frac{\varphi(x, \theta)}{r^{m-1}}\right] u(y)\, dy - u(x) \int_S \varphi(x, \theta) \cos(r, x_k)\, dS. \tag{9.12}$$

If $u(y)$ satisfies a Lipschitz condition with exponent α, $0 < \alpha < 1$, in $\bar{D}$, then the derivatives $\partial\omega/\partial x_k$ are continuous in D.

§2 The symbol

2.1 *The singular operator*

For the present, we regard singular integrals as operators defined on a given Banach space of functions in E_m. Below, we extend the notion of a singular operator. Let E be a Banach space. The *singular operator A* in E is defined by

$$(Au)(x) = a(x)\, u(x) + \int_{E_m} \frac{f(x, \theta)}{r^m} u(y)\, dy + (Tu)(x), \tag{9.13}$$

where T is a compact operator in E.

2.2 *The symbol of a singular operator*

The *symbol* of the singular operator A is the function $\Phi_A(x, \theta)$ {with $x \in E_m$ and $\theta \in S$} which satisfies, with respect to appropriate restrictions on $a(x)$ and $f(x. \theta)$, the following three conditions:
1) the symbol of any compact operator is equal to zero;
2) the symbol of the sum of two singular operators is equal to the sum of their symbols;
3) the symbol of the product of two singular operators is equal to the product of their symbols.

2.3 *Formulas for the determination of the symbol*

The symbol of the operator A can be determined as

$$\Phi_A(x, \theta) = a(x) + \lim_{\varepsilon \to 0} \int_{E_m} \frac{f\left(x, \frac{y}{|y|}\right)}{|y|^m} e^{-i(y,z)}\, dy; \qquad \theta = \frac{z}{|z|}. \tag{9.14}$$

The limit on the r.h.s. of (9.14) exists for rather general assumptions about the characteristic. For example, it is sufficient that

$$\int_S |f(x, \theta)|^2\, dS < \infty. \tag{9.15}$$

We shall mention two other formulas for the determination of the symbol, which are simpler in structure than (9.14). One of them is

$$\Phi_A(x, \theta) = a(x) + \int_S \left[\ln \frac{1}{\cos \gamma} - \frac{i\pi}{2} \operatorname{signcos} \gamma \right] f(x, \theta')\, dS', \tag{9.16}$$

with γ the angle between vectors drawn from the origin to the points θ and θ'. The integration in (9.16) is performed with respect to the variable θ'.

We denote by $Y_{n,m}^{(k)}(\theta)$ the system of m-dimensional spherical harmonics which are orthonormal with respect to the metric of $L_2(S)$. Here, n is the order of the spherical harmonics and varies between zero and infinity; k varies between unity and $k_{n,m} = (2n + m - 2)\,(n + m - 3)!/(m - 2)!\, n!$, with $k_{m,n}$ the number of linearly independent m-dimensional spherical harmonics of order n {see [42]}. In terms of spherical harmonics, the characteristic has the series representation

$$f(x, \theta) = \sum_{m=1}^{\infty} \sum_{k=1}^{k_{n,m}} a_n^{(k)}(x)\; Y_{n,m}^{(k)}(\theta). \tag{9.17}$$

The term with $n = 0$ is absent because of the identity (9.5). The symbol of the operator (9.13) can also be determined from the series

$$\Phi(x, \theta) = a(x) + \sum_{n=1}^{\infty} \sum_{k=1}^{k_{n,m}} \gamma_{n,m}\, a_n^{(k)}(x) Y_{n,m}^{(k)} F(\theta), \tag{9.18}$$

where

$$\gamma_{n,m} = \frac{i^n \pi^{m/2} \Gamma(n/2)}{\Gamma\left(\frac{m+n}{2}\right)}. \tag{9.19}$$

The following estimates, which are based on Stirling's formula for large n, are important in the study of properties of the symbol:

$$\gamma_{n,m} = O(n^{-m/2}), \qquad \frac{1}{\gamma_{n,m}} = O(n^{m/2}). \tag{9.20}$$

From (9.18) it clearly follows that: if the symbol is given by

$$\Phi(x, \theta) = \sum_{n=0}^{\infty} \sum_{k=1}^{k_{m,n}} b_n^{(k)}(x)\; Y_{n,m}^{(k)}(\theta),$$

then the corresponding characteristic is determined by the series

$$f(x, \theta) = \sum_{n=1}^{\infty} \sum_{k=1}^{k_{m,n}} \frac{b_n^{(k)}(x)}{\gamma_{n,m}}\; Y_{n,m}^{(k)}(\theta).$$

For $m = 2$, the sum of this series is given in [10], so it follows that

$$f(x, \theta) = -\frac{1}{4\pi i} \frac{d}{d\theta} [\Phi(x, \theta + \pi) - \Phi(x, \theta)]$$

$$- \frac{1}{4\pi^2} \int_0^{\pi} ctg(\theta - \psi) \frac{d}{d\psi} [\Phi(x, \psi + \pi) + \Phi(x, \psi)]\, d\psi. \tag{9.21}$$

The symbol, determined by (9.14), obviously satisfies conditions 1) and 2) formulated above. Condition 3), which we call the *multiplication rule for symbols* is satisfied for certain restrictions on $a(x)$ and $f(x, \theta)$. These restrictions depend in their turn on the Banach space in which the singular operator A is considered. The multiplication rule for symbols will be considered in detail in §3.

2.4 *Some new concepts*

Let $g(x, \theta)$ be an arbitrary function of x and θ, where x varies on some set M and θ on the unit sphere S. We assume that $g(x, \theta)$, considered as a function of θ, is extended to the whole space E_m, except for the origin and the point at infinity, so that it takes a constant value on any ray passing through the origin. Usually, we regard such an extension as a function with domain as the spherical layer Ω, defined by the inequalities $\rho_1 \leqslant |\theta| \leqslant \rho_2$, where ρ_1 and ρ_2 are arbitrary fixed positive numbers.

We say that $g(x, \theta)$ belongs to the space $W_p^{(l)}(S)$ for fixed x, if its extension, as defined above, belongs to the space $W_p^{(l)}(\Omega)$, where $W_p^{(l)}(\Omega)$ denotes a Sobolev space (see [24], [25]) for integer l and a Slobodezky space (see [23]) for non-integer l. We define the norm in $W_p^{(l)}(S)$ by

$$\|g\|_{W_p^{(l)}(S)} = \|g\|_{W_p^{(l)}(\Omega)}. \tag{9.22}$$

In addition, we say that $g(x, \theta)$ belongs uniformly (on the set M) to the space $W_p^{(l)}(S)$ and denote this symbolically as $g \mathrel{\hat{\in}} W_p^{(l)}(S)$, if $g \in W_p^{(l)}(S)$ for any $x \in M$ and if

$$\|g\|_{W_p^{(l)}(S)} \leqslant C,$$

where C does not depend on x.

Henceforth, we shall assume that the coefficient $a(x)$ of the operator A is bounded.

THEOREM 2.1. *If the operator* (9.13) *is such that the characteristic* $f \mathrel{\hat{\in}} W_2^{(l)}(S)$, $l \geqslant 0$, *then the symbol* $\Phi_A \mathrel{\hat{\in}} W_2^{(l+m/2)}(S)$. *Conversely, if* $\Phi_A \mathrel{\hat{\in}} W_2^{(\chi)}(S)$, $\chi \geqslant m/2$, *then* $f \mathrel{\hat{\in}} W_2^{(\chi-m/2)}(S)$.

This form of the theorem is given in [1]. The corresponding theorem for integer indices is given in [18].

2.5 *Variation in the symbol with respect to changes in the independent variables*

In E_m, let the coordinates x_k be replaced by x'_k in accordance with

$$x_k = x_k(x'_1, x'_2, \ldots, x'_n). \tag{9.23}$$

We denote the Jacobian of the transform (9.23), calculated at the fixed point x, by $\mathfrak{A}$. As above, we assume that θ is an arbitrary point of E_m, which differs from the origin and the point at infinity, and that the functions of θ under consideration are constant on rays passing through the origin. Thus, denoting the symbol of a singular operator before and after the transform (9.23) by $\Phi(x, \theta)$ and $\check{\Phi}(x', \theta)$, we obtain

$$\check{\Phi}(x', \theta) = \Phi(x, (\mathfrak{A}^*)^{-1}\theta). \tag{9.24}$$

This formula becomes simpler, if the matrix $\mathfrak{A}$ is orthogonal at the given point x. In such case,

$$\check{\Phi}(x', \theta) = \Phi(x, \mathfrak{A}\theta). \tag{9.25}$$

2.6 *An integral representation for the operator **A** in terms of its symbol*

The concept of symbol enables the concept of a 'singular operator' to be extended in the following way. Consider an operator of the form

$$(\mathbf{A}u)(x) = a(x)\, u(x) + \int_{E_m} \frac{f(x, \theta)}{r^m}\, u(y)\, dy. \tag{9.26}$$

For certain assumptions about the symbol of this operator, it can be shown that the operator (9.26) has a representation of the form

$$(\mathbf{A}u)(x) = \int_{\Pi} \Phi(x, \theta)\, dE(\theta) u. \tag{9.27}$$

The notation of the formula has the following sense:

$$dE(\theta) = \prod_{k=1}^{m-1} dE_k(\vartheta_k),$$

where $\vartheta_1, \vartheta_2, \ldots, \vartheta_{m-1}$ are the angular coordinates of the point θ. Further

$$dE_k(\vartheta_k) = \mathscr{F}^{-1}\, dE'_k(\vartheta_k)\mathscr{F},$$

where $\mathscr{F}$ denotes the Fourier transform, and $E'_k(\vartheta_k)$ is the spectral function for the operation of multiplication by ϑ_k. Finally, Π is the parallelepiped defined by the inequalities

$$0 \leqslant \vartheta_k \leqslant \pi, \qquad 1 \leqslant k \leqslant m - 2, \qquad -\pi \leqslant \vartheta_{m-1} \leqslant \pi. \tag{9.28}$$

The operator, defined by (9.27), is called *the simplest singular operator* with symbol $\Phi(x, \theta)$. In the general case, a singular operator means the sum of two operators: the simplest singular operator and a compact operator.

§3 Singular operators in $L_p(E_m)$

3.1 *Boundedness conditions for singular operators*

THEOREM 3.1. *The operator* (9.27) *is bounded in* $L_p(E_m)$, $1 < p < \infty$, *if its symbol* $\Phi(x, \theta) \mathrel{\hat{\in}} W_2^{(l)}(S)$, *where*

$$l \geqslant \frac{m-1}{p} + 1, \text{ if } 1 < p < 2, \text{ and } l \geqslant \frac{m+1}{2}, \text{ if } 2 \leqslant p < \infty.$$

We note that a theorem regarding the boundedness of a singular operator in $L_p(E_m)$ {formulated in terms of the characteristic but not the symbol} was given in §1 {Theorem 1.4}.

The following two theorems are valid for $p = 2$.

THEOREM 3.2. *If the symbol* $\Phi(x, \theta)$ *and its derivatives*

$$\frac{\partial \Phi}{\partial \vartheta_1}, \frac{\partial \Phi}{\partial \vartheta_2}, \ldots, \frac{\partial^2 \Phi}{\partial \vartheta_1\, \partial \vartheta_2}, \ldots, \frac{\partial^{m-2}}{\partial \vartheta_1\, \partial \vartheta_2 \ldots \partial \vartheta_{m-2}}$$

are measurable with respect to the totality of points (x, θ), *are continuous for fixed* x *and bounded independently of* x, *and the derivative*

$$\frac{\partial^{m-1} \Phi}{\partial \vartheta_1\, \partial \vartheta_2 \ldots \partial \vartheta_{m-1}}$$

exists in the sense of a Sobolev generalized derivative and satisfies the inequality

$$\int_{\Pi} \left| \frac{\partial^{m-1}\Phi}{\partial\vartheta_1\,\partial\vartheta_2\ldots\partial\vartheta_{m-1}} \right|^2 d\vartheta_1\,d\vartheta_2\ldots d\vartheta_{m-1} \leqslant C = \text{const},$$

then its singular operator is bounded in $L_2(E_m)$.

THEOREM 3.3. *If the symbol* $\Phi(\theta)$ *does not contain a pole and is essentially bounded, then the singular operator* $\boldsymbol{A} = \mathscr{F}^{-1}\,\Phi(\theta)\mathscr{F}$ *is bounded in* $L_2(E_2)$ *and*

$$\|\boldsymbol{A}\| = \sup \text{ess } |\Phi(\theta)|. \tag{9.29}$$

Note. It is well known that a function is *essentially bounded*, if it can be bounded after changing it on a set of measure zero. The greatest lower bound of the supremums of the functions obtained from all possible changes is denoted by 'sup ess'.

We also note a relationship which connects in a certain sense the norm of a singular operator with its symbol. Let $\boldsymbol{A}$ be a singular operator with symbol $\Phi_A(x, \theta)$ and $\boldsymbol{T}$ be an arbitrary compact operator in $L_2(E_m)$. We denote by ξ, the image of a point x under the stereographic transform of E_m onto the sphere Σ, and let the symbol be continuous as a function of both points ξ and θ, as ξ ranges over the sphere Σ and θ over the sphere S. Then (see [8], [14])

$$\inf \|\boldsymbol{A} + \boldsymbol{T}\|_{L_2(E_m)} = \max_{\substack{x \in E_m \\ \theta \in S}} |\Phi_A(x, \theta|, \tag{9.30}$$

where the 'inf' is taken over all operators $\boldsymbol{T}$ which are compact in $L_2(E_m)$.

For generalizations of the equality (9.30) see [1].

We now state a theorem (see [41]) concerning the boundedness of a singular operator in a space of functions, some power of which is summable with respect to a given weight. We denote the space of functions, the p-th power of which is summable with respect to the weight $\rho(x)$, by $L_p(\rho, E_m)$; it is well known that the norm in this space is defined by

$$\|u\|^p = \int_{E_m} \rho(x)\,|u(x)|^p\,dx.$$

The properties of singular operators in such spaces are also studied in [6].

THEOREM 3.4. *Let the symbol of a singular operator* $\boldsymbol{A}$ *satisfy the conditions of Theorem* 3.1 (*or the conditions of Theorem* 3.2 *or* 3.3 *for* $p = 2$) *and, in addition, let the characteristic of* $\boldsymbol{A}$ *be bounded. If a constant* β *exists such that* $-m < \beta < m(p-1)$, *then the operator* $\boldsymbol{A}$ *is bounded in the space* $L_p(|x|^\beta, E_m)$.

We note, without entering into details, that the boundedness of multidimensional singular operators in Orlicz spaces is studied in [22] and [36].

3.2 *The decomposition of the simplest singular operator into a series*

The following theorem states the conditions under which the simplest singular operator can be decomposed into a series with respect to singular operators, the characteristics of which are spherical harmonics.

THEOREM 3.5. *Let* $\boldsymbol{A}$ *be the simplest singular operator with symbol* $\Phi_A(x, \theta) \mathrel{\hat{\in}} W_2^{(l)}(S)$, *where*

$$l \geqslant \frac{m-1}{p} + 2, \text{ if } 1 < p < 2, \text{ and } l \geqslant \frac{m+3}{2}, \text{ if } 2 \leqslant p < \infty.$$

If the symbol is decomposed into a series with respect to the spherical harmonics

$$\Phi_A(x, \theta) = \sum_{n=0}^{\infty} \sum_{k=1}^{k_{n,m}} a_n^{(k)}(x)\, Y_{n,m}^{(k)}(\theta), \tag{9.31}$$

then the corresponding decomposition of the operator $\boldsymbol{A}$ *into a series converging with respect to the norm* $L_p(E_m)$ *has the form*

$$\boldsymbol{A} = a_0^{(l)}(x)\boldsymbol{I} + \sum_{n=1}^{\infty} \sum_{k=1}^{k_{n,m}} a_n^{(k)}(x)\boldsymbol{A}_n^{(k)}, \tag{9.32}$$

where $\boldsymbol{I}$ *is the identity operator and* $\boldsymbol{A}_n^{(k)}$ *is the simplest singular operator with symbol* $Y_{n,m}^{(k)}(\theta)$:

$$(\boldsymbol{A}_n^{(k)}u)(x) = \frac{1}{\gamma_{n,m}} \int_{E_m} \frac{Y_{n,m}^{(k)}(\theta)}{r^m}\, u(y)\, dy.$$

3.3 *The multiplication rule for symbols*

THEOREM 3.6. *Let* $\boldsymbol{A}$ *and* $\boldsymbol{B}$ *be singular operators with symbols* $\Phi_A(x, \theta)$ *and* $\Phi_B(x, \theta)$ *which satisfy the conditions of Theorem* 3.4. *Let these symbols be uniformly continuous with respect to* θ *on the Riemannian sphere. Then the operator* $\boldsymbol{AB} - \boldsymbol{BA}$ *is compact in* $L_p(E_m)$ *and the symbol of the product* $\boldsymbol{AB}$ *is equal to the product* $\Phi_A(x, \theta)\, \Phi_B(x, \theta)$.

3.4 *The operator conjugate to a singular operator*

If the singular operator $\boldsymbol{A}$ is bounded in $L_p(E_m)$, then there exists a conjugate operator $\boldsymbol{A}^*$, acting in $L_{p'}(E_m)$, $1/p + 1/p' = 1$, and bounded in this space. The conditions under which $\boldsymbol{A}^*$ is also a singular operator in the sense of the definition, given in §2, are of interest. These conditions are stated in the following theorems.

THEOREM 3.7. *Let* $\boldsymbol{A}$ *be the simplest singular operator, with symbol* $\Phi(\theta)$, *which does not contain a pole. Then the conjugate operator* $\boldsymbol{A}^*$ *is the simplest singular operator with symbol* $\Phi(\theta)$.

THEOREM 3.8. *Let* $\boldsymbol{A}$ *be a singular operator with symbol* $\Phi_A(x, \theta)$ *which satisfies the following conditions*: (1) $\Phi_A \mathrel{\hat{\in}} W_2^{(l)}(S)$, *where* $l \geqslant (m - 1)/q + 2$, $q = \min(p, p')$; (2) Φ_A *is uniformly continuous with respect to* θ *on the Riemannian sphere. Then the conjugate operator* $\boldsymbol{A}^*$ *is a singular operator with symbol* $\Phi_{A^*}(x, \theta) = \Phi_A(x, \theta)$.

§4 Singular integrals over a manifold

4.1 *The definition of a singular operator and its symbol*

Ler Γ be a Liapunov compact manifold of m-dimensions; ξ and η be points in Γ; ρ be the distance between ξ and η, defined with respect to the internal metric of the manifold; $d_\eta \Gamma$ be an element of measure in Γ. We consider integrals of the form

$$\int_\Gamma K(\xi, \eta)\, u(\eta)\, d_\eta \Gamma, \tag{9.33}$$

where the kernel $K(\xi, \eta)$ satisfies conditions formulated below.

Given a number $\varepsilon > 0$, we fix a point $\xi \in \Gamma$ and consider the set Γ_ε of points $\eta \in \Gamma$ such that $\rho < \varepsilon$. We assume first of all that the kernel $K(\xi, \eta)$ is summable on the set $\Gamma \backslash \Gamma_\varepsilon$.

If ε is sufficiently small, then there exists a one-to-one mapping between Γ_ε and a region D of the Euclidean space E_m. We can stipulate that the Jacobian of this mapping should satisfy a Lipschitz condition with some positive exponent. Finally, we stipulate that the mapping be conformal at the point ξ. For instance, this mapping can be performed in the following manner: we construct a tangent plane to Γ; we take the orthogonal projection of a point $\eta \in \Gamma_\varepsilon$ onto this plane as y. Below, we assume that the transform $\eta \to y$ has been chosen in this way.

We denote by x and y the images of points ξ and η under the above transform. We introduce the notation

$$K(\xi, \eta)\, d_\eta \Gamma = L(x, y)\, dy, \qquad u(\eta) = v(y),$$

and assume that the kernel $L(x, y)$ has the representation

$$L(x, y) = \frac{f(x, \theta)}{r^m} + \frac{M(x, y)}{r^\mu}, \qquad \mu = \text{const.} < m,$$

where $M(x, y)$ is a bounded function and r, θ are the same as in §1. We note that θ ranges over the unit sphere S, situated in the tangent plane to Γ.

On the basis of the above definitions, we obtain

$$\int_\Gamma K(\xi, \eta)\, u(\eta)\, d_\eta \Gamma = \int_{\Gamma \backslash \Gamma_\varepsilon} K(\xi, \eta)\, u(\eta)\, d_\eta \Gamma + \int_D \frac{f(x, \theta)}{r^m} v(y)\, dy + \int_D \frac{M(x, y)}{r^\mu} v(y)\, dy. \tag{9.34}$$

We define the symbol of the integral (9.33) at the point ξ {in accordance with (9.16) and (9.18)} with the help of the characteristic $f(x, \theta)$. We give the following defining relationships: the symbol of the operator for multiplication by a function is equal to that function; the symbol of a compact operator is equal to zero. All the theorems of §§1–3 for manifolds without boundaries are valid under this definition. In all theorems, the continuity conditions on the Riemannian sphere are changed to analogous conditions on Γ.

The definition of symbol we have given is local: in order to define the symbol, we must fix a point $\xi \in \Gamma$, then construct a unit sphere with centre at

ξ in the tangent plane passing through ξ and then choose a coordinate system on this sphere. We can construct the symbol globally, i.e. on the whole manifold, only when we can introduce a regular coordinate system on Γ. In such cases, we can indicate how to choose a coordinate system on S which does not depend on the position of the point ξ, and the symbol can be considered as a function of the two independent points $\xi \in \Gamma$ and $\theta \in S$. Besides the Euclidean space considered above, the same situation occurs, if for example Γ is a torus. In general, a regular coordinate system cannot be constructed on a manifold and ξ and θ cannot be regarded as independent points. In this case, the symbol on the whole manifold is defined as a function of its linear elements. We recall that a linear element of a manifold is the union of its points and an arbitrary ray, lying in the tangent plane, which passes through a given point.

4.2 *Another definition for the singular operator*

The above definition of a singular integral on a manifold is equivalent to the definition given by Seeley (see [38] and also [17]).

Let B be a Banach space of functions defined on a manifold Γ. Let $\varphi(\xi)$ and $\psi(\xi)$ be arbitrary smooth functions which satisfy the following condition: the support (i.e. the closure of the set of points on which the given function is not equal to zero) of any of these functions is contained inside a Liapunov sphere and, hence, can be mapped onto some closed region of the Euclidean space E_m in the way discussed in §4.1. Let $\boldsymbol{A}$ be an operator, the domain of which is a set of functions, each of which is defined on Γ. We call $\boldsymbol{A}$ singular, if it satisfies two conditions:

i) if the supports of φ and ψ do not overlap, then $\varphi \boldsymbol{A} \psi$ is compact;

ii) if the supports of φ and ψ overlap and both supports lie inside one and the same Liapunov sphere, then $\varphi \boldsymbol{A} \psi = \tilde{\boldsymbol{A}} + \boldsymbol{T}$, where $\boldsymbol{T}$ is a compact operator in B and $\tilde{\boldsymbol{A}}$ becomes a singular operator of the form (9.26) after transformation to the Euclidean space.

§5 Regularization and Fredholm theorems

In this section, Γ is either a Euclidean space E_m or a compact Liapunov m-dimensional manifold without boundaries. Let τ be an arbitrary linear element of the manifold Γ; if a regular coordinate system can be introduced on this manifold, then τ can be considered as a pair of independent points $\xi \in \Gamma$ and $\theta \in S$. Let $\boldsymbol{A}$ be a singular operator, the symbol $\Phi_A(\tau)$ of which satisfies the conditions of Theorem 3.8. Assume that $\Phi_A(\tau)$ does not vanish anywhere, then we way that the symbol is not *degenerate*, and it can be proved that $[\Phi_A(\tau)]^{-1}$ also satisfies the conditions of Theorem 3.7, when $l \geqslant m/2 + 1$. We construct a singular operator $\boldsymbol{B}$ with symbol $\Phi_B(\tau) = [\Phi_A(\tau)]^{-1}$: $\boldsymbol{A}$ and $\boldsymbol{B}$ are considered as operators in $L_p(\Gamma)$. According to Theorem 3.1, the operator $\boldsymbol{B}$ (as well as $\boldsymbol{A}$) is bounded in $L_p(\Gamma)$, and according to Theorem 2.5, $\boldsymbol{BA} = \boldsymbol{I} + \boldsymbol{T}'$, $\boldsymbol{AB} = \boldsymbol{I} + \boldsymbol{T}''$, where $\boldsymbol{T}'$ and $\boldsymbol{T}''$ are compact operators in $L_p(\Gamma)$. This implies that $\boldsymbol{B}$ is a double-side regularizator for $\boldsymbol{A}$, and the Nöther Theorem {Theorem 6.2, Chapter VI} is valid for the singular equation

$$\boldsymbol{A}u = g(\xi), \qquad g \in L_p(\Gamma). \tag{9.35}$$

It can be proved then that *the index of* $\boldsymbol{A}$ *is equal to zero* and, hence, the Fredholm theorems (see §1, Chapter II) are valid for (9.35). In the present context, they become: $\boldsymbol{A}u = 0$ and $\boldsymbol{A}^*u = 0$ have one and the same finite number of linearly independent solutions; the non-homogeneous equation (9.35) is solvable if and only if $g(\xi)$ is orthogonal to all the solutions of $\boldsymbol{A}^*u = 0$. From this fact it follows, in particular, that (9.35) is uniquely solvable for any $g \in L_p(\Gamma)$, if the corresponding homogeneous equation $\boldsymbol{A}u = 0$ has no non-trivial solutions.

For $p = 2$, Gohberg [7], [8] has proved that a non-degenerate symbol is not only sufficient but also necessary for the singular operator to admit regularization. For arbitrary p, this theorem has been proved in [14].

§6 Systems of singular equations

6.1 *Matrix singular operators*

We consider systems of singular equations of the form

$$\sum_{k=1}^{h} \boldsymbol{A}_{jk} u_k = g_j(x), \tag{9.36}$$

where $g_j \in L_p(\Gamma)$ and $\boldsymbol{A}_{jk}$ are singular operators.

We assume that either $\Gamma = E_m$ or Γ is an m-dimensional manifold, satisfying the conditions of §4. The system (9.36) can be rewritten in the form

$$\boldsymbol{A}u = g, \tag{9.37}$$

if we introduce *the matrix singular operator*

$$A = \begin{pmatrix} A_{11} & A_{12} & \dots & A_{1n} \\ A_{21} & A_{22} & \dots & A_{2n} \\ \dots & \dots & \dots & \dots \\ A_{n1} & A_{n2} & \dots & A_{nn} \end{pmatrix} \tag{9.38}$$

and vector-columns u and g with components $u_1, u_2, \dots, u_n$ and $g_1, g_2, \dots, g_n$, respectively.

6.2 *The symbol matrix*

We denote the symbol of the operator $\boldsymbol{A}_{jk}$ by $\Phi_{jk}(\tau)$, where τ is a linear element of the manifold Γ. We assume that the $\Phi_{jk}(\tau)$ satisfy the conditions of Theorem 3.8. The matrix

$$\Phi_A(\tau) = \begin{pmatrix} \Phi_{11}(\tau) & \Phi_{12}(\tau) & \dots & \Phi_{1n}(\tau) \\ \Phi_{21}(\tau) & \Phi_{22}(\tau) & \dots & \Phi_{2n}(\tau) \\ \dots & \dots & \dots & \dots \\ \Phi_{n1}(\tau) & \Phi_{n2}(\tau) & \dots & \Phi_{nn}(\tau) \end{pmatrix} \tag{9.39}$$

is called *the symbol matrix* or simply the symbol of the operator $\boldsymbol{A}$. The sum and product of symbols correspond to the sum and product (in the same order) of matrix singular operators of the form (9.38).

Consider now the case when the symbol matrix (9.39) is not degenerate, so that

$$\inf |\det \Phi_A(\tau)| > 0. \tag{9.40}$$

Then a singular operator $\boldsymbol{B}$ with symbol matrix $\Phi_B(\tau) = [\Phi_A(\tau)]^{-1}$ is a double-sided regularizator for $\boldsymbol{A}$; hence, the Nöther Theorem is valid for the equation (9.37) {or, equivalently, for the system (9.36)}.

6.3 *The index*

In contrast to the case of one equation, the index of a singular multi-dimensional system is not in general equal to zero. The index of multi-dimensional systems is studied in many papers: see [2], [4], [5], [19], [20], [27], [39].

Rather general and complete results concerning systems of a more general type than that of singular integral equations are given in the remarkable paper of Atiyah and Singer [26]. The proofs are not elementary: thus, two volumes of the Seminaire Henri Cartan (see [40]) are dedicated to the analysis of the basic Theorem of Atiyah and Singer. We cite the basic results of the mentioned works, and consider the cases when the index of a system is equal to zero.

We consider the system (9.36). If $n > m$, i.e. the order of the system is higher than the dimension of the manifold, then we can construct a new singular system of order m, the index of which is equal to the index of the given system. If $n < m$, then introducing additional equations

$$u_j = 0, \qquad j = n + 1, n + 2, \ldots, m,$$

we obtain a system of order m, the index of which coincides with the index of the given system. Thus, it is sufficient to study systems the order of which coincides with the dimension of the manifold of integration.

So, let $n = m$. The symbol matrix of the system has the form

$$\Phi_A(\tau) = \begin{pmatrix} \Phi_{11}(\tau) & \Phi_{12}(\tau) & \ldots & \Phi_{1m}(\tau) \\ \Phi_{21}(\tau) & \Phi_{22}(\tau) & \ldots & \Phi_{2m}(\tau) \\ \cdots & \cdots & \cdots & \cdots \\ \Phi_{m1}(\tau) & \Phi_{m2}(\tau) & \ldots & \Phi_{mm}(\tau) \end{pmatrix}, \tag{9.41}$$

and satisfies the inequality (9.40). Consider, for example, the first line of the matrix (9.41). Its elements are, generally speaking, complex-valued functions; we separate their real and imaginary parts and set

$$\Phi_{1k}(\tau) = \Phi'_{1k}(\tau) + i\Phi''_{1k}(\tau),$$

where the functions $\Phi'_{1k}(\tau)$ and $\Phi''_{1k}(\tau)$ are real. The sum

$$\Phi_1^2(\tau) = \sum_{k=1}^{m} \{\Phi'^2_{1k}(\tau) + \Phi''^2_{1k}(\tau)\} = \sum_{k=1}^{m} |\Phi_{1k}(\tau)|^2$$

is positive. In the opposite case, (9.40) does not hold. Let us set

$$\varphi_{2k-1}(\tau) = \frac{\Phi'_{1k}(\tau)}{\Phi_1(\tau)}, \qquad \varphi_{2k}(\tau) = \frac{\Phi''_{1k}(\tau)}{\Phi_1(\tau)} \tag{9.42}$$

and examine the vector-function

$$\varphi(\tau) = \{\varphi_1(\tau), \varphi_2(\tau), \ldots, \varphi_{2m}(\tau)\}. \tag{9.43}$$

It defines a continuous mapping of $(2m - 1)$-dimensional space of the linear elements of the manifold Γ onto a $(2m - 1)$-dimensional sphere. *If Γ is a surface, bounding a region of Euclidean E_{m+1} space, then the index of the system* (9.36) *is equal to the order of the mapping* (9.43). It follows from this statement that the index of a singular system is equal to zero, if its order is less than the dimension of the manifold of integration.

We now examine some cases when the index of a singular system (we shall also refer to it as the index of the corresponding symbol matrix) is equal to zero. Below we do not assume that $n = m$.

1) $\Gamma = E_m$ and the symbol matrix does not depend on x.
2) The symbol matrix is Hermitian or skew-symmetric.
3) The symbol matrix has the form

$$\Phi(\tau) = I - \psi(\tau),$$

where I is the unit matrix, and the eigenvalues of the matrix $\psi(\tau)$ are less than unity for all τ.

4) The lower bound for the moduli of the minors

$$\delta_1 = \Phi_{11}, \quad \delta_2 = \begin{vmatrix} \Phi_{11} & \Phi_{12} \\ \Phi_{21} & \Phi_{22} \end{vmatrix}, \quad \ldots, \quad \delta_n = \begin{vmatrix} \Phi_{11} & \Phi_{12} & \ldots & \Phi_{1n} \\ \Phi_{21} & \Phi_{22} & \ldots & \Phi_{2n} \\ \ldots & \ldots & \ldots & \ldots \\ \Phi_{n1} & \Phi_{n2} & \ldots & \Phi_{nn} \end{vmatrix} \tag{9.44}$$

is positive.

5) There exists a smooth curve, connecting points $\zeta = 0$ and $\zeta = \infty$ in the complex ζ-plane, which has no common points with the set of eigenvalues of the symbol matrix (9.39).

6) Let Γ be either a two-dimensional plane, or a two-dimensional torus such that we can introduce a regular coordinate system on Γ. Then the space $\tilde{\Gamma}$ of linear elements is the Cartesian product of the surface Γ and the unit circle; each linear element is defined by a point $\zeta = e^{i\theta}$ on the unit circle and a point $x \in \Gamma$; the symbol matrix of the system (9.36) can be written in the form $\Phi(x, \zeta)$. Once x has been fixed, we can apply *the left standard factorization* to the matrix $\Phi(x, \zeta)$, i.e. express it in the form

$$\Phi(x, \zeta) = \Phi_+(x, \zeta)\, \Phi_0(x, \zeta)\, \Phi_-(x, \zeta), \tag{9.45}$$

where $\Phi_+(x, \zeta)$ and $\Phi_-(x, \zeta)$ admit analytic continuations inside and outside of the unit circle of the ζ-plane, respectively, and have determinants which do not vanish in the mentioned regions; and $\Phi_0(x, \zeta)$ is the diagonal matrix

$$\Phi_0(x, \zeta) = (\zeta^{\kappa_j}\delta_{jk})_{j,k=1}^{j,k=n},$$

where the values κ_j are called *left partial indices* of the matrix $\Phi(x, \zeta)$. They are integers and do not depend on ζ. *The right standard factorization* and *right partial indices* can be introduced in an analogous manner. If the left (or right) partial indices of a matrix do not depend on x, then its index is equal to zero. In particular, the index of a symbol matrix, which does not depend on the pole, is equal to zero on a torus.

7) Sometimes, factorization can be applied to the symbol matrix of a multi-dimensional system. Let a regular coordinate system be introduced on Γ. The symbol matrix has the form $\Phi(x, \theta)$, where $x \in \Gamma$ and $\theta = (\vartheta_1, \vartheta_2, \ldots, \vartheta_{m-1})$ is a point of the unit $(m - 1)$ dimensional sphere and $\vartheta_{m-1} \in [0, 2\pi]$. Putting $\zeta = e^{i\vartheta_{m-1}}$ and fixing other arguments, we can apply the standard factorization to the matrix $\Phi(x, \theta)$. If all left (or right) partial indices are equal to zero, then $\operatorname{Ind} \Phi = 0$.

8) Let τ be an arbitrary linear element of the manifold Γ of arbitrary dimension and $\Phi(\tau)$ be the symbol matrix of the system (9.36). We set $\Phi(\tau) = \Phi_1(\tau) + i\Phi_2(\tau)$, where the matrices Φ_1 and Φ_2 are selfadjoint. If at least one of these matrices is (positive or negative) definite, then $\operatorname{Ind} \Phi = 0$.

§7 Singular equations in Lipschitz spaces

The results of §§5–6 enable us to assert, under certain conditions, that a singular equation (or a system) has a solution which is summable with degree $p > 1$. The cases when a solution has some smoothness properties {for example, satisfies a Lipschitz condition with a positive exponent} are also of interest. Results of this kind are presented in the book [18]; they are extended in [17], which we shall mainly refer to below.

We consider an m-dimensional closed Liapunov manifold Γ and the space $\mathrm{Lip}_\alpha(\Gamma)$ of functions, defined on Γ and satisfying a Lipschitz condition with exponent α, where $0 < \alpha < 1$. The norm in $\mathrm{Lip}_\alpha(\Gamma)$ is defined by

$$\|u\| = \max_{x \in \Gamma} |u(x)| + \sup_{x,y \in \Gamma} \frac{|u(x) - u(y)|}{|x - y|^\alpha}. \tag{9.46}$$

Let A be a singular operator with characteristic $f(\tau)$, coefficient $a(x)$ in the non-integral term, and symbol $\Phi_A(\tau)$ with τ a linear element of the manifold Γ. We assume that $a(x) \in \mathrm{Lip}_\alpha(\Gamma)$. Below, we always consider $f(\tau)$ and $\Phi_A(\tau)$ on a part of the manifold Γ, which can be mapped in a one-to-one and sufficiently smooth manner onto some region of Euclidean space. Thus, τ can be regarded as a pair of two independent points $x \in \Gamma$ and $\theta \in S$. Accordingly, we denote the characteristic and the symbol by $f(x, \theta)$ and $\Phi_A(x, \theta)$, respectively.

THEOREM 7.1. *If* $f(x, \theta) \hat{\in} W_1^{(l)}(S)$ *and*

$$\| f(y, \theta) - f(x, \theta)\|_{L_1(S)} \leqslant B\,|x - y|^\alpha, \qquad B = \text{const.},$$

then the operator $\boldsymbol{A}$ *is bounded in* $\mathrm{Lip}_\alpha(\Gamma)$.

Theorem 7.1 can be sharpened for $m = 2$.

THEOREM 7.2. *If* $\Phi_A(x, \theta) \hat{\in} W_2^{(l)}(S)$, $l \geqslant m/2 + 1$ *and*

$$\|\Phi_A(y, \theta) - \Phi_A(x, \theta)\|_{W_2^{l-1}(S)} \leqslant B\,|x - y|^\alpha, \qquad B = \text{const.},$$

then the operator $\boldsymbol{A}$ *is bounded in* $\mathrm{Lip}_\alpha(\Gamma)$.

THEOREM 7.3. *Let* $\boldsymbol{A}_1$ *and* $\boldsymbol{A}_2$ *be singular operators on* Γ. *If* $\Phi_{A_j}(x, \theta) \hat{\in} W_2^{(l)}(S)$, $l \geqslant (m + 3)/2$, $(j = 1, 2)$ *and*

$$\|\Phi_{A_j}(y, \theta) - \Phi_{A_j}(x, \theta)\|_{W_2^{l-1}(S)} \leqslant B\,|x - y|^\alpha, \qquad j = 1, 2,$$

then the symbol of the product $\boldsymbol{A}_1\boldsymbol{A}_2$ *is equal to the product* $\Phi_{A_1}(x, \theta) \times \Phi_{A_2}(x, \theta)$.

THEOREM 7.4. *Let* $g \in \mathrm{Lip}_\alpha(\Gamma)$, $\inf |\Phi_A(x, \theta)| > 0$ *and*

$$f(x, \theta) \hat{\in} W_2^{(1+\varepsilon)}(S) \cap C^{(1)}(S), \qquad \varepsilon > 0.$$

In addition, assume

$$|f(y, \theta) - f(x, \theta)| \leqslant B\,|x - y|^\beta, \qquad \beta > \alpha,$$
$$\|f(y, \theta) - f(x, \theta)\|_{W_2^{(1+\varepsilon)}(S)} \leqslant B_h\,|h|^\alpha, \qquad B_h \to 0, \qquad h \to 0.$$

Then any square summable solution of (9.35) *belongs to the space* $\mathrm{Lip}_\alpha(\Gamma)$.

All the results of this section can be extended naturally to systems of singular equations.

We also note that singular equations for classes of Lipschitz functions were studied by Giraud (see [31]) for rather special assumptions on the characteristic. The method of Giraud was extended to some classes of singular systems by Kupradze (see [15]).

§8 **Singular equations on a cylinder** (see [21])

We consider an m-dimensional cylinder, the evolute of which is the strip Ω in m-dimensional Euclidean space, defined by the inequalities

$$|x_j| \leqslant \tfrac{1}{2}, \qquad j = 1, 2, \dots, k, \qquad -\infty < x_j < \infty, \qquad j = k+1, k+2, \dots, m. \tag{9.47}$$

We introduce the notation

$$x^{(1)} = (x_1\ x_2, \dots, x_k, 0, 0, \dots, 0)$$

$$x^{(2)} = (0, 0, \dots, 0, x_{k+1}, \dots, x_m),$$

$$n = (n_1, n_2, \dots, n_k, 0, 0, \dots, 0),$$

where the n_j are integers. We denote by Ω_1 the k-dimensional cube $|x_j| \leqslant \frac{1}{2}$, $j = 1, 2, \dots, k$, and by Ω_2 the $(m-k)$-dimensional Euclidean space.

We introduce the space $L_2(\Omega_2)$ of vector-valued functions $v = \{v_n(x^{(2)})\}$ with norm

$$\|v\|^2 = \sum_a \int_{\Omega_2} |v_n(x^{(2)})|^2 \, dx^{(2)}.$$

The Fourier transform in Ω is defined by $\mathscr{F}v \equiv \{\hat{v}_n(x^{(2)})\}$, where

$$\hat{v}_n(x^{(2)}) = (2\pi)^{-[(m-k)/2]} \int_\Omega v(y^{(1)}, y^{(2)}) \exp\{-2\pi i(n, y^{(1)}) - i(x^{(2)}, y^{(2)})\} \, dy$$

The following properties of the Fourier transform can be proved: 1) It is isometric. 2) The inversion formula is

$$v(x^{(1)}, x^{(2)}) = (2\pi)^{-(m-k)/2} \sum_n \int_{\Omega_2} \hat{v}_n(y^{(2)}) \exp(2\pi i(n, x^{(1)}) + i(x^{(2)}, y^{(2)})) \, dy^{(2)}.$$

3) Let a star, viz. $*$, denote the convolution of two functions with respect to the point $x^{(2)}$, and let $u(x^{(1)}, x^{(2)})$ and $v(x^{(1)}, x^{(2)})$ be periodic with unit period with respect to any of the coordinates $x_1, x_2, \ldots, x_k$; then

$$\mathscr{F}(u * v) = \{(2\pi)^{(m-k)/2}\, \tilde{u}_n(x^{(2)})\, \tilde{v}_n(x^{(2)})\}.$$

Let $\Phi(x) = \Phi(x^{(1)}, x^{(2)})$ be a homogeneous function of order zero such that $\Phi(x) = \Phi(\theta)$, $\theta = x/|x|$. We introduce the following definition for a singular operator A with a symbol $\Phi(\theta)$ which does not contain the pole. We denote the Fourier transform in $\Omega_2 = E_{m-k}$ by $\mathscr{F}_2$. If $u \in L_2(\Omega)$, then we write

$$u_n(x^{(2)}) = \int_{\Omega_1} u(x^{(1)}, x^{(2)}) e^{-2\pi i (n, x^{(1)})}\, dx^{(1)}.$$

Finally, we set $\Phi_n(x_2) = \Phi(2\pi n, x_2)$. Now, we define the singular operator A by the formula

$$Au = \sum_n e^{2\pi i (n, x^{(1)})} \mathscr{F}_2^{-1} \Phi_n \mathscr{F}_2 u_n. \tag{9.48}$$

It can be proved that the operator (9.48) is bounded in $L_p(\Omega)$ for any p, $1 < p < \infty$, if the symbol $\Phi(\theta)$ has continuous derivatives up to order m inclusive with respect to the cartesian coordinates of the point θ. For $p = 2$, the operator A is bounded, only if the symbol is bounded, and $\|A\| =$ supess $|\Phi(\theta)|$.

Next, we give a definition for a singular operator with a symbol which contains the pole. Let $\Phi(x, \theta)$, $x \in \Omega$, $\theta \in S$, be periodic with unit period with respect to all the variables $x_1, x_2, \ldots, x_k$, and let $\Phi(x, \theta) \hat{\in} w_2^{(l)}(S)$, where $l \geqslant (5m/2) - 1$ with l an integer. By $A_{\nu,m}^{(k)}$ we denote the singular operator with symbol $Y_{\nu,m}^{(k)}$. We decompose Φ into the series

$$\Phi(x, \theta) = \sum_{\nu=1}^{\infty} \sum_{k=1}^{k_{\nu,m}} a_\nu^{(k)}(x)\, Y_{\nu,m}^{(k)}(\theta). \tag{9.49}$$

The operator A, defined by

$$A = \sum_{\nu=0}^{\infty} \sum_{k=1}^{k_{\nu,m}} a_\nu^{(k)}(x)\, A_\nu^{(k)} + T, \tag{9.50}$$

where T is any compact operator in $L_p(\Omega)$, is called the singular operator with symbol $\Phi(x, \theta)$. It can be proved, under the above assumptions about the symbol, that the series (9.50) converges in norm.

The multiplication rule for symbols remains valid for this definition of a singular operator. Since this rule holds, the whole theory remains valid (see §5).

§9 Singular equations in spaces of generalized functions

Multidimensional singular integral equations have been studied by Horvath (see [32]–[35]), Malgrange (see [37]) and Krupnik (see [11]–[14]). Below we present some results of these authors.

Horvath [35] considers integrals of the form

$$\boldsymbol{K}u = \int_{E_m} \frac{f(\theta)}{r^m} u(y)\, dy, \tag{9.51}$$

where $f \in L^{(p)}(S)$, $p > 1$.

He has proved that this operator $\boldsymbol{K}$ maps any infinitely differentiable function $u(y)$ with a compact support into some generalized function. For arbitrary q, $1 < q < \infty$, its convolution with any generalized function of D'_{L^q} is again a generalized function from D'_{L^q}. We recall that

$$D'_{L^q} = \bigcup_{k=1}^{\infty} W_q^{(-k)}(E_m),$$

where $W_q^{(-k)}$ is the dual space of the Sobolev space $W_{q'}^{(,k)}$ with $1/q + 1/q' = 1$. For details see [37].

The results of [37] show that the operator (9.51) maps the space $W_p^{(k)}(E_m)$ into the same space for any k, $0 \leqslant k \leqslant \infty$, if $f \in L^{(p)}(S)$. Here $W_p^{(0)} \equiv L_p$ and $W_p^{(\infty)}$ is the intersection

$$W_p^{(\infty)} = \bigcap_{k=1}^{\infty} W_p^{(k)}.$$

Krupnik [11]–[13] introduces a new space M of basis functions. It is the normalized space of complex valued infinitely differentiable functions in E_m, for which each of the products

$$(1 + |x|)^{m - 1/(1+k)} D^q u(x), \qquad |q| \leqslant k \quad (k = 0, 1, 2, \ldots), \tag{9.52}$$

is bounded. Here, we use the following notation: q is a multi-index, {i.e., an m-tuple of non-negative integers: $q = (q_1, q_2, \ldots, q_m)$}; $|q| = q_1 + q_2 + \ldots + q_m$ and

$$D^q u = \frac{\partial^{|q|} u}{\partial x_1^{q_1}\, \partial x_2^{q_2} \ldots \partial x_m^{q_m}}.$$

The system of norms in M is defined by

$$\|u\|_k = \sup_{x \in E_m,\, |q| \leqslant k} (1 + |x|)^{m - 1/(1+k)}\, |D^{|q|}\, u(x)| \quad (k = 0, 1, 2, \ldots). \tag{9.53}$$

We shall denote the dual of M by M'.

The following singular operators are considered

$$\boldsymbol{A} u = a(x)\, u(x) + \int_{E_m} \frac{f(x, \theta)}{r^m}\, u(y)\, dy \tag{9.54}$$

and

$$\boldsymbol{B} u = a(x)\, u(x) + \int_{E_m} \frac{f(y, \theta)}{r^m}\, u(y)\, dy. \tag{9.55}$$

It is necessary to note that the difference $\boldsymbol{A} - \boldsymbol{B}$ is not compact in the spaces M and M' even if the characteristic f is infinitely differentiable.

It can be proved that each of operators $\boldsymbol{A}$ and $\boldsymbol{B}$ is continuous in M, if $a(x)$ and $f(x, \theta)$ are infinitely differentiable with respect to x and each of the derivatives is bounded independently of θ. Then, the function $\Phi(x, \theta)$, defined by (9.16), is called the symbol of the operators $\boldsymbol{A}$ and $\boldsymbol{B}$. The following theorem is valid.

THEOREM 9.1. *Let, for any $\theta \in S$, the symbol $\Phi(x, \theta)$ be infinitely differentiable with respect to x and $D_x^q \Phi \,\hat{\in}\, W_2^{(l)}(S)$, $1 \geqslant m + [m/2] + 2$, $|q| > 2$, and the functions $D_x^q \Phi$ have continuous derivatives of order $m + [m/2] + 1$ with respect to the Cartesian coordinates of the point θ for $|q| \leqslant 2$. Further, let every derivative of $D_x^q \Phi$ be continuous with respect to the totality of variables θ and ξ* (ξ is the image of the point x under the stereographic mapping of E_m onto the Riemannian sphere). *If the symbol does not vanish everywhere, then the operators $\boldsymbol{A}$ and $\boldsymbol{B}$ are normally solvable and have zero index in M and M'.*

An analogous theorem is valid for matrix singular operators with the difference that the index is not necessarily equal to zero. In this case, it can be proved that the index of the operators $\boldsymbol{A}$ and $\boldsymbol{B}$ in the spaces M and M' coincides with the index of the operator $\boldsymbol{A}$ in L_2.

§10 Equations with degenerate symbol

Multidimensional singular integral equations with degenerate symbol have not been studied extensively. However, it is clear that the general solution must be sought in a class of generalized functions. In this section, we present some results from [16a], which is one of the first papers to deal with problems of this kind.

We consider a singular equation, defined on the Euclidean plane E_2, and introduce the polar coordinates ρ, θ: $0 \leqslant \rho < \infty$, $0 < \theta < 2\pi$. Let l be a non-negative integer. By $L_2^{(l)}(S)$, we denote the Hilbert space of functions which are defined on the unit circle S, are periodic with period 2π with respect to θ, are absolutely continuous on the unit circle with all their derivatives or order $\leqslant l - 1$, and have square summable l-th derivatives. We define the norm in $L_2^{(l)}(S)$ by

$$|\varphi|_l^2 = \int_0^{2\pi} \sum_{k=0}^{l} \left| \frac{\partial^k \varphi}{\partial \theta^k} \right|^2 d\theta, \tag{9.56}$$

with the scalar product denoted by square brackets:

$$[\varphi, \psi]_l = \int_0^{2\pi} \sum_{k=0}^{l} \frac{\partial^k \varphi}{\partial \theta^k} \frac{\partial^k \bar{\psi}}{\partial \theta^k} d\theta. \tag{9.57}$$

As usual, the dual of $L_2^{(l)}(S)$ with respect to the scalar product $[\ ,\ ]_0$ will be denoted by $L_2^{(-l)}(S)$.

Functions, defined on the plane E_2, will be considered as abstract functions of the variable ρ with their range of values in this or that functional space. In particular, consider the Hilbert space $H_2^{(l)}$ of functions of the variable ρ with their range of values in $L_2^{(l)}(S)$, the scalar product and the norm of which are

$$(u, v)_l = \int_0^\infty [u, v]_l \rho \, d\sigma, \qquad \|u\|_l^2 = \int_0^\infty |u|_l^2 \, \rho \, d\rho, \tag{9.58}$$

where l is an arbitrary integer.

It is easy to see that the spaces $H_2^{(l)}$ and $H_2^{(-l)}$ are dual with respect to the scalar product $(u, v)_0$. Let $\Phi(\theta) \in L_2^{(l)}(S)$, $l \geqslant 0$. Then the singular operator $\boldsymbol{A} = \mathscr{F}^{-1}\Phi\mathscr{F}$ is defined and bounded in any of the spaces $H_2^{(l)}$ and $H_2^{(-l)}$, where $\mathscr{F}$ is the Fourier transform.

Let the points θ_j $(j = 1, 2, \ldots, N)$ be zeros of the symbol $\Phi(\theta)$ with multiplicities l_j, and $\Phi(\theta) \neq 0$ for $\theta \neq \theta_j$. We set $l = \max l_j$.

Consider the equation

$$\boldsymbol{A}u = f. \tag{9.59}$$

If $f \in L_2(E_2)$, then it can be proved that this equation has a solution $u \in H_2^{(-l)}$. This solution is not unique: the general solution of the homogeneous equation

$$\boldsymbol{A}u = 0 \tag{9.60}$$

has the form

$$u = \sum_{j=0}^{N} \sum_{k=0}^{l_j - 1} \mathscr{F}^{-1}[c_{k_j}(\rho)\, \delta^{(k)}\, (\theta - \theta_j)]. \tag{9.61}$$

Here δ is the Dirac function, and the $c_{k_j}(\rho)$ are arbitrary square summable functions on $(0, \infty)$.

Since the solution of the equation (9.59) is not unique, it can be subjected to some additional conditions. We introduce arbitrary symbols $\Phi_k(\theta) \in L_2^{(l)}(S)$ and arbitrary functions $\psi_k(\rho) \in L_2(0, \infty)$ $(k = 1, 2, \ldots, s)$, where $s = \sum_{j=1}^{N} l_j$. We denote by $\boldsymbol{A}_k$ the singular operators with symbols $\Phi_k(\theta)$, so that $\boldsymbol{A}_k = \mathscr{F}^{-1}\Phi_k\mathscr{F}$, and seek the solution of (9.59), satisfying the conditions

$$[\boldsymbol{A}_k u, 1]_0 = \psi_k(\rho) \qquad (k = 1, 2, \ldots, s). \tag{9.62}$$

We denote by B the square matrix

$$B = (B_1, B_2, \ldots, B_N), \tag{9.63}$$

where B_j is the rectangular matrix of order $(l_j \times s)$:

$$B_j = \begin{bmatrix} \Phi_1(\theta_j) & \ldots & \Phi_1^{(l_j-1)}(\theta_j) \\ \Phi_2(\theta_j) & \ldots & \Phi_2^{(l_j-1)}(\theta_j) \\ \cdots & \cdots & \cdots \\ \Phi_s(\theta_j) & \ldots & \Phi_s^{(l_j-1)}(\theta_j) \end{bmatrix}. \tag{9.64}$$

The equation (9.59) has a unique solution under the additional conditions (9.62), if and only if $\det B \neq 0$. If this condition is satisfied, then the solution depends continuously on f and ψ_k.

The paper [16] also contains some results concerning properly posed problems for the equation (9.59) with degenerate symbol, and the case when the symbol depends 'weakly' on x.

§11 Singular integro-differential equations

The theory of multi-dimensional singular integrodifferential (SID) equations was developed intensively during recent years. Initially, the notion of a SID-operator and its symbol appeared in the paper of Calderon and Zygmund [30]. The first paper to contain more or less general results was Dinin [9]. Contemporary results and problems are considered in the review paper of Agranovich [1] and in the paper of Seeley [39]. Below we refer to some of the results of [1].

Let, as in §9, q be a multi-index, $q = (q_1, q_2, \ldots, q_m)$, and $\boldsymbol{D}^q$ the differential operator

$$\boldsymbol{D}^q = \frac{\partial^{|q|}}{\partial x_1^{q_1}\, \partial x_2^{q_2} \ldots \partial x_m^{q_m}}.$$

Further, let $\boldsymbol{A}_q$ be a matrix singular operator. A SID-operator of order s is the operator (a more general definition is given in [1]):

$$\boldsymbol{A} = \sum_{|q|=s} \boldsymbol{A}_q \boldsymbol{D} + \boldsymbol{T}, \tag{9.65}$$

where $\boldsymbol{T}$ is an operator 'almost of the order $s - 1$' in $M_2^{(s)}(E_m)$. The last has the following sense: given an arbitrary $\varepsilon > 0$, there exists a decomposition $\boldsymbol{T} = \boldsymbol{T}'_\varepsilon + \boldsymbol{T}''_\varepsilon$, where $\boldsymbol{T}'_\varepsilon$ is a bounded operator acting from $W_2^{(s)}(E_m)$ into $L_2(E_m)$ with norm less than ε, and T''_ε is a bounded operator acting from $W_2^{(s)}(E_m)$ into $W_2^{(s-1)}(E_m)$.

Let $\theta_1, \theta_2, \ldots, \theta_m$ be the Cartesian coordinates of the point θ and $\theta^q = \theta_1^{q_1}\theta_2^{q_2}\ldots\theta_m^{q_m}$. If $\Phi_q(x, \theta)$ is the symbol of the singular integral operator $\boldsymbol{A}_q$, then the symbol of the SID-operator (9.65) is defined by

$$\Phi(x, \theta) = \sum_{|q|=s} \Phi_q(x, \theta)\theta^q. \tag{9.66}$$

It is worth noting that the symbol (9.66) is a homogeneous function of degree s with respect to $\theta_1, \theta_2, \ldots, \theta_m$. The symbol of the singular integral operator is, evidently, a homogeneous function of order zero with respect to $\theta_1, \theta_2, \ldots, \theta_m$. It is easy to prove that the operator (9.65) admits the representation

$$\boldsymbol{A} = \boldsymbol{A}_0(-\Delta)^{s/2} + \boldsymbol{T}, \tag{9.67}$$

where Δ is the Laplace operator in $\boldsymbol{E}_m$ and $\boldsymbol{A}_0$ is a singular integral operator with symbol $\Phi(x, \theta)\,|\theta|^{-s}$; conversely, if $\boldsymbol{A}_0$ is a singular operator with symbol $\Phi_0(x, \theta)$, then (9.67) defines the SID-operator with symbol $\Phi_0(x, \theta)\,|\theta|^s$. It is also easy to prove, under suitable smoothness assumptions on $\Phi_q(x, \theta)$, that the operator (9.65) is bounded as an operator acting from $W_2^{(l)}(E_m)$ into $W_2^{(l-s)}(E_m)$, where l is an integer which depends in a definite way on the smoothness of the symbol $\Phi_q(x, \theta)$. A SID-operator on a compact sufficiently smooth manifold without boundary can be defined in a manner which is analogous to that of §4 for a singular integral operator (for details see [1], page 51). We do not consider these details.

A SID-operator is *elliptic*, if its symbol is not degenerate.

Let $\boldsymbol{A}$ be a SID-operator of order s on a compact sufficiently smooth manifold Γ without boundary and let the symbol of $\boldsymbol{A}$ satisfy certain smoothness conditions. We choose the integer l which is connected in a definite way with the smoothness of the symbol. The basic result of the theory of SID-operators is the following: *in order that a SID-operator be a Nöther operator from $W_2^{(l)}(\Gamma)$ into $W_2^{(l-s)}(\Gamma)$, it is necessary and sufficient that it be elliptic.*

§12 Singular equations on a manifold with boundary

This problem has been studied in [1] and [3]. Below, we state the simplest results of [3].

Let Γ be a finite region of the space E_m, bounded by a sufficiently smooth surface $\partial\Gamma$. Obviously, Γ can be considered as an m-dimensional manifold with boundary $\partial\Gamma$. Consider an operator of the form

$$\boldsymbol{A}u = \int_{\Gamma} A(x, x - y)\, u(y)\, dy, \tag{9.68}$$

where the kernel $A(x, \theta)$ (it can be a generalized function) is defined for $x \in \Gamma$ and $\theta \in E_m$. We complete the definition of the function u by setting it equal to zero outside Γ. Then we can consider the integral (9.68) over the whole space E_m. We assume that the Fourier transform $\Phi_A(x, \theta)$ of the kernel $A(x, \theta)$ with respect to the variable θ is a positive homogeneous function of θ of some real order α:

$$\Phi_A(x, \lambda\theta) = \lambda^\alpha\, \Phi_A(x, \theta), \qquad \lambda > 0. \tag{9.69}$$

The function $\Phi_A(x, \theta)$ is called the symbol of the operator $\boldsymbol{A}$ and α is the order of $\boldsymbol{A}$.

We also consider sums of the form $\boldsymbol{A} + \boldsymbol{T}$, where $\boldsymbol{A}$ is the operator just described, and $\boldsymbol{T}$ is an operator which is in some sense 'subordinate' to $\boldsymbol{A}$; for example, we can consider $\boldsymbol{T}$ as an operator of the same type as $\boldsymbol{A}$ but of order $\alpha' < \alpha$.

We adopt the following 'rule of summation for symbols'. If we sum two operators of the type (9.68) of the same order, then we sum their symbols; and if the item $\boldsymbol{T}$ is subordinate to $\boldsymbol{A}$, then the symbol of the sum $\boldsymbol{A} + \boldsymbol{T}$ is equal to the symbol of the 'leading' item $\boldsymbol{A}$.

Consider now the equation

$$(\boldsymbol{A} + \boldsymbol{T})u = f(x) \tag{9.70}$$

and assume that $u \in W_2^{(s)}(\Gamma)$, $f \in W_2^{(s-\alpha)}(\Gamma)$, where s is some fixed number.

The factorization of the symbol in (9.70), i.e. its decomposition into multipliers with the properties described below, plays a crucial role in the study of (9.70).

Let the Cartesian coordinates of the point θ be $\theta_1, \theta_2, \ldots, \theta_m$. We denote by θ' a point of $(m - 1)$-dimensional Euclidean space. We write $\theta = (\theta', \theta_m)$ and $f(x, \theta) = f(x, \theta', \theta_m)$, where f is an arbitrary function. We assume that the symbol Φ does not vanish for $\theta \neq 0$ and for any $x \in \bar{\Gamma} = \Gamma \cup \partial\Gamma$. We fix a point $x \in \partial\Gamma$ and introduce a local coordinate system in which the tangent plane to $\partial\Gamma$ at the point x has the equation $x_m = 0$. It can be proved that the following decomposition of the symbol holds

$$\Phi(x, \theta) = \Phi_+(x, \theta', \theta_m)\, \Phi_-(x, \theta', \theta_m), \tag{9.71}$$

where $\Phi_+(x, \theta, \theta_m)$ and $\Phi_-(x, \theta, \theta_m)$ are homogeneous functions with respect to θ, and $\Phi_+(x, \theta, \theta_m)$ admits an analytic continuation in the upper half-plane $Im\theta_m > 0$, and $\Phi_-(x, \theta, \theta_m)$ admits an analytic continuation in the lower half-plane $Im\theta_m < 0$. Here, the mentioned functions do not vanish in the corresponding half-planes.

In the general case, the order of homogeneity of any of the functions Φ_+ and Φ_- depends on x. The case when the order κ of Φ_- is constant and an integer is considered in [3]. Then the order of homogeneity of the function Φ_-, which is equal to $\alpha - \kappa$, is also constant (but not necessarily an integer).

The rest depends on the sign of κ. It is assumed that the symbol satisfies some smoothness conditions and the multipliers of the decomposition (9.71) have some analytic properties, which we shall not formulate because of their awkwardness. If $\kappa > 0$, then we can add κ boundary conditions on the boundary $\partial\Gamma$ to (9.70). The operators introduced by these conditions must satisfy an algebraic condition, which is analogous to the well-known Lopatinsky condition. Then the problem is Nöther in a properly chosen pair of spaces.

If $\kappa = 0$, then no boundary conditions are required: the operator (9.68) is Nöther.

If $\kappa < 0$, then Nöther problems are formed by adding to the left hand side of (9.70) a sum of the form

$$\sum_{k=1}^{|\kappa|} G_k \rho_k,$$

where the ρ_k are new unknown functions and the G_k are convolution operators satisfying a Lopatinsky type condition.

Chapter X

NON-LINEAR INTEGRAL EQUATIONS

At the present time, the theory of non-linear integral equations is treated as a part of the general theory of non-linear operator equations. All the statements relating to non-linear integral equations are simple consequences of general theorems. The specific character of a given equation is expressed only in terms of the general properties of the integral operators, which are contained in this equation. Such properties are continuity, compactness, potentiality and others. For this reason, §1 deals with the general properties of integral operators. The same comments are valid for systems of integral equations.

The theory of non-linear operator equations in functional spaces makes up a large part of non-linear functional analysis. It is impossible to state even the basic facts of this theory in a short chapter. For this reason, the choice of material was limited to that which is not bulky but which gives a clear notion of basic problems and methods of solution in the theory of integral equations. In particular, it was necessary to do without results which involve cumbersome calculations.

In §2, we present basic theorems concerning the existence of solutions; in §3, we study the dependence of solutions on a parameter.

§1 Non-linear integral operators

1.1 *The basic concepts of the theory of non-linear operators*

In this section, we study different classes of non-linear integral operators which act in spaces of functions defined on some bounded closed set of a finite-dimensional space.

The simplest and the most important non-linear integral operators are *the Urison operators*:

$$\boldsymbol{A}u(x) = \int_{\Omega} K[x, t, u(t)]\, dt. \tag{10.1}$$

Here, $K(x, t, u)$, the *kernel* of the operator (10.1), is a function, which is defined for $x, t \in \Omega$, $-\infty < u < \infty$, is continuous with respect to u for almost all $x, t \in \Omega$, and is measurable with respect to the totality of variables x, t on $\Omega \times \Omega$ for almost all $u \in (-\infty, \infty)$. The domain of the Urison operator consists of measurable functions $u(x)$, for which the function $K[x, t, u(t)]$ is summable with respect to t on Ω for almost all $x \in \Omega$. For every such $u(x)$, the value $\boldsymbol{A}u(x)$ is a measurable function.

A particular case of a Urison operator is the *Hammerstein operator*:

$$\boldsymbol{A}u(x) = \int_\Omega K(x, t) f[t, u(t)]\, dt, \tag{10.2}$$

where $f(t, u)$ is a function, which is defined for $t \in \Omega$, $-\infty < u < \infty$, is continuous with respect to u for almost all $t \in \Omega$, and is measurable with respect to t on Ω for almost all $u \in (-\infty, \infty)$; and $K(x, t)$ is a function, which is measurable with respect to the totality of variables on $\Omega \times \Omega$. The Hammerstein operator can be expressed in the form of a product

$$\boldsymbol{A} = \boldsymbol{K}\boldsymbol{f} \tag{10.3}$$

of the non-linear *superposition operator*

$$\boldsymbol{f}u(t) = f[t, u(t)] \tag{10.4}$$

and the linear integral operator

$$\boldsymbol{K}u(x) = \int_\Omega K(x, t)\, u(t)\, dt \tag{10.5}$$

with kernel $K(x, t)$.

A more complex class of integral operators is defined by:

$$\boldsymbol{A}u(t) = \boldsymbol{F}\left\{x, u(x), \int_\Omega K_1[x, t, u(t)]\, dt, \ldots, \int_\Omega K_n[x, t, u(t)]\, dt\right\}. \tag{10.6}$$

This operator is the superposition of the Urison operators $\boldsymbol{A}_1, \ldots, \boldsymbol{A}_n$ (with kernels $K_1(x, t, u), \ldots, K_n(x, t, u)$) along with the non-linear operator

$$\boldsymbol{F}(u, u_1, \ldots, u_n)(x) = F(x, u(x), u_1(x)), \ldots, u_n(x)]. \tag{10.7}$$

Among the other types of non-linear integral operators, we mention *the Liapunov-Lichtenstein operators:*

$$Au(x) = \sum_{k_1,\dots,k_l} \int_\Omega \dots \int_\Omega K_{[k_1,\dots,k_l]}(x, t_1, \dots, t_l)\, u(t_1)^{k_1} \dots u(t_l)^{k_l}\, dt_1 \dots dt_l; \tag{10.8}$$

Here $[k_1, \dots, k_l]$ denotes a collection of non-negative integers, and $K_{[k_1,\dots,k_l]}(x, t_1, \dots, t_l)$ functions, which are measurable with respect to the totality of variables.

The study of non-linear integral operator equations is simplified, if we know pairs $\{E_1, E_2\}$ of Banach spaces such that the non-linear operator maps from E_1 into E_2 and has 'good' properties (continuity, compactness, etc.). Just as for linear operators, the determination of such pairs is extremely difficult. In the present section, we give conditions under which a given integral operator defines a mapping from an L_p into an L_q space, and is continuous, compact, etc.

We recall some definitions concerning non-linear operators. For simplicity, we restrict attention to operators defined on the whole space.

The operator A, mapping from E_1 into E_2, is *continuous at the point* $u_0 \in E_1$, if

$$\lim_{\|u-u_0\|_{E_1}\to 0} \|Au - Au_0\|_{E_2} = 0.$$

The operator A is *continuous*, if it is continuous at every point of the space E_1. It is natural that the continuity of a non-linear operator on the whole space does not follow from its continuity at a single point.

The operator A, mapping from E_1 into E_2, is called *bounded*, if it maps bounded sets into bounded sets; *compact*, if it maps bounded sets of E_1 into compact sets of E_2. Consequently, compact operators are bounded.

The properties of continuity and compactness are not related for non-linear operators: an operator can be compact, but not continuous; and continuous, but not compact. The operator A, mapping from E_1 into E_2, is called *completely continuous*, if it is simultaneously continuous and compact.

We introduce an additional definition for operators with range in L_q, $q < \infty$. The operator A, mapping from L_p into L_q ($q < \infty$) is *absolutely bounded*, if it maps bounded sets of L_p into sets of L_q with equi-continuous norms. We recall that the norms of functions of a set $\mathfrak{M}$ from L_q ($q < \infty$) are *absolutely equicontinuous*, if

$$\lim_{\operatorname{mes} D\to 0} \sup_{u\in\mathfrak{M}} \|P_D u\|_{L_q} = 0,$$

where P_D is the operator of multiplication by the characteristic function of the set $D \subset \Omega$. Compact operators, mapping from L_p into L_q, $q < \infty$, are absolutely bounded. The contrary is, generally speaking, not valid.

1.2 *Urison operators with their range in the space C*

We begin the study of non-linear integral operators with a study of Urison operators with range in the space of continuous functions C.

THEOREM 1.1. *Let $K(x, t, u)(x, t \in \Omega, -\infty < u < \infty)$ be continuous with respect to u for almost all $x, t \in \Omega$ and be measurable with respect to the totality of variables x, t on $Q \times Q$ for all u. Let, for any $h > 0$, the following inequality hold*

$$|K(x, t, u)| \leqslant R_h(x, t) \qquad (|u| \leqslant h), \tag{10.9}$$

where $R_h(x, t)$ is a measurable function with respect to the totality of variables which satisfies the condition

$$\int_\Omega R_h(x, t)\, dt \leqslant k(h) < \infty \qquad (0 < h < \infty). \tag{10.10}$$

Further, let

$$\lim_{\delta \to 0} \sup_{x \in \Omega} \int_\Omega \max_{|u_i| \leqslant h,\, |u_1 - u_2| \leqslant \delta} |K(x, t, u_1) - K(x, t, u_2)|\, dt = 0. \tag{10.11}$$

Finally, we assume that the following equality holds for any measurable set D and any $x_0 \in \Omega$

$$\lim_{x \to x_0} \int_D K(x, t, u)\, dt = \int_D K(x_0, t, u)\, dt. \tag{10.12}$$

*Then the Urison operator **A** with kernel $K(x, t, u)$ maps from L_∞ into C and is continuous.*

THEOREM 1.2. *Let $K(x, t, u)(x, t \in \Omega, -\infty < u < \infty)$ be continuous with respect to u for almost all $x, t \in \Omega$ and be measurable with respect to the totality of variables x, t in $\Omega \times \Omega$ for all u. Let, for any h,*

$$|K(x, t, u)| \leqslant R_h(x, t) \qquad (|u| \leqslant h), \tag{10.13}$$

where $R_h(x, t)$ is a measurable function with respect to the totality of variables which satisfies the condition

$$\int_\Omega R_h(x, t)\, dt \leqslant k(h) < \infty \qquad (0 < h < \infty). \tag{10.14}$$

Finally, we assume that for any $h > 0$ and any $x_0 \in \Omega$

$$\lim_{x \to x_0} \int_\Omega \max_{|u| \leqslant h} |K(x, t, u) - K(x_0, t, u)|\, dt = 0. \tag{10.15}$$

Then the Urison operator A with kernel $K(x, t, u)$ maps from L_∞ into C and is completely continuous.

The conditions of Theorems 1.1–1.2 are satisfied, if $K(x, t, u)$ is continuous with respect to the totality of variables. Thus, a Urison operator with a continuous kernel maps from L_∞ to C and is completely continuous.

THEOREM 1.3. *Let $K(x, t, u)(x, t \in \Omega, -\infty < u < \infty)$ be continuous with respect to u for almost all $x, t \in \Omega \times \Omega$ and be measurable with respect to the totality of variables x, t on $\Omega \times \Omega$ for all u. Let $K(x, t, u)$ satisfy the inequality*

$$|K(x, t, u)| \leqslant R(x, t) + k\, |u|^p, \tag{10.16}$$

where $R(x, t)$ is a measurable function with respect to the totality of variables which satisfies the condition

$$\int_\Omega R(x, t)\, dt \leqslant k_1 < \infty, \tag{10.17}$$

and for any $h > 0$

$$\lim_{\delta \to 0} \sup_{x \in \Omega} \int_\Omega \max_{|u_1| \leqslant h,\, |u_1 - u_2| \leqslant \delta} |K(x, t, u_1) - K(x, t, u_2)|\, dt = 0. \tag{10.18}$$

Further, let, for any measurable subset $D \subset \Omega$, any $u \in (-\infty, \infty)$ and each $x_0 \in \Omega$

$$\lim_{x \to x_0} \int_D K(x, t, u)\, dt = \int_D K(x_0, t, u)\, dt.$$

Finally, we assume that for any non-negative function z from L_p

$$\lim_{\text{mes } D \to 0} \sup_{|u| \leqslant z} \sup_{x \in \Omega} \left| \int_D K[x, t, u(t)]\, dt \right| = 0. \tag{10.19}$$

*Then the Urison operator **A** with kernel K(x, t, u) maps from L_p into C and is continuous.*

THEOREM 1.4. *Let $K(x, t, u)(x, t \in \Omega, -\infty < u < \infty)$ be continuous with respect to u for almost all $x, t \in \Omega$, and be measurable with respect to the totality of variables x, t on $\Omega \times \Omega$ for all u. Let K(x, t, u) satisfy the inequality*

$$|K(x, t, u)| \leqslant R(x, t) + k\, |u|^p, \tag{10.20}$$

where R(x, t) is a measurable function with respect to the totality of variables which satisfies the condition

$$\int_\Omega R(x, t)\, dt \leqslant k_1 < \infty, \tag{10.21}$$

and let for any $h > 0$

$$\lim_{x \to x_0} \int_\Omega \max_{|u| \leqslant h} |K(x, t, u) - K(x_0, t, u)|\, dt = 0. \tag{10.22}$$

Finally, we assume that

$$\lim_{\text{mes } D \to 0} \sup_{\|u\|_{L_p} \leqslant 1} \sup_{x \in \Omega} \left| \int_D K[x, t, u(t)]\, dt \right| = 0. \tag{10.23}$$

*Then the Urison operator **A** with kernel K(x, t, u) maps from L_p into C and is completely continuous.*

Now, we assume that

$$|K(x, t, u)| \leqslant \sum_{k=0}^{\infty} R_k(x, t)\, |u|^{\delta_k}, \tag{10.24}$$

where $0 = \delta_0 < \delta_1 < \ldots \delta_n = p$ and the $R_k(x, t)(x, t \in \Omega)$ are non-negative measurable functions with respect to the totality of variables. Then $K(x, t, u)$ obviously satisfies the relationships (10.16)–(10.18), if

$$\int_\Omega R_k(x, t)^{p/(p-\delta_k)}\, dt \leqslant c_k < \infty \qquad (k = 0, \ldots, n-1),$$

$$\operatorname*{vrai\,max}_{t \in \Omega} |R_n(x, t)| \leqslant c_n < \infty. \tag{10.25}$$

Consequently, the condition (10.19) is satisfied, if

$$\lim_{\text{mes}\, D \to 0} \sup_{x \in \Omega} \int_D R_0(x, t)\, dt = 0, \tag{10.26}$$

and the condition (10.23) is satisfied, if

$$\lim_{\text{mes}\, D \to 0} \sup_{x \in \Omega} \int_D R_k(x, t)^{p/(p-\delta_k)}\, dt = 0 \qquad (k = 0, \ldots, n-1) \tag{10.27}$$

and if $R_n(x, t) \equiv 0$.

1.3 *Hammerstein operators with range in L_q*

It is convenient to study the Hammerstein operator

$$\boldsymbol{A}u(x) = \int_\Omega K(x, t)\, f[t, u(t)]\, dt \tag{10.28}$$

using the splitting (10.3) for $\boldsymbol{A}$. Different conditions for the continuity and complete continuity of $\boldsymbol{A}$ can be obtained as the union of conditions for the continuity of the non-linear operator $\boldsymbol{f}$ along with conditions for the continuity and complete continuity of a linear operator $\boldsymbol{K}$. More delicate statements can be obtained by using this method.

We present basic results concerning superposition operators.

The function $f(t, u)(t \in \Omega, -\infty < u < \infty)$ is said to satisfy the *Caratheodory conditions*, if it is continuous with respect to u for almost all $t \in \Omega$ and is measurable with respect to $t \in \Omega$ for all $u \in (-\infty, \infty)$.

THEOREM 1.5. *Let $f(t, u)$ satisfy the Caratheodory conditions. Then the superposition operator*

$$\boldsymbol{f}u(t) = f[t, u(t)] \tag{10.29}$$

maps from L_p into L_q if and only if $f(t, u)$ satisfies the inequality (for $p < \infty$)

$$|f(t, u)| \leqslant a(t) + b\,|u|^{p/q}, \tag{10.30}$$

where $a(t) \in L_q$, b is a constant, or satisfies the inequality (for $p = \infty$)

$$|f(t, u)| \leqslant a_h(t) \qquad (|u| \leqslant h), \tag{10.31}$$

where $a_h(t) \in L_q$, $0 < h < \infty$.

THEOREM 1.6. *Let $f(t, u)$ satisfy the Caratheodory conditions and let the superposition operator $\boldsymbol{f}u = f[t, u(t)]$ map from L_p into L_q.*

Then the operator $\boldsymbol{f}$ is continuous, if and only if one of the following conditions is satisfied:

a) $q < \infty$;
b) $q = \infty, p < \infty, f(t, u) \equiv a(t)$;
c) $q = p = \infty$, *$f(t, u)$ satisfies the inequalities*

$$|f(t, u) - f(t, v)| \leqslant \varphi_h(u - v) \qquad (|u|, |v| \leqslant h), \tag{10.32}$$

where $\varphi_h(z)$ is a continuous function with respect to z for any $h > 0$ with $\varphi_h(0) = 0$.

THEOREM 1.7. *Let $f(t, u)$ satisfy the Caratheodory conditions and let the superposition operator $\boldsymbol{f}u = f(t, u(t))$ map from L_p into L_q, $q < \infty$.*

Then the operator $\boldsymbol{f}$ is absolutely bounded, if and only if $f(t, u)$ satisfies the inequality

$$M[f(t, u)] \leqslant a(t) + b\,|u|^{p/q}, \tag{10.33}$$

where $a(t) \in L_q$, b is a constant, and $M(u)$ is a continuous function, satisfying the condition

$$\lim_{u \to \infty} \frac{M(u)}{u} = \infty. \tag{10.34}$$

Different conditions for the continuity, regularity and complete continuity of linear integral operators are given in §1, Chapter V.

We now state general results about Hammerstein operators.

THEOREM 1.8. *Let $f(t, u)$ satisfy Caratheodory conditions and $K(x, t)$ be measurable with respect to the totality of its variables. Let the operator $fu = f[t, u(t)]$ map from L_p into L_r and the integral operator $\mathbf{K}$ with kernel $K(x, t)$ map from L_r into L_q.*

Then:

I) *The Hammerstein operator $\mathbf{A} = \mathbf{K}\boldsymbol{f}$ maps from L_p into L_q;*

II) *The operator $\mathbf{A}$ is continuous, if one of the following conditions is satisfied*

a) $r < \infty$,

b) $r = \infty$, *and $\mathbf{K}$ is regular;*

III) *The operator $\mathbf{A}$ is completely continuous, if one of the following conditions is satisfied:*

a) $r < \infty$, *and $\mathbf{K}$ is completely continuous;*

b) $r < \infty$, *$\boldsymbol{f}$ is absolutely bounded, and $\mathbf{K}$ is regular;*

c) $r = \infty$, *and $\mathbf{K}$ is regular.*

1.4 *The continuity of Urison operators with range in L_q*

The continuity of a Urison operator is not a consequence of the fact that it maps from L_p into L_q (compare with §1.4 of Chapter V). Some simple and convenient sufficient conditions for the continuity of non-linear integral operators, mapping from L_p into L_q, are known. We state the simplest two of them:

THEOREM 1.9. *Let $K(x, t, u)(x, t \in \Omega, -\infty < u < \infty)$ be continuous with respect to u for almost all $x, t \in \Omega \times \Omega$ and be measurable with respect to the totality of variables x, t on $\Omega \times \Omega$ for all u. Let the Urison operator $\mathbf{A}_0$ with kernel*

$$K_0(x, t, u) = \max_{|v| \leqslant u} |K(x, t, v)| \tag{10.35}$$

map from L_p into L_q, $q < \infty$.

Then the Urison operator $\mathbf{A}$ with kernel $K(x, t, u)$ maps from L_p into L_q and is continuous.

In particular, it follows from this theorem that the Urison operator $\boldsymbol{A}$ with kernel $K(x, t, u)$, mapping from L_p into L_q ($q < \infty$), is continuous, if $K(x, t, u)$ is non-negative and monotonic with respect to u.

THEOREM 1.10. *Let $K(x, t, u)(x, t \in \Omega, -\infty < u < \infty)$ be continuous with respect to u for almost all $x, t \in \Omega$ and be measurable with respect to the totality of variables x, t on $\Omega \times \Omega$ for all u. Let the Urison operator $\mathbf{A}$ with* *kernel* $\mathbf{K}(x, t, u)$ *map from* L_p *into* L_q $(q < \infty)$ *and let the following conditions be satisfied:*

a) *for any* $h > 0$ *the function*

$$R_h(x, t) = \max_{|u| \leqslant h} |K(x, t, u)| \tag{10.36}$$

is summable with respect to t for all $x \in \Omega$*;*

b) *for any* $h > 0$

$$\lim_{\text{mes } D \to h} \sup_{|u| \leqslant h} \|\boldsymbol{P}_D \boldsymbol{A} u\|_{L_q} = 0; \tag{10.37}$$

c) *for any non-negative function* $z(x)$ *from* L_p

$$\lim_{\text{mes } D \to 0} \sup_{|u| \leqslant z} \left\| \int_D K[x, t, u(t)]\, dt \right\|_{L_q} = 0. \tag{10.38}$$

Then $\boldsymbol{A}$ *is continuous.*

The conditions a) and b) are satisfied, if, for any $h > 0$, the function (10.36) satisfies the condition

$$\int_\Omega R_h(x, t)\, dt \in L_q. \tag{10.39}$$

We state, finally, an additional theorem concerning the continuity of Urison operators—*a majorant principle.*

THEOREM 1.11. *Let* $K(x, t, u)$ *and* $K_0(x, t, u)(x, t \in \Omega, -\infty < u < \infty)$ *be continuous with respect to u for almost all* $x, t \in \Omega \times \Omega$, *be measurable with respect to the totality of variables on* $\Omega \times \Omega$ *for all u and satisfy*

$$|K(x, t, u)| \leqslant K_0(x, t, u). \tag{10.40}$$

Let the Urison operator $\boldsymbol{B}$ *with kernel* $K_0(x, t, u)$ *map from* L_p *into* L_q $(q < \infty)$ *and be continuous.*

Then the operator $\boldsymbol{A}$ *with kernel* $K(x, t, u)$ *also maps from* L_p *into* L_q *and is continuous.*

1.5 *The complete continuity of Urison operators with range in* L_q

THEOREM 1.12. *Let* $K(x, t, u)$ *be continuous with respect to* u *for almost all* $x, t \in \Omega \times \Omega$ *and be measurable with respect to the totality of variables* x, t *on* $\Omega \times \Omega$ *for all* u. *Let the Urison operator* $\boldsymbol{A}$ *with kernel* $K(x, t, u)$ *map from* L_p *into* L_q $(q < \infty)$ *and let the following conditions be satisfied:*

a) *for any* $h > 0$, *the function*

$$R_h(x, t) = \max_{|u| \leqslant h} |R(x, t, u)| \tag{10.41}$$

is summable with respect to t *for almost all* $x \in \Omega$;

b) *for any* $h > 0$

$$\lim_{\text{mes}\, D \to 0} \sup_{|u| \leqslant h} \|\boldsymbol{P}_D \boldsymbol{A} u\|_{L_q} = 0; \tag{10.42}$$

c) *for some* $a > 0$

$$\lim_{\text{mes}\, D \to 0} \sup_{\|u\|_{L_q} \leqslant a} \left\| \int_D K[x, t, u(t)]\, dt \right\|_{L_q} = 0. \tag{10.43}$$

Then the operator $\boldsymbol{A}$ *is completely continuous.*

Unfortunately, the majorant principle {see Theorem 1.11} cannot be applied to prove complete continuity.

1.6 *Some special conditions*

General conditions for the continuity and complete continuity of Urison operators were given in §§1.4 and 1.5. These conditions reduce the study of continuity and complete continuity to the verification that a given Urison operator maps from L_p into L_q. Now we shall present one particular, more simply formulated statement.

THEOREM 1.13. *Let* $K(x, t, u)(x, t \in \Omega, -\infty < u < \infty)$ *be continuous with respect to* u *for almost all* $x, t \in \Omega \times \Omega$, *be measurable with respect to the totality of variables on* $\Omega \times \Omega$ *for all* u *and satisfy the inequality*

$$|K(x, t, u)| \leqslant \sum_{k=0}^{n} R_k(x, t)\, f_k(t, u), \tag{10.44}$$

where the $f_k(t, u)$ $(k = 0, \ldots, n)$ *are non-negative functions, satisfying the*

Caratheodory conditions, and the $R_k(x, t)$ $(k = 0, 1, \ldots, n)$ *are non-negative, measurable functions, with respect to all the variables. Let the superposition operator* $\boldsymbol{f}_k u = f_k[t, u(t)]$ *map from* L_p *into* L_{r_k}, *where* r_k *is a number from the interval* $[1, \infty]$, *and the linear integral operator* $\boldsymbol{R}_k$ *with kernel* $R_k(x, t)$ *maps from* L_{r_k} *into* L_q $(q < \infty)$.

Then:

I) *The Urison operator* $\boldsymbol{A}$ *with kernel* $K(x, t, u)$ *maps from* L_p *into* L_q *and is continuous;*

II) *The operator* $\boldsymbol{A}$ *is completely continuous, if for every* $k = 0, 1, \ldots, n$ *one of the following conditions is satisfied:*

a) *the operators* $\boldsymbol{f}_k$ *are absolutely bounded;*

b) $r_k > 1$, *and the* $\boldsymbol{R}_k$ *are completely continuous.*

1.7 *Operators with range in* L_∞

The results of §§1.4–1.6 are not valid for operators with their range in L_∞. We state simple conditions for the continuity and complete continuity of such operators.

THEOREM 1.14. *Let* $K(x, t, u)(x, t \in \Omega, -\infty < u < \infty)$ *be continuous with respect to u for almost all* $x, t \in \Omega$, *be measurable with respect to the totality of variables on* $\Omega \times \Omega$ *for all u and, for any* $h > 0$, *let*

$$\lim_{\delta \to 0} \left\| \int_\Omega \max_{|u_i| \leq h,\, |u_1 - u_2| \leq \delta} |K(x, t, u_1) - K(x, t, u_2)| \, dt \right\|_{L_\infty} = 0. \tag{10.45}$$

Let the Urison operator $\boldsymbol{A}$ *with kernel* $K(x, t, u)$ *map from* L_p *into* L_∞ *with either* $p = \infty$ *or* $p < \infty$, *and, for any non-negative function* $z(x)$ *from* L_p, *let*

$$\lim_{\text{mes}\, D \to 0} \sup_{|u| \leq z} \left\| \int_D K[x, t, u(t)] \, dt \right\|_{L_\infty} = 0. \tag{10.46}$$

Then $\boldsymbol{A}$ *is continuous.*

THEOREM 1.15. *Let* $K(x, t, u)(x, t \in \Omega, -\infty < u < \infty)$ *be continuous with respect to u for almost all* $x, t \in \Omega \times \Omega$, *be measurable with respect to the totality of variables on* $\Omega \times \Omega$ *for all u, and for given arbitrary* $\varepsilon, h > 0$, *let there exist a partition* $D_0 \cup D_1 \cup \ldots \cup D_l$ *of* Ω *such that* $\operatorname{mes} D_0 = 0$ *and*

$$\int_\Omega \max_{|u| \leqslant h} |K(x', t, u) - K(x'', t, u)|\, dt < \varepsilon \tag{10.47}$$

$$(x', x'' \in D_k, k = 1, \ldots, l).$$

Let the Urison operator $\boldsymbol{A}$ *with kernel* $K(x, t, u)$ *map from* L_p *into* L_∞ *with either* $p = \infty$ *or* $p < \infty$, *and, for some 'a', let*

$$\lim_{\operatorname{mes} D \to 0} \sup_{\|u\|_{L_p} \leqslant a} \left\| \int_D K[x, t, u(t)]\, dt \right\|_{L_\infty} = 0. \tag{10.48}$$

Then $\boldsymbol{A}$ *is completely continuous.*

We now introduce a special condition.

THEOREM 1.16. *Let* $K(x, t, u)(x, t \in \Omega \times \Omega, -\infty < u < \infty)$ *be continuous with respect to* u *for almost all* $x, t \in \Omega \times \Omega$, *be measurable with respect to all the variables on* $\boldsymbol{\Omega} \times \Omega$ *for all* u *and satisfy the inequality*

$$|K(x, t, u)| \leqslant \sum_{k=0}^{n} R_k(x, t)\, f_k(t, u), \tag{10.49}$$

where the $f_k(t, u)\,(k = 0, 1, \ldots, n)$ *are non-negative functions, satisfying the Caratheodory conditions, and the* $R_k(x, t)\,(k = 0, 1, \ldots, n)$ *are non-negative functions measurable with respect to all the variables. Let the superposition operator,* $\boldsymbol{f}_k u = f_k[t, u(t)]$ *map from* L_p *into* L_{r_k}, *where* r_k *is an integer from* $[1, \infty]$, *and the linear operator* $\boldsymbol{R}_k$ *with kernel* $R_k(x, t)$ *map from* L_{r_k} *into* L_∞.

Then:

I) *The Urison operator* $\boldsymbol{A}$ *with kernel* $K(x, t, u)$ *maps from* L_p *into* L_∞.

II) *The operator* $\boldsymbol{A}$ *is continuous, if* $K(x, t, u)$ *satisfies*

$$\lim_{\delta \to 0} \left\| \int_\Omega \max_{|u_i| \leqslant h, |u_1 - u_2| \leqslant \delta} |K(x, t, u_1) - K(x, t, u_2)|\, dt \right\|_{L_\infty} = 0 \tag{10.50}$$

with either $p = \infty$ *or* $p < \infty$, *and for any* k *such that* $r_k = \infty$

$$\lim_{\operatorname{mes} D \to 0} \left\| \int_D R_k(x, t)\, dt \right\|_{L_\infty} = 0. \tag{10.51}$$

III) *The operator* $\boldsymbol{A}$ *is completely continuous if, given arbitrary* ε, $h > 0$, *there exists a partition* $D_0 \cup D_1 \cup \ldots \cup D_l$ *of* Ω, *such that* $\operatorname{mes} D_0 = 0$ *and*

$$\int_\Omega \max_{|u| \leqslant h} |K(x', t, u) - K(x'', t, u)| \, dt < \varepsilon \tag{10.52}$$

$$(x', x'' \in D_k, k = 1, \ldots, l),$$

and if, for every $k = 0, 1, \ldots, n$, *one of the following conditions is satisfied:*

a) *the operator* $\boldsymbol{f}_k$ *is absolutely bounded;*

b) $r_k > 1$, *and* $\boldsymbol{R}_k$ *is completely continuous.*

1.8 *Continuity and complete continuity for other types of integral operators*

The study of non-linear integral operators of a more general form than the Urison type can often be reduced to a study of a superposition of Urison operators and non-linear superposition operators. Different continuity and compactness conditions for such operators can be obtained as the union of corresponding conditions for Urison and superposition operators.

Consider, for example, the operator

$$\boldsymbol{A}u(x) = F\left[x, \int_\Omega K(x, t)\, u(t)\, dt\right]. \tag{10.53}$$

We assume that the linear integral operator $\boldsymbol{K}$ with kernel $K(x, t)$ maps from L_p into L_r and that the non-linear superposition operator $\mathscr{F}u(x) = F[x, u(x)]$ maps from L_r into L_q. Then, the operator $\boldsymbol{A} = \mathscr{F}\boldsymbol{K}$ is continuous, if $\boldsymbol{K}$ and $\mathscr{F}$ are continuous, and is completely continuous, if $\boldsymbol{K}$ is completely continuous and $\mathscr{F}$ is continuous or if $\boldsymbol{K}$ is regular and $\mathscr{F}$ is absolutely bounded (see §1.1 of Chapter V and §1.1 of Chapter VI).

Analogous results can be formulated for non-linear integral operators more general than (10.53):

$$\boldsymbol{A}u(x) = \boldsymbol{F}\left\{x, \int_\Omega K_1[x, t, u(t)]\, dt, \ldots, \int_\Omega K_n[x, t, u(t)]\, dt\right\}. \tag{10.54}$$

The study of continuity and complete continuity of Liapunov operators is not difficult. It reduces to the study of continuity and complete continuity of the integral operators

$$\boldsymbol{K}_{[k_1,\ldots,k_l]}u(x) = \int_\Omega \ldots \int_\Omega \boldsymbol{K}_{[k_1,\ldots,k_l]}(x, t_1, \ldots, t_l)\, u(t_1, \ldots, t_l)\, dt_1 \ldots dt_l \tag{10.55}$$

(it is convenient to study these operators using the Cartesian product of L_p spaces: for example, the space L_p of functions of l variables) and to the study of convergence for the series

$$\boldsymbol{A}u(x) = \sum_{k_1,\ldots,k_l} \boldsymbol{K}_{[k_1,\ldots,k_l]}\{u(t_1)^{k_1} \ldots u(t_l)^{k_l}\}(t). \tag{10.56}$$

1.9 *Derivatives of non-linear operators*

Let $\boldsymbol{A}$ be a non-linear operator, mapping from E_1 into E_2. The operator $\boldsymbol{A}$ is called *Fréchet differentiable at a point* u_0, if the increment $\boldsymbol{A}(u_0 + h) - \boldsymbol{A}u_0$ of the operator $\boldsymbol{A}$ can be expressed in the form

$$\boldsymbol{A}(u_0 + h) - \boldsymbol{A}u_0 = \boldsymbol{B}h + \omega(u_0, h), \tag{10.57}$$

where $\boldsymbol{B}$ is a linear continuous operator mapping from E_1 into E_2 and $\omega(u_0, h)$ is an operator satisfying the condition

$$\lim_{\|h\| \to 0} \frac{\|\omega(u_0, h)\|_{E_2}}{\|h\|_{E_1}} = 0. \tag{10.58}$$

The operator $\boldsymbol{B}$ is called the (Fréchet) *derivative* of $\boldsymbol{A}$ at the point u_0 and is, usually, denoted by $\boldsymbol{A}'(u_0)$.

If an operator $\boldsymbol{A}$ is differentiable at every point of E_1, then $\boldsymbol{A}'(u)$ can be considered as the operator with domain E_1 and range the space of linear continuous operators, mapping from E_1 into E_2. If $\boldsymbol{A}'(u)$ is continuous, then $\boldsymbol{A}$ is said to be continuously differentiable.

The derivative of the non-linear integral operator

$$\boldsymbol{A}u(x) = \int_\Omega K[x, t, u(t)]\, dt \tag{10.59}$$

at a point u_0 is defined by

$$\boldsymbol{A}'(u_0)\, h(x) = \int_\Omega K_0(x, t)\, h(t)\, dt, \tag{10.60}$$

for all known differentiability conditions, where

$$K_0(x, t) = \lim_{u \to 0} \text{as} \frac{K[x, t, u_0(t) + u] - K[x, t, u_0(t)]}{u}, \tag{10.61}$$

and we denote by 'lim as' the measure limit.

We note, by the way, that whatever be the smoothness of the kernel $K(x, t, u)$ of a Urison operator A, mapping from L_p into L_q, the differentiability of A is not guaranteed (even if the linear integral operator with kernel $K'_u[s, t, u_0(t)]$ maps from L_p into L_q). Some essential additional conditions are necessary.

1.10 *Derivatives of the Hammerstein operator*

The study of the differentiability of the Hammerstein operator

$$Ax(t) = \int_\Omega K(x, t)\, f[t, u(t)]\, dt \tag{10.62}$$

reduces to a study of the differentiability of the superposition operator

$$fu = f[t, u(t)]. \tag{10.63}$$

Suppose that f maps from L_p into L_r with $r < \infty$. Then the derivative of f (if it exists) has the form

$$f'(u)h = a(t)\, h(t), \tag{10.64}$$

where

$$a(t) = \lim_{u \to 0} \text{as} \frac{f[t, u_0(t) + u] - f[t, u_0(t)]}{u}. \tag{10.65}$$

Thus, if the superposition operator (10.63) maps from L_p into L_r and is differentiable at a point u_0, then the limit (10.65) exists with either $p < r$ and $a(t) \equiv 0$, or $p \geqslant r$ and $a(t) \in L_{pr/(p-r)}$. What is more, if $f(t, u)$ depends linearly on u, then $p > r$. The mentioned conditions are not sufficient for differentiability at the point u_0.

THEOREM 1.17. *Let $f(t, u)(t \in \Omega, -\infty < u < \infty)$ satisfy the Caratheodory conditions. Let $p > r$, $u_0 \in L_p$, $f[t, u_0(t)] \in L_r$ and the limit*

$$a(t) = \lim_{u \to 0} \text{as} \frac{f[t, u_0(t) + u] - f[t, u_0(t)]}{u}, \tag{10.66}$$

exist with $a(t) \in L_{pr/(p-r)}$. Further, let either $p = \infty$, or $p < \infty$ with the inequality

$$|f[t, u_0(t) + u] - f[t, u_0(t)]| \leqslant a_0(t)\,|u| + b_0\,|u|^{p/r}, \tag{10.67}$$

satisfied where $a_0(t) \in L_{pr/(p-r)}$ and b_0 is a constant.

Then the superposition operator $\boldsymbol{f}u = f[t, u(t)]$ maps from L_p into L_r, is continuous, and differentiable at the point u_0 with

$$f'(u_0)h = a(t)\,h(t). \tag{10.68}$$

THEOREM 1.18. *Let $f(t, u)(t \in \Omega, -\infty < u < \infty)$ have a continuous derivative $f'_u(t, u)$ with respect to u for almost all $t \in \Omega$ and be measurable with respect to t for all u $(-\infty, \infty)$. Further, let the superposition operator $\boldsymbol{f}u = f[t, u(t)]$ map from L_p into L_r, $p > r$.*

Then f is differentiable at every point of the space L_p, if and only if the superposition operator $\boldsymbol{g}u = f'_u[t, u(t)]$ maps from L_p into $L_{pr/(p-r)}$. Here, the operator $\boldsymbol{f}$ is continuously differentiable.

The last two theorems concerned superposition operators with range in L_r, $r < \infty$. Differentiability conditions are simpler for operators with range in L_∞. We present the simplest results.

THEOREM 1.19. *Let $f(t, u)$ $(t \in \Omega, -\infty < u < \infty)$ satisfy the Caratheodory conditions and, for any $h > 0$, let $a_h(t) = \max_{|u| \geqslant h} |f(t, u)|$ belong to L_∞. Let $u_0 \in L_\infty$ and the limit*

$$a(t) = \lim_{u \to 0} \frac{f[t, x_0(t) + u] - f[t, x_0(t)]}{u} \tag{10.69}$$

exist uniformly with respect to $t \in \Omega$.

Then the superposition operator $\boldsymbol{f}u = f[t, u(t)]$ maps from L_∞ into L_∞, and is differentiable at the point x_0 with

$$f'(u_0)h = a(t)\,h(t). \tag{10.70}$$

THEOREM 1.20. *Let $f(t, u)$ $(t \in \Omega, \ -\infty < u < \infty)$ have a derivative with respect to u, $f'_u(t, u)$, which is uniformly continuous with respect to $t \in \Omega$ for almost all $t \in \Omega$, and be measurable with respect to t for all $u \in (-\infty, \infty)$. Let $f(t, 0) \in L_\infty$.*

Then the superposition operator $\boldsymbol{f}u = f[t, u(t)]$ maps from L_∞ into L_∞, and is continuously differentiable with

$$f'(u)h = f'_u[t, u(t)]\, h(t). \tag{10.71}$$

Theorems 1.19–1.20 can certainly be applied to the study of superposition operators mapping from C into C, as such operators define mappings from L_∞ into L_∞.

We obtain different differentiability conditions for Hammerstein operators, by unifying the differentiability conditions for superposition operators with continuity conditions for linear integral operators. Some subtler conditions are:

THEOREM 1.21. *Let $f(t, u)$ satisfy the Caratheodory conditions, $u_0 \in L_p$, $f[(t, u_0(t)] \in L_r$ $(p \geqslant r)$ and the limit*

$$a_0(t) = \lim_{u \to 0} \frac{f[t, u_0(t) + u] - f[t, u_0(t)]}{u}. \tag{10.72}$$

exist. Let the superposition operator $\boldsymbol{a}h = a[t, h(t)]$, where

$$a(t, u) = \begin{cases} \dfrac{f[t, u_0(t) + u] - f[t, u_0(t)]}{u}, & \text{if} \quad u \neq 0, \\ \quad a_0(t) & \text{if} \quad u = 0, \end{cases} \tag{10.73}$$

maps from L_p into $L_{pr/(p-r)}$. Let a linear integral operator with kernel $K(x, t)$ map from L_r into L_q, and let one of the following conditions hold:

a) $p > r$;

b) $1 < p = r < \infty$, $\boldsymbol{K}$ *is regular and completely continuous*;

c) $p = r = \infty$, $\boldsymbol{K}$ *is regular and completely continuous.*

Then the Hammerstein operator $\boldsymbol{A} = \boldsymbol{K}\boldsymbol{f}$ maps from L_p into L_q, is continuous, and is differentiable at the point u_0 with

$$\boldsymbol{A}'(u_0)\, h(x) = \int_\Omega K(x, t)\, f'_u[t, u_0(t)]\, h(t)\, dt. \tag{10.74}$$

THEOREM 1.22. *Let $f(t, u)$ have a derivative with respect to u, $f'_u(t, u)$, which is continuous for almost all $t \in \Omega$, and be measurable with respect to t for all $u \in (-\infty, \infty)$; let a given function $K(x, t)(x, t \in \Omega)$ be measurable with respect to all its variables. Let the superposition operators $\boldsymbol{f}u = f[t, u(t)]$ map from L_p into L_r, $p \geqslant r$, the superposition operator $\boldsymbol{g}u = f'_u[t, u(t)]$ map from L_p into $L_{pr/(p-r)}$ and the linear integral operator $\boldsymbol{K}$ with kernel $K(x, t)$ map from L_r into L_q. Let, finally, one of the following conditions hold:*

a) $r < p$;
b) $1 < r = p < \infty$, *$\boldsymbol{K}$ is regular and completely continuous;*
c) $r = p = \infty$, *$\boldsymbol{K}$ is regular and completely continuous.*

Then the Hammerstein operator $\boldsymbol{A} = \boldsymbol{Kf}$ maps from L_p into L_q, is continuous, and is continuously differentiable with

$$\boldsymbol{A}'(u)\, h(x) = \int_{\Omega} K(x, t)\, f'_u[t, u(t)]\, h(t)\, dt. \tag{10.75}$$

1.11 *Auxiliary theorems about superposition operators*

The study of differentiability and continuous differentiability of Urison operators usually reduces to a study of the continuity of certain constructed superposition operators, which map in special spaces of continuous functions of two variables.

Let $p \in [1, \infty]$. We denote by N_p the space of functions $u(x, t)(x, t \in \Omega)$ of two variables, for which the norm

$$\|u(x, t)\|_{N_p} = \left\| \operatorname*{vrai\,max}_{x \in \Omega} |u(x, t)| \right\|_{L_p}, \tag{10.76}$$

is finite and has sense and N_p is a Banach space. It is convenient to consider functions $u(t)$ from the space L_p as functions from N_p which are independent of x.

Let $p, q \in [1, \infty]$. We denote by $B_{p,q}$ (see §2.3 of Chapter V) the space of functions $k(x, t)(x, t \in \Omega)$ of two variables, for which the norm

$$\|k(x, t)\|_{B_{p,q}} = \sup_{\|u\|_{L_p},\, \|v\|_{L_{q'}} \leqslant 1} \int_{\Omega}\int_{\Omega} |k(x, t)\, u(t)\, v(x)|\, dt\, dx. \tag{10.77}$$

has sense and is finite. By $B^0_{p,q}$ we denote the subspace of $B_{p,q}$ which consists of functions $k(x, t)$ such that

$$\lim_{\text{mes } D \to 0} \sup_{\|u\|_{L_p}, \|v\|_{L_{q'}} \leqslant 1} \int_\Omega \int_D |k(x, t)\, u(t)\, v(x)|\, dt\, dx$$

$$= \lim_{\text{mes } D \to 0} \sup_{\|u\|_{L_p}, \|v\|_{L_{q'}} \leqslant 1} \int_D \int_\Omega |k(x, t)\, u(t)\, v(x)|\, dt\, dx = 0. \quad (10.78)$$

We recall that the functions $k(x, t)$ from $B_{p,q}$ are kernels of regular integral operators, mapping from L_p into L_q. The subspace $B^0_{p,q}$ contains only the zero function if $p = 1$ or $q = \infty$. If $p > 1$ and $q < \infty$, then the functions of $B^0_{p,q}$ are kernels of regular integral operators $\boldsymbol{K}$, for which $\boldsymbol{K}$ and $|\boldsymbol{K}|$ are completely continuous.

The theorems stated below contain different sufficiency conditions for the continuity of a superposition operator

$$\boldsymbol{Q}u(x, t) = Q[x, t, u(x, t)] \quad (10.79)$$

mapping from N_p into $B_{p,q}$.

THEOREM 1.23. *Let $Q(x, t, u)$ be continuous with respect to u for almost all $x, t \in \Omega \times \Omega$, and be measurable with respect to the totality of variables x, t for all u. Let $Q(x, t, 0) \in B_{p,q}$ and*

$$|Q(x, t, u) - Q(x, t, v)| \leqslant \sum_{j=1}^{m} S_j(x, t)\, g_j(t, w, u - v) \quad (10.80)$$

$$(|\boldsymbol{u}|, |u| \leqslant w),$$

where $S_j \in B_{pr_j/(p+r_j),q}$; the functions $g_j(t, w, z)$ be continuous with respect to the totality of variables w, z for almost all $t \in \Omega$, and be measurable with respect to t for all w, z satisfying the condition $g_j(t, w, 0) = 0$; the superposition operator $\boldsymbol{g}_j(w, h) = g_j[t, w(t)\, h(t)]$ be defined for $w, h \in L_p$ and its range belong to L_{r_j}.

Finally, for every j, let one of the following conditions hold:

a) $r_j < \infty$;

b) $r_j = \infty$, $S_j \in B^0_{p,q}$.

Then the operator $\boldsymbol{Q}u = Q[x, t, u(x, t)]$ maps from N_p into $B_{p,q}$ and is continuous.

We note that the operator $\boldsymbol{g}_j(w, h) = g_j[t, w(t), h(t)]$ is defined for $w, h \in L_p$ and its range belongs to L_{r_j}, if and only if the functions $g_j(t, w, h)$ satisfy the inequality

$$|g_j(t, w, h)| \leqslant g_j(t) + a\,|w|^{p/r_j} + b\,|h|^{p/r_j}, \tag{10.81}$$

where $g_j(t) \in L_{r_j}$ and a, b are constants.

In many cases it is difficult to verify the conditions of the Theorem 1.23. We now present a particular but simple statement of verification.

THEOREM 1.24. *Let $Q(x, t, u)$ have a derivative with respect to u, $Q'_u(x, t, u)$, which is continuous for almost all $(x, t) \in \Omega \times \Omega$, and be measurable with respect to the totality of variables x, t for all u. Let $Q(x, t, 0) \in B_{p,q}$; the function $Q'_u(x, t, u)$ satisfy the inequality*

$$|Q'_u(x, t, u)| \leqslant \sum_{j=1}^{m} S_j(x, t)\, g_j(t, u), \tag{10.82}$$

where $S_j \in B_{pr_j/(p+r_j),q}$; the functions $g_j(t, u)$ satisfy the Caratheodory conditions; and every superposition operator $\boldsymbol{g}_j u = g_j[t, u(t)]$ map from L_p into $L_{pr_j/(p-r_j)}$, $r_j \leqslant p$.

Then the operator $\boldsymbol{Q}u = Q[(x, t, u(x, t)]$ maps from N_p into $B_{p,q}$ and is continuous.

The continuity conditions for the superposition operator (10.79), which do not contain estimates for the rate of change of $Q(x, t, u)$, are of greater interest. Below, we state some of these conditions.

THEOREM 1.25. *Let $Q(x, t, u)$ be continuous with respect to u for almost all $x, t \in \Omega \times \Omega$, and be measurable with respect to the totality of variables x, t for all u. Let the superposition operator $\boldsymbol{Q}u = Q[x, t, u(x, t)]$ map from N_p into $B^0_{p,q}$, $p > 1, q < \infty$.*

Then the operator $\boldsymbol{Q}$ is continuous.

THEOREM 1.26. *Let $Q(x, t, u)$ be continuous with respect to u for almost all $x, t \in \Omega \times \Omega$, be measurable with respect to the totality of variables x, t for all u, and satisfy the inequality*

$$|Q(x, t, u)| \leqslant \sum_{j=1}^{m} S_j(x, t)\, g_j(t, u), \tag{10.83}$$

where $S_j(x, t) \in B_{pr_j/(p+r_j),q}$; *and let the functions* $g_j(t, u)$ $(j = 1, \ldots, m)$ *satisfy the Caratheodory conditions, and the superposition operator* $\boldsymbol{g}_j u = g_j[t, u(t)]$ *map from* L_p *into* L_{r_j}. *Finally, let* $p > 1$, $q < \infty$ *and for every* $j = 1, \ldots, m$ *one of the following conditions hold:*

a) $r_j < \infty$;

b) $r_j = \infty$, $S_j \in B^0_{p,q}$.

Then the superposition operator $\boldsymbol{Q}u = Q[x, t, u(t)]$ *maps from* N_p *into* $B_{p,q}$ *and is continuous.*

Theorems 1.25–1.26 are not applicable if $p = 1$ or $q = \infty$. It turns out that the superposition operator (10.79), mapping from N_1 into $B_{1,q}$, is continuous, only when the function $Q(x, t, u)$ does not depend on u. Analogous theorems to Theorems 1.25–1.26 are also not valid for superposition operators with their range in $B_{p,\infty}$. The general continuity condition for such operators is given by:

THEOREM 1.27. *Let* $Q(x, t, u)$ *be continuous with respect to u for almost all* $x, t \in \Omega \times \Omega$ *and be measurable with respect to the totality of variables* x, t *on* $\Omega \times \Omega$ *for all* $u \in (-\infty, \infty)$; *and, for any* $h > 0$, *let the function*

$$L_h(x, t) = \max_{|u| \leqslant h} |Q(x, t, u)|$$

belong to $B_{1,\infty}$. *Besides, let for any* $h > 0$,

$$\lim_{\delta \to 0} \operatorname{vrai\,max}_{x \in \Omega} \int_\Omega \max_{|u_i| \leqslant h,\, |u_1 - u_2| \leqslant \delta} |K(x, t, u_1) - K(x, t, u_2)|\, dt = 0.$$

Then the operator $\boldsymbol{Q}$ *maps from* N_∞ *into* $B_{\infty,\infty}$ *and is continuous.*

THEOREM 1.28. *Let* $Q(x, t, u)$ *be continuous with respect to u for almost all* $x, t \in \Omega \times \Omega$, *be measurable with respect to the totality of variables* x, t *on* $\Omega \times \Omega$ *for all* $u \in (-\infty, \infty)$ *and satisfy the inequality*

$$|Q(x, t, u)| \leqslant R(x, t) + b\,|u|^{p-1}, \tag{10.84}$$

where $R(x, t) \in B_{p,\infty}$, *b is a constant*, $1 < p < \infty$. *Besides, let the following conditions be satisfied:*

a) *for any* $h > 0$

$$\lim_{\delta \to 0} \operatorname{vrai\,max}_{x \in \Omega} \int_\Omega \max_{|u_i| \leqslant h,\, |u_1 - u_2| \leqslant \delta} |Q(x, t, u_1) - Q(x, t, u_2)|\, dt = 0;$$

b) *the following equality holds*

$$\lim_{\text{mes}\, D \to 0} \left\| \int_D |R(x, t)^{p'}| \, dt \right\|_{L\infty} = 0.$$

Then the operator $\boldsymbol{Q}$ *maps from* N_p *into* $B_{p,\infty}$ *and is continuous.*

1.12 *The differentiability of Urison operators*

THEOREM 1.29. *Let* $K(x, t, u)$ *be continuous with respect to* u *for almost all* $x, t \in \Omega \times \Omega$, *and be measurable with respect to the totality of variables* x, t *on* $\Omega \times \Omega$ *for all* $u \in (-\infty, \infty)$. *Let* $u_0 \in L_p$,

$$\int_\Omega K[x, t, u_0(x, t)] \, dt \in L_q$$

and for almost $x, t \in \Omega \times \Omega$ *the following limit exist*

$$K_0(x, t) = \lim_{u \to 0} \text{as} \frac{K[x, t, u_0(t) + u] - K[x, t, u_0(t)]}{u}.$$

Let the superposition operator $Hu = H[x, t, u(x, t)]$ *where*

$$H(x, t, u) = \begin{cases} \dfrac{K[x, t, u_0(t) + u] - K[x, t, u_0(t)]}{u}, & \text{if} \quad u \neq 0, \\ K_0(x, t), & \text{if} \quad u = 0, \end{cases}$$

map from L_p *into* $B_{p,q}$ *and be continuous at the point* 0.

Then the Urison operator $\boldsymbol{A}$ *with kernel* $K(x, t, u)$ *map from* L_p *into* L_q, *is differentiable at the point* u_0 *with*

$$A'(u_0)\, h(x) = \int_\Omega K_0(x, t)\, h(t) \, dt. \tag{10.85}$$

Applying the different continuity conditions for superposition operators to the operator $\boldsymbol{H}$, we can obtain different sufficiency conditions for the differentiability of Urison operators.

1.13 *Continuous differentiability of Urison operators*

THEOREM 1.30. *Let $K(x, t, u)$ have a derivative with respect to u, $K'_u(x, t, u)$, which is continuous for almost all $x, t \in \Omega \times \Omega$, and be measurable with respect to the totality of variables x, t on $\Omega \times \Omega$ for all $u \in (-\infty, \infty)$. Let the operator $\boldsymbol{A}$ with kernel $K(x, t, u)$ map from L_p into L_q and the superposition operator $\boldsymbol{Q}u = Q[x, t, u(x, t)]$ map from N_p into $B_{p,q}$ and be continuous.*

Then $\boldsymbol{A}$ is continuously differentiable and

$$\boldsymbol{A}'(u)\, h(x) = \int_\Omega K'_u[x, t, u(t)]\, h(t)\, dt. \tag{10.86}$$

1.14 *Higher order derivatives*

Let $\boldsymbol{A}$ be a non-linear operator mapping from a Banach space E_1 into another Banach space E_2. We assume that the increment $\boldsymbol{A}(u_0 + h) - \boldsymbol{A}(u_0)$ can be expressed in the form

$$\boldsymbol{A}(u_0 + h) - \boldsymbol{A}u_0 = \sum_{k=1}^{n} \boldsymbol{B}_k h + \omega(u_0, h), \tag{10.87}$$

where

$$\boldsymbol{B}_k h = \hat{\boldsymbol{B}}_k (h_1, \ldots, h_k), \tag{10.88}$$

and $\hat{\boldsymbol{B}}_k (h_1, \ldots, h_k)$ is an operator which is linear with respect to each of the variables $h_1, \ldots, h_k$, and is continuous with respect to the totality of variables; and $\omega(u_0, h)$ is an operator, satisfying the condition

$$\lim_{\|h\|_{E_1} \to 0} \frac{\|\omega(u_0, h)\|_{E_1}}{\|h\|^n_{E_1}} = 0. \tag{10.89}$$

Then we say, that $\boldsymbol{A}$ is *n times differentiable at the point u_0*; the operator $k!\,\boldsymbol{B}_k$ is called the *derivative* of $\boldsymbol{A}$ at u_0 and is denoted by $\boldsymbol{A}^{(k)}(u_0)$.

We state two theorems concerning high order derivatives of integral operators:

THEOREM 1.31. *Let $f(t, u)$ satisfy the Caratheodory conditions. Let $u_0 \in L_p$, $f[t, u_0(t)] \in L_r$ and*

$$|f[t, u_0(t) + u] - f[t, u_0(t)]| \leqslant \sum_{k=1}^{n} a_k(t)\, u^k + w(t, u)\, u^n, \tag{10.90}$$

where $a_k(t) \in L_{pr/(p-r_k)}$, $w(t, 0) = 0$ *and the superposition operator* $\boldsymbol{w}u = w[t, u(t)]$ *maps from* L_p *into* $L_{pr/(p-rn)}$. *Let a given linear integral operator* $\boldsymbol{K}$ *with kernel* $K(x, t)$ *map from* L_r *into* L_q. *Finally, let one of the following conditions hold:*

a) $p > nr$;

b) $p = nr$, *the operator* $\boldsymbol{K}$ *is regular and compact.*

Then the Hammerstein operator $\boldsymbol{A} = \boldsymbol{Kf}$ *maps from* L_p *into* L_q, *is continuous, has derivatives at* u_0 *of order up to* n, *and*

$$A^{(k)}(u_0)\, h(x) = \int_{\Omega} K(x, t)\, a_k(t)\, h^k(t)\, dt \qquad (k = 1, \ldots, n). \tag{10.91}$$

THEOREM 1.32. *Let* $K(x, t, u)\,(x, t \in \Omega, -\infty < u < \infty)$ *be measurable with respect to* (x, t) *on* $\Omega \times \Omega$ *for all* $u \in (\infty, \infty)$ *and continuous with respect to* u *for all* $(x, t) \in \Omega \times \Omega$. *Let* $u_0 \in L_p$,

$$\int_{\Omega} K[x, t, u_0(t)]\, dt \in L_q \tag{10.92}$$

and

$$K[x, t, u_0(t) + u] - K[x, t, u_0(t)] = \sum_{k=1}^{n} K_k(x, t)\, u^k + W(x, t, u)\, u^n, \tag{10.93}$$

where $W(x, t, 0) \equiv 0$. *Let* $K_k(x, t) \in B_{(p/k),q}$ $(k = 1, \ldots, n)$ *and the superposition operator* $\boldsymbol{W}u = W[x, t, u(t)]$ *map from* L_p *into* $B_{p,q}$ *and be continuous.*

Then the Urison operator $\boldsymbol{A}$ *with kernel* $K(x, t, u)$ *maps from* L_p *into* L_q, *is continuous, and has a derivative at* u_0 *of order up to* n *with*

$$A^{(k)}(u_0)\, h(x) = \int_{\Omega} K_k(x, t)\, h^{(k)}(t)\, dt \qquad (k = 1, \ldots, n). \tag{10.94}$$

1.15 *Analytic operators*

An operator $\boldsymbol{A}$, mapping from a space E_1 into a space E_2, is called *analytic at a point* u_0, if its increment can be represented in the form

$$A(u_0 + h) - Au_0 = \sum_{k=1}^{\infty} B_k h, \tag{10.95}$$

where the $\boldsymbol{B}_k$ are operators of the form (10.38) and the series on the r.h.s. converges with respect to the norm uniformly in some ball $\|h\| \leqslant \rho$. An analytic operator $\boldsymbol{A}$ at a point u_0 has derivatives of all orders $\boldsymbol{A}^{(k)}(u_0)$ with $\boldsymbol{A}^{(k)}(u_0) = k!\, \boldsymbol{B}_k$. It is easy to see that an analytic operator $\boldsymbol{A}$ at a point u_0 is analytic in a neighbourhood of the point u_0.

All the definitions and statements of this section apply to both real and complex spaces. We now present an important result which is not valid for real valued spaces: In a complex valued space, the analyticity of an operator $\boldsymbol{A}$ at a point u_0 follows from its differentiability at u_0 and in a neighbourhood of u_0.

As an example, consider the Urison operator

$$\boldsymbol{A}u(x) = \int_\Omega K[x, t, u(t)]\, dt. \tag{10.96}$$

Let

$$K(x, t, u) = \sum_{k=0}^{\infty} K_k(x, t)\, u^k, \tag{10.97}$$

where $K_k(x, t)$ is measurable with respect to the totality of variables and

$$\overline{\lim_{k\to\infty}} \sqrt[k]{\left(\sup_{x\in\Omega} \int_\Omega |K(x, t)|\, dt\right)} < 1/a. \tag{10.98}$$

Then the operator $\boldsymbol{A}$ is defined on the ball $\|u\| < a$ of the space L_∞ and is analytic. The Urison operator is analytic in spaces L_p only when its kernel is a polynomial with respect to u of order not more than p.

The Liapunov-Lichtenschtein operator

$$\boldsymbol{A}u(x) = \sum_{k_1,\ldots,k_l} \int_\Omega \cdots \int_\Omega K_{[k_1,\ldots,k_l]}(x, t_1, \ldots, t_l)\, u(t_1)^{k_1} \ldots u(t_l)^{k_l}\, dt_1 \ldots dt_l \tag{10.99}$$

is analytic under some natural conditions.

1.16 *Asymptotically linear operators*

A non-linear operator $\boldsymbol{A}$, mapping from E_1 into E_2, is called *asymptotically linear*, if it can be expressed in the form

$$\boldsymbol{A}u = \boldsymbol{B}u + \omega(u), \tag{10.100}$$

where $\boldsymbol{B}$ is a linear continuous operator, mapping from E_1 into E_2, and the operator $\omega(u)$ satisfies the condition

$$\lim_{\|u\| \to \infty} \frac{\|\omega(u)\|_{E_2}}{\|u\|_{E_1}} = 0. \tag{10.101}$$

The operator $\boldsymbol{B}$ is called *the asymptotic derivative* or *the derivative at infinity* and is denoted by $\boldsymbol{A}'(\infty)$.

We state conditions for the asymptotic linearity of superposition and integral operators.

THEOREM 1.33. *Let $f(t, u)$ satisfy the Caratheodory conditions and let the limit*

$$g(t) = \lim_{u \to \infty} \frac{f(t, u)}{u} \tag{10.102}$$

exist. Further, let one of the following conditions be satisfied:
a) $p = r$, $g(t) \in L_\infty$ *and*

$$|f(t, u) - g(t)u| \leqslant a(t) + b(t)\,|u|^k, \tag{10.103}$$

where $a(t) \in L_r$, $b(t) \in L_{p/(1-k)}$, $0 < k < 1$;
b) $p > r$, $g(t) \in L_{pr/(p-r)}$ *and*

$$|f(t, u) - g(t)u| \leqslant a(t) + b(t)\,|u|, \tag{10.104}$$

where $a(t) \in L_r$, $b(t) \in L_{pr/(p-r)}$.

Then the superposition operator $\boldsymbol{f}u = f[t, u(t)]$ *is asymptotically linear and*

$$\boldsymbol{f}'(\infty)\, h(t) = q(t)\, h(t). \tag{10.105}$$

THEOREM 1.34. *Let $f(t, u)$ satisfy the Caratheodory conditions and the limit*

$$g(t) = \lim_{u \to \infty} \frac{f(t, u)}{u} \tag{10.106}$$

exist. Let $K(x, t)$ be measurable with respect to the totality of variables. Let *the superposition operator* $fu = f[t, u(t)]$ *map from* L_p *into* L_r, $r \leqslant p$, $g(t) \in L_{pr/(p-r)}$ *and the integral operator* $\boldsymbol{K}$ *map from* L_r *into* L_q. *Finally, let one of the following conditions be satisfied:*

a) $r < p$ *and the function* $f(t, u)$ *satisfies the inequality*

$$|f(t, u) - q(t)u| \leqslant a(t) + b(t)\,|u|, \tag{10.107}$$

where

$$a(t) \in L_r, \qquad b(t) \in L_{pr/(p-r)};$$

b) $r = p = \infty$ *and*

$$|f(t, u) - q(t)u| \leqslant a(t) + b(t)\,|u|, \tag{10.108}$$

where $a(t) \in L_r$, $b(t) \in L_\infty$ *and the operator* $\boldsymbol{K}$ *is regular and compact;*

c) $r = p$ *and*

$$|f(t, u) - q(t)u| \leqslant a(t) + b(t)\,|u|^k, \tag{10.109}$$

where $a(t) \in L_r$, $b(t) \in L_{p/(1-k)}$, $0 < k < 1$.

Then the Hammerstein operator $\boldsymbol{A} = \boldsymbol{Kf}$ *maps from* L_p *into* L_q, *and is asymptotically linear with*

$$\boldsymbol{A}'(\infty)\, u(x) = \int_\Omega K(x, t)\, g(t)\, u(t)\, dt. \tag{10.110}$$

The study of asymptotic linearity of Urison operators is a more difficult problem. In applications, it is usually sufficient to obtain an inequality of the form

$$|K(x, t, u) - K_\infty(x, t)u| \leqslant \sum_{j=1}^{m} S_j(x, t)\, g_j(t, u). \tag{10.111}$$

If the Hammerstein operator

$$\boldsymbol{B}_j u(x) = \int_\Omega S_j(x, i)\, g_j[t, u(t)]\, dt \tag{10.112}$$

as an operator mapping from L_p into L_q has an asymptotic derivative, which is not equal to zero, and if $K_\infty(x, t) \in B_{p,q}$, then the Urison operator with kernel $K(x, t, u)$ maps from L_p into L_q, is asymptotically linear and

$$A'(\infty)\, u(x) = \int_{\Omega} K_{\infty}(x, t)\, u(t)\, dt. \tag{10.113}$$

1.17 *Supplementary notes*

For a detailed treatment of continuity, complete continuity, differentiability and other properties of non-linear integral operators see the monograph [11]; some properties are also studied in the monographs [2], [8], [9], [10], [12].

The theorems of this section, which concerned operators acting in L_p $(1 \leqslant p \leqslant \infty)$, are applicable only to operators with non-linearities of the power type. The study of operators with other types of non-linearity (for example, exponential) require spaces other than L_p, such as Orlicz, Lorentz, Marcinkievicz. The theorems of this section, which concerned continuity, complete continuity and differentiability of Hammerstein and Urison operators are particular cases of general theorems, which concern operators, acting in general spaces of functions. These theorems are presented in [6].

Sometimes, we deal with operators, defined not on the whole space but on some part of it (for example, on the set of all non-negative functions or on the set of functions, satisfying the inequality $u_1(t) \leqslant u(t) \leqslant u_2(t)$). Theorems, analogous to those stated in this section, are valid for such operators (see, for instance, [6], [12]).

§2 The existence and uniqueness of solutions

2.1 *The formulation of the problem*

In this section, we mainly study equations of the form

$$u(x) = \mu \int_{\Omega} K[x, t, u(t)]\, dt + f(x), \tag{10.114}$$

where Ω is a bounded closed set of a finite-dimensional space of non-zero Lebesgue measure, $K[x, t, u]$, $f(x)$ $(x, t \in \Omega, u \in (a, b))$ are given functions, μ is a parameter, and $u(x)$ an unknown function.

As a rule, (see §1) the non-linear operator

$$Au(x) = \int_{\Omega} K[x, t, u(t)]\, dt, \tag{10.115}$$

defined by the first term of (10.114), acts in some Banach spaces E of measurable functions on Ω and has 'good' properties (it is continuous, completely continuous, differentiable etc.). If $f(x) \in E$, then it is convenient to consider (10.114) as a non-linear operator equation

$$u = \mu A u + f \tag{10.116}$$

in E. It is necessary to remember that, when considering (10.116) in E, we 'lose' solutions which do not belong to E (if such solutions exist). If the operator equation (10.116) can be examined in different spaces, then the choice of the space E depends on the type of solutions in which we are interested.

In this section, we study existence and uniqueness conditions for the solutions of (10.116). If the trivial solution satisfies (10.116), we study the existence of a second non-zero solution. We also study the existence of non-negative solutions.

The solutions of (10.116) are the fixed points of the operator $A_1 u = \mu A u + f$. So fixed-point theorems will be the main tools used when proving the existence of solutions of (10.116) and, hence (10.114). The most widely used tools in the analysis of fixed points are the Banach Principle of contraction mapping (see §2.2) and the Schauder Principle (see §2.3). In this section, we formulate some existence theorems for the solutions of (10.114), which are based on contraction mapping principles.

2.2 *Equations with operators satisfying a Lipschitz condition*

An operator A, acting in a Banach space E, is said to satisfy *a Lipschitz condition* with constant q on a set $M \subset E$, if

$$\|Au - Av\| \leqslant q \|u - v\| \qquad (u, v \in M). \tag{10.117}$$

If $q < 1$, then A is called *a contraction operator*. We recall that a linear bounded operator satisfies (10.117) with $q = \|A\|$.

THE CONTRACTION MAPPING PRINCIPLE: *If the contraction operator A maps a closed set $M \subset E$ into itself ($AM \subset M$), then it has a unique fixed-point*

u^* in M. This fixed point can be obtained as the limit of the sequence

$$u_n = Au_{n-1} \qquad (n = 1, 2, \ldots), \tag{10.118}$$

where u_0 is an arbitrary element of M. The rate of convergence of the sequence (10.118) *is given by the inequality*

$$\|u_n - u^*\| \leqslant \frac{q^n}{1-q}\|u_1 - u_0\| \qquad (n = 1, 2, \ldots). \tag{10.119}$$

Applying the contraction mapping principle, it is easy to derive the following results:

THEOREM 2.1. *Let $K(x, t, u)$ be continuous with respect to the totality of variables $x, t \in \Omega$, $|u| \leqslant \rho$, and let*

$$\left|\frac{\partial K(x, t, u)}{\partial u}\right| \leqslant C \qquad (x, t \in \Omega, |u| \leqslant \rho).$$

Then the equation

$$u(x) = \mu\int K[x, t, u(t)]\,dt \tag{10.120}$$

has a unique continuous solution $u^*(x)$ $(x \in \Omega)$, satisfying the inequality $|u(x)| \leqslant \rho$,

if $|\mu| \cdot C \cdot \operatorname{mes}\Omega < 1$,

and

$$|\mu| \max_{x\in\Omega}\int_{\Omega}\max_{|u|\leqslant\rho}|K(x, t, u)|\,dt \leqslant \rho. \tag{10.121}$$

If $u_0(x)$ is an arbitrary continuous function, satisfying the inequality $|u_0(x)| \leqslant \rho$ $(x \in \Omega)$, then the successive approximations

$$u_n(x) = \mu\int K[x, t, u_{n-1}(t)]\,dt \qquad (n = 1, 2, \ldots) \tag{10.122}$$

converge to $u^*(t)$ uniformly on Ω.

THEOREM 2.2. *Let the operator* (10.115) *act in the space* $L_p(\Omega)$ $(p > 1)$. *Also, let, for all* u_1, u_2 *and* $x, t \in \Omega$, *the following inequality hold*

$$|K(x, t, u_1) - K(x, t, u_2)| \leqslant K_1(x, t)\,|u_1 - u_2|,$$

where $K_1(x, t)$ *is measurable and*

$$\Lambda^k = \int_\Omega \left(\int_\Omega K_1^{p/(p-1)}(x, t)\,dt \right)^{p-1} dx < \infty.$$

Then (10.114) *has a unique solution in* $L_p(\Omega)$ *for* $|\mu| < (1/\Lambda)$ *and* $f \in L_p$.

2.3 *Equations with completely continuous operators*

THE SCHAUDER PRINCIPLE: *If a completely continuous operator* **A** *maps a bounded convex closed set* $M \subset E$ *into itself then* **A** *has at least one fixed point in* M.

When applying the Schauder Principle, it is important to:
i) prove the complete continuity of the operator (see §1); and,
ii) prove the existence of convex sets which are invariant for the given operator.

From the Schauder Principle, it immediately follows that (10.114) has at least one solution in the space E for small μ, if the operator (10.115) is completely continuous in E and $f \in E$. More explicit theorems contain estimates of the admissible values for μ.

THEOREM 2.3 *Let* $K(x, t, u)$ $(x, t \in \Omega, |u| < \infty)$ *be continuous with respect to u and satisfy the inequality*

$$|K(x, t, u)| \leqslant K_1(x, t)(a + b\,|u|^\alpha) \qquad (x, t \in \Omega, |u| < \infty),$$

where $a, b, \alpha > 0$ *and*

$$\int_\Omega \int_\Omega |K_1(x, t)|^{\alpha+1}\,dx\,dt < \infty.$$

Then (10.114) *has a solution* $u_\mu^*(x) \in L_{\alpha+1}$ *for all sufficiently small (in absolute magnitude) values of* μ *and* $f(x) \in L_{\alpha+1}$. *If* $\alpha < 1$, *then solutions exist for all* μ.

THEOREM 2.4. *Let $K(x, t, u)$ be continuous with respect to the totality of variables $x, t \in \Omega, |u| \leqslant a$.*
Then for

$$|\mu| \leqslant \frac{a}{\operatorname{mes} \Omega \max\limits_{\substack{x, t \in \Omega \\ |u| \leqslant a}} |K(x, t, u)|}$$

the integral equation (10.120) *has at least one continuous solution $u_\mu^*(x)$, satisfying the inequality $|u_\mu^*(x)| \leqslant a$.*

THEOREM 2.5. *Let $K(x, t, u)$ be continuous with respect to the totality of variables $x, t \in \Omega, |u| < \infty$. We set*

$$\varphi(r) = \max_{x,\, t \in \Omega,\, u \leqslant r} K\,|(x, t, u)|,$$

$$\Lambda = \sup_{0 < r < \infty} \frac{r}{\varphi(r) \cdot \operatorname{mes} \Omega}.$$

Then (10.120) *has at least one continuous solution $u_\mu^*(x)$ for $\mu < \Lambda$.*

If, in particular,

$$\lim_{r \to \infty} \frac{\varphi(r)}{r} = 0, \tag{10.123}$$

then (10.120) has a solution for any μ. In this case, (10.114) has a solution for any μ and any continuous $f(x)$.

We note in conclusion that the following fixed-point principle is often useful when proving existence theorems.

An operator $A(u, \lambda)(u \in E, 0 \leqslant \lambda \leqslant 1)$ is called *completely continuous*, if it is continuous with respect to the totality of its variables and if every set $\{v \colon v = A(u, \lambda), \|u\| \leqslant r, 0 \leqslant \lambda \leqslant 1\}$ is compact in E.

Assume that, for all solutions $u(\lambda)$ of the equations

$$u = A(u, \lambda) \qquad (0 \leqslant \lambda \leqslant 1)$$

with completely continuous operator $A(u, \lambda)$, the following general *a priori* estimate has been established:

$$\|u(\lambda)\| \leqslant R_0 \qquad (0 \leqslant \lambda \leqslant 1).$$

Also, let the operator $A(u, 1)$ satisfy the condition

$$\|A(u, 1)\| < R_0 \qquad (\|u\| = R_0).$$

Then the equation

$$u = A(u, 0)$$

has at least one solution.

2.4 *The use of upper estimates*

The following principle is a generalization of the Schauder Principle:

Let a completely continuous operator A *satisfy the condition*

$$(Au, u) \leqslant (u, u) \qquad (\|u\| = R). \tag{10.124}$$

Then there exist at least one solution of the equation $u = Au$ *in the ball* $\|u\| \leqslant R$.

We illustrate this principle by establishing an existence theorem for a system of non-linear integral equations

$$u_i(x) = \mu \int_{\Omega} K_i(x, t)\, f_i[t, u_1(t), \ldots, u_n(t)]\, dt \qquad (i = 1, \ldots, n) \tag{10.125}$$

with continuous symmetric positive-definite kernels $K_i(x, t)$ and functions $f_i(t, u_1, \ldots, u_n)$ continuous with respect to all their variables. Let the functions $f_i(t, u_1, \ldots, u_n)$ satisfy the inequalities

$$\sum_{i=1}^{n} u_i f_i(t, u_1, \ldots, u_n) \leqslant a \sum_{i=1}^{n} u_i^2 + b \qquad (i = 1, 2, \ldots, n),$$

where $a < 1/\lambda_0$, and λ_0 is the largest eigenvalue of the kernels $K_i(x, t)$.

Then the system (10.125) has at least one continuous solution.

2.5 *Equations with asymptotically linear operators*

We assume that the non-linear operator (10.115) is completely continuous in a functional space E, and is asymptotically linear (see §1.16). Let its asymptotic derivative $A'(\infty)$ be the integral operator

$$A'(\infty)\, u(x) = \int_{\Omega} B(x, t)\, u(t)\, dt. \tag{10.126}$$

THEOREM 2.6. *If* $1/\mu$ *is not an eigenvalue of the operator* (10.126), *then* (10.114) *has at least one solution in* E.

For example, let $K(x, t, u)$ satisfy the inequality

$$|K(x, t, u) - K_0(x, t)u| \leqslant a + b\,|u|^{1-\varepsilon}. \tag{10.127}$$

Then the equation (10.114) has at least one continuous solution for any continuous $f(x)$ and for μ such that $1/\mu$ is not an eigenvalue of the kernel $K_0(x, t)$.

2.6 *Variational methods*

Let a functional $F(u)$ be defined on L_2 and be *differentiable* at any point $u \in L_2$ in the sense that its increment can be expressed in the form

$$F(u + h) - F(u) = (v, h) + o(\|h\|).$$

The operator Γ, defined by the equality

$$\Gamma u = v,$$

is called the *gradient of the functional* F, $\Gamma = \operatorname{grad} F$. An operator, which is the gradient of a functional, is called a *potential operator*.

It is convenient to prove existence theorems for equations with potential operators using the variational method. This method consists in examining the solutions of

$$\Gamma u = \theta, \tag{10.128}$$

as extreme points of the functional $F(u)$. So every theorem concerning the existence of an extremum of a functional $F(u)$ is, simultaneously, an existence theorem for (10.128).

We state a condition under which the non-linear functional $F(u)$ attains a minimum. Let

$$\lim_{||u|| \to \infty} F(u) = \infty, \tag{10.129}$$

and

$$F(u) = \beta(u, u) + F_1(u) \tag{10.130}$$

where β is a non-negative integer and $F_1(u)$ is a weakly continuous functional (this means, that $F_1(u_n) \to F_1(u^*)$, if u_n converges weakly to u^*); then the minimum value of the functional $F(u)$ is attained at some point.

The main difficulty with this study of integral equations is the reduction of a given integral equation to one with a potential operator.

Consider the Hammerstein equation

$$u(x) = \int_{\Omega} K(x, t)\, f[t, u(t)]\, dt \tag{10.131}$$

with a continuous, positive-definite kernel $K(x, t)$ and a function $f(t, u)$ continuous with respect to all its variables.

Let

$$K(x, t) = \sum_{i=1}^{\infty} \frac{1}{\lambda_i} e_i(x) e_i(t).$$

We define the operator $\boldsymbol{H}$, acting in L_2

$$\boldsymbol{H}u(x) = \sum_{i=1}^{\infty} \frac{1}{\sqrt{\lambda_i}} e_i(x) \int_{\Omega} e_i(t)\, u(t)\, dt.$$

It can be proved that the functional

$$F(u) = \tfrac{1}{2}(u, u) - \int_{\Omega} \left[\int_0^{\boldsymbol{H}u(x)} f(t, u)\, du \right] dt \tag{10.132}$$

can be expressed in the form (10.130). So grad F vanishes at some point, if the condition (10.129) is satisfied. This holds, if $f(t, u)$ satisfies the inequality

$$\int_0^x f(t, u)\, du \leqslant \frac{a}{2} u^2 + b \qquad (t \in \Omega, |u| < \infty), \tag{10.133}$$

where $a < 1/\lambda_0$ with λ_0 the largest eigenvalue of the kernel $K(x, t)$.

Thus, if (10.133) is satisfied, then there exists at least one solution $u^* = u^*(x)$ of the equation

$$\operatorname{grad} F(u) \equiv u - \boldsymbol{H} f(t, \boldsymbol{H}u) = 0 \tag{10.134}$$

in L_2. The continuous function

$$v(x) = \boldsymbol{H}u(x)$$

is a solution of the Hammerstein equation (10.131).

Thus, we have proved

THEOREM 2.7. *If the condition* (10.133) *is satisfied, then* (10.131) *has at least one continuous solution.*

The analysis of (10.131) is more complicated, if the positive-definite, symmetric kernel $K(x, t)$ is not bounded.

THEOREM 2.8. *Let the kernel* $K(x, t)$ *of* (10.131) *satisfy the condition*

$$\int_\Omega \int_\Omega |K(x, t)|^p\, dx\, dt < \infty,$$

where $p \geqslant 2$ *and let the function* $f(t, u)$ *satisfy the inequality* (10.133) *and the condition*

$$|f(t, u)| \leqslant a + b\,|u|^{p-1} \qquad (t \in \Omega, -\infty < u < \infty).$$

Then the equation (10.131) *has at least one solution.*

Variational methods are applicable when the symmetric kernel $K(x, t)$ has a finite number of negative eigenvalues. We denote the smallest absolute magnitude of them by λ_-.

THEOREM 2.9. *Let the symmetric kernel* $K(x, t)$ *be continuous and have a finite number of negative eigenvalues. Let* $f(t, u)$ *satisfy the inequality*

$$-\int_0^u f(t, u)\, du \geqslant au^2 - c \qquad (t \in \Omega,\ -\infty < u < \infty),$$

where $a\,|\lambda_-| > 1$.

Then the equation (10.131) *has at least one continuous solution.*

2.7 *The existence of non-zero solutions*

Most of the theorems formulated above stated that non-linear integral equations have at least one solution. But the uniqueness of a solution is an exceptional fact for many non-linear problems (such as in the loss of stability of elastic systems, and in hydrodynamics). It often happens that one solution is known and it is necessary to find another. In such cases, the theorems formulated above are not applicable.

We assume for simplicity that the equation under consideration, viz. $u = \boldsymbol{A}u$, has a non-zero solution. Some existence conditions of non-zero solutions can be obtained using topological methods, but a discussion of these methods is beyond the scope of this reference book. We limit considerations to the formulation of a simple general theorem.

THEOREM 2.10. *Let* $\boldsymbol{A}$ *be a completely continuous operator and* $\boldsymbol{A}T \subset T$, *where* $T = \{u, \|u\| \leqslant r\}$. *Let* $\boldsymbol{A}\theta = \theta$ *and* $\boldsymbol{A}$ *be differentiable at the zero function* θ. *Finally, assume that unity is not an eigenvalue of the operator* $A'(\theta)$ *and that the sum of the multiplicities of the eigenvalues of* $A'(\theta)$ *which exceed unity is odd.*

Then the operator $\boldsymbol{A}$ *has at least one non-zero fixed point in the ball* T.

It is convenient to use this theorem together with conditions for complete continuity and differentiability of non-linear integral operators (see §1). We now present the simplest corollaries of Theorem 2.10.

THEOREM 2.11. *Let* $K(x, t, u)$ *be a continuous function with respect to the totality of variables and*

$$|K(x, t, u)| \leqslant a + b\,|u|^{1-\varepsilon} \qquad (x, t \in \Omega,\ -\infty < u < \infty),$$

Let $K(x, t, u)$ *be differentiable with respect to* u *for small values of* u. *Assume that unity is not an eigenvalue of the kernel* $K'_u(x, t, 0)$ *and that the sum of the multiplicities of the eigenvalues which exceed unity is odd.*

Then the equation

$$u(x) = \int_{\Omega} K[x, t, u(t)]\, dt$$

has at least one non-zero continuous solution.

2.8 *The existence of a positive solution*

In this subsection, we limit consideration to the integral equation

$$u(x) = \int_{\Omega} K[x, t, u(t)]\, dt, \tag{10.135}$$

the kernel $K(x, t, u)$ of which is continuous with respect to all its variables and satisfies the condition

$$K(x, t, u) \geqslant 0 \qquad (x, t \in \Omega, u \geqslant 0). \tag{10.136}$$

In this case, the integral operator $\boldsymbol{A}$, defined by the r.h.s. of (10.135), transforms the cone K of non-negative functions of the space $C(\Omega)$ into itself and is completely continuous.

General methods of the theory of non-linear operators which leave a cone in a Banach space invariant can be applied to the study of (10.135).

THEOREM 2.12. *Let*

$$K(x, t, u) \leqslant a + G(x, t)u \qquad (x, t \in \Omega, u \geqslant 0), \tag{10.137}$$

and the largest eigenvalue λ *of the non-negative kernel* $G(x, t)$ *be less than unity.*

Then the equation (10.135) *has at least one continuous non-negative solution.*

We now assume that the equation (10.135) has a zero solution. Then the problem of the existence of non-negative solutions, which are not identically equal to zero (see §2.7), arises. In order to prove the existence of such solutions, it is convenient to apply the fixed-point principles for operators which define a contraction and an expansion of a cone. We state here simplified formulations for these principles.

Let an operator A leave a cone K in a Banach space E invariant (see §4 of Chapter V). We say that A is a *contraction of the cone* K, if $u - Au \notin K$ for non-zero $u \in K$ with small norms, and $Au - u \notin K$ for $u \in K$ with large norms. Analogously, A is an *expansion of the cone* K, if $Au - u \notin K$ for non-zero $u \in K$ having small norms and $u - Au \notin K$ for $u \in K$ with large norms.

THEOREM 2.13. *If a completely continuous positive operator A is the contraction or the extension of a cone K, then it has at least one non-zero fixed point in K.*

We state the direct corollaries of this theorem for the equation (10.135).

THEOREM 2.14. *Let $K(x, t, u)$ satisfy the conditions of Theorem* 2.12. *Let*

$$K(x, t, u) \geqslant G_1(x, t)u \qquad (0 \leqslant u \leqslant r),$$

and the continuous kernel $G_1(x, t)$ be positive and its largest eigenvalue exceed unity.

Then (10.135) *has at least one positive solution.*

THEOREM 2.15. *Let*

$$K(x, t, u) \leqslant G_2(x, t)u \qquad (0 \leqslant u \leqslant r),$$

and all the eigenvalues of the kernel $G_2(x, t)$ be less than unity. Let

$$K(x, t, u) \geqslant G_3(x, t)u - a \qquad (x, t \in \Omega, u \geqslant 0),$$

and the kernel $G_3(x, t)$ be positive and its largest eigenvalue exceed unity. Finally, let the kernel $K(x, t, u)$ satisfy the inequality

$$\sup_{x,y,t\in\Omega,u>0} \frac{K(x, t, u)}{K(y, t, u)} < \infty. \qquad (10.138)$$

Then the equation (10.135) *has at least one positive solution.*

We note that the condition (10.138) is satisfied, if $K(x, t, u) = K(x, t)\, f(t, u)$ and the kernel $K(x, t)$ is continuous and positive.

2.9 *Equations with concave non-linearities*

The theory of the so-called *concave* operators is an important part of non-linear analysis.

Applied to integral equations, this theory enables one to obtain simple uniqueness conditions for non-zero, non-negative solutions. We state the basic theorem of the theory of concave operators in one of its simplest forms.

Let a non-linear operator A be positive and monotone in the sense that it leaves a cone K invariant and that $v - u \in K$ implies $Av - Au \in K$.

Let u_0 be some fixed non-zero element from K. The operator A is called u_0-*concave*, if, given any non-zero $u \in K$, there exist positive α and β such that

$$\alpha u_0 \leqslant Au \leqslant \beta u_0,$$

and if, for any non-zero $u \in K$ and any $\gamma \in (0, 1)$, the following inequality holds:

$$A(\gamma u) \geqslant \gamma_1 Au,$$

where $\gamma_1 > \gamma$.

THEOREM 2.16. *A positive monotone u_0-concave operator can have no more than one fixed point in a cone K.*

In the following theorem, we assume that a cone K is *normal* in the sense that $\theta \leqslant u \leqslant v$ implies $\|u\| \leqslant C\|v\|$ where C is a constant. We recall that the cones of non-negative functions in the spaces L_p and C are normal.

THEOREM 2.17. *Let the equation $u = Au$ with a positive monotone u_0-concave operator A have a non-zero solution u^* in K.*

Then the successive approximations

$$u_{n+1} = Au_n \qquad (n = 1, 2, \ldots)$$

tend to u^* for any non-zero initial approximation $u_1 \in K$.

An application of these theorems to non-linear integral equations requires the conditions of u_0-concavity for integral operators.

We again limit consideration to the simplest results.

THEOREM 2.18. *Let $K(x, t, u)$ and its derivative $K'_u(x, t, u)$ be continuous with respect to the totality of variables ($x, t \in \Omega, 0 \leqslant u \leqslant \infty$). Let K and K'_u be non-negative, and $K'_u(x, t, u)$ decrease as u increases. Then the integral operator* (10.115) *is u_0-concave on the cone of non-negative functions, where $u_0(x) \equiv 1$.*

In conclusion, we state a general theorem concerning equations with non-linearities of the type u^α, where $\alpha \in (-1, 1)$.

THEOREM 2.19. *Let $K(x, t)$ be a positive continuous function with respect to the variables $x, t \in \Omega$.*

Let $f(t, u)$ be a continuous function with respect to the variables $t \in \Omega$, $u \geqslant 0$, which is non-negative for $u \geqslant 0$ and strictly positive for $u > 0$ for almost all t. Let at least one of the following two conditions be satisfied:
1) *$f(t, u)$ be non-decreasing with respect to u and $f(t, u)/u^\alpha$ be non-increasing with respect to u for $\alpha \in (0, 1)$;*
2) *$f(t, u)$ be non-increasing for increasing u and $u^{-\alpha}f(t, u)$ be increasing with respect to u for $\alpha \in (0, 1)$.*

Then (10.135) *has a unique positive solution u^*. This solution is the uniform limit of the successive approximations*

$$u_n(x) = \int_\Omega K(x, t) f[t, u_{n-1}(t)]\, dt \qquad (n = 1, 2, \ldots),$$

where $u_0(t)$ is an arbitrary non-zero, non-negative function.

2.10 *Equations with a parameter*

As we pointed out above, equations with a completely continuous operator A of the form

$$\lambda u = A u \tag{10.139}$$

have a solution for large values of λ. In some problems only non-zero solutions are of interest. This is the case, for example, for non-linear integral equations in the theory of vibrations. By analogy with the theory for linear integral equations, non-zero solutions of the non-linear equation (10.139) are called *eigenfunctions* of the non-linear operator A and the corresponding values of λ are called its *eigenvalues*.

An immediate corollary of the Schauder principle is

THEOREM 2.20. *Let $A\theta \neq \theta$ where A is a completely continuous operator. Then A has a continuum of eigenfunctions.*

This theorem can be supplemented with the remark that the condition $A\theta \neq \theta$ implies the existence of eigenfunctions, corresponding to all λ of sufficiently large absolute magnitude.

The problem is much more complicated, if $A\theta = \theta$. Some statements, concerning this case, were given in §§2.7 and 2.8, where we have been dealing with the non-zero solutions of equations. Now, we state existence theorems for eigenfunctions, which correspond to certain unknown eigenvalues λ.

THEOREM 2.21. *Let a completely continuous operator A be differentiable at the zero-function θ and let the linear operator $A'(\theta)$ have at least one real eigenvalue of odd multiplicity.*

Then A has a continuum of eigenfunctions.

This theorem will be examined in detail in the next section.

The application of the last two theorems requires the verification of the complete continuity and differentiability of integral operators at a single point. These conditions were presented in §1.

In order to prove the existence of eigenfunctions of potential operators, we can apply the variational principle, which is based on the fact that *the gradient operator of a functional, which is differentiable in a neighbourhood of the zero-function of a Hilbert space and is weakly continuous, has a continuum of eigenfunctions.* This principle, when applied to the integral operator

$$Au(x) = \int_{\Omega} K(x, t)\, f[t, u(t)]\, dt \tag{10.140}$$

leads to the following results:

THEOREM 2.22. *Let $K(x, t)$ be symmetric, continuous and positive-definite, and let $f(t, u)$ be continuous with respect to all its variables.*

Then the operator (10.140) *has a continuum of eigenfunctions.*

THEOREM 2.23. *Let a symmetric, positive-definite kernel $K(x, t)$ satisfy the inequality*

$$\int_{\Omega}\int_{\Omega} |K(x, t)|^p\, dx\, dt < \infty,$$

where $p \geqslant 2$. *Let a continuous function* $f(t, u)$ *satisfy the inequality*

$$|f(t, u)| \leqslant a + b\,|u|^{p-1}.$$

Then the operator (10.140) *has a continuum of eigenfunctions.*

Variational principles are also applicable to the study of the operators $\boldsymbol{A}$ with kernels, which have eigenvalues of different signs.

THEOREM 2.24. *Let* $K(x, t)$ *be continuous, symmetric and have a finite number of negative eigenvalues. Let* $f(t, u)$ *be continuous and satisfy the inequality*

$$uf(t, u) \geqslant \alpha u^2 - \beta \qquad (t \in \Omega, \quad -\infty < u < \infty),$$

where $\alpha > 0$.

Then the operator (10.140) *has a continuum of eigenfunctions.*

Different theorems can be proved for positive operators. We state the simplest of them for the operator (10.115).

THEOREM 2.25. *Let* $K(x, t, u)$ *be continuous with respect to all its variables and satisfy the inequality*

$$K(x, t, u) \geqslant G(x, t)u \qquad (x, t \in \Omega, u > 0),$$

where $G(x, t)$ *is a positive continuous kernel.*

Then the operator (10.115) *has a continuum of eigenfunctions.*

The assumption of positiveness for $G(x, t)$ can be made a slightly more complicated but less restrictive assumption. For example, for the validity of the last theorem it is sufficient to stipulate that, for any set $\Omega_0 \subset \Omega$, with mes $\Omega_0 > 0$, the following inequality holds

$$\int_{\Omega} G(x, t)\,dt \leqslant \alpha(\Omega_0) \int_{\Omega_0} G(x, t)\,dt \qquad (x \in \Omega).$$

From Theorem 2.16 it follows that not more than one positive eigenfunction corresponds to any eigenvalue of a u_0-concave positive operator. Hence, eigenvalues, corresponding to positive eigenfunctions of u_0-concave positive operators, form an interval.

2.11 *Supplementary notes*

The simplest existence theorems for non-linear integral equations are presented in the monograph of Tricomi [21]. For general theorems see [8]. Topological and variational methods of proof of existence theorems for non-linear integral equations are given in the book of Krasnoselskii [9]. The book of Vainberg [2] is also devoted to variational methods. The study of positive solutions of non-linear equations is presented in the book of Krasnoselskii [10] {see also Krein and Rutman [13]}.

§3 The extension and bifurcation of solutions of non-linear integral equations

3.1 *The formulation of the problem*

In this section, we study the dependence of the solutions of non-linear integral equations on different parameters which are contained in them. In order to illustrate the basic results, we consider

$$u(x) = \int_{\Omega} K[x, t, u(t); \lambda]\, dt, \tag{10.141}$$

where Ω is a bounded closed set of a finite-dimensional space and λ is a scalar or vector parameter.

The equation (10.141), as well as many other equations, can be rewritten in the form

$$f(u, \lambda) = 0, \tag{10.142}$$

where u is an element of a Banach space E, the parameter λ is an element of a Banach space Λ and $f(u, \lambda)$ is an operator with range in a Banach space F.

Let

$$f(u_0, \lambda_0) = 0, \tag{10.143}$$

i.e. u_0 is a solution of (10.142) for $\lambda = \lambda_0$. We study the solutions $u(\lambda)$ of (10.142), which are near to u_0 for values of λ which are near to λ_0. If such solutions $u(\lambda)$ exist, then they are called *implicit functions* (or *abstract implicit functions*) defined by (10.142).

As will be seen below, an implicit function, defined in a neighbourhood of the point λ_0, is single valued, continuous and has 'good' properties in basic non-degenerate situations. But different degenerate cases are also important in applications: it can happen that an implicit function is many-valued (even infinitely many-valued) for all values of λ near to λ_0; an implicit function may not be defined for all values of λ, which are near to λ_0; and so on.

Below, we present basic theorems connected with the existence and properties of implicit functions. The construction of an analytic implicit function is examined.

Most results of this section relate to equations in both real and complex Banach spaces.

3.2 *The basic theorem for implicit functions*

Henceforth in this section, we assume that the operator $f(u, \lambda)$ is defined for $\|u - u_0\|_E \leqslant r_1$, $\|\lambda - \lambda_0\| \leqslant r_2$, where r_1 and r_2 are positive numbers, $f(u, \lambda)$ and $f'_u(u, \lambda)$ are continuous with respect to all the variables. We recall that the derivative $f'_u(u, \lambda)$ is a linear continuous operator, mapping from E into F. Consequently, the continuity of the derivative $f'_u(u, \lambda)$ must be understood in the sense that the operator $f'_u(u, \lambda)$ must be continuous with respect to the norm

$$\lim_{\lambda \to \lambda_1, u \to u_1} \sup_{\|h\|_E \leqslant 1} \|f'_u(u, \lambda)h - f'_u(u_1, \lambda_1)h\|_F = 0.$$

The derivative $f'_u(u_0, \lambda_0)$ plays an important role in the study of implicit functions. The *non-degeneracy* of the implicit function problem implies that the operator $f'_u(u_0, \lambda_0)$ has a continuous inverse

$$\Gamma = [f'_u(u_0, \lambda_0)]^{-1} \tag{10.144}$$

which is defined on the whole of F.

THEOREM 3.1. *Let the condition* (10.143) *be satisfied. Let the operator* (10.144) *exist. Then* (10.142) *defines a single-valued implicit function* $u(\lambda)$ *in a neighbourhood* $\|\lambda - \lambda_0\| \leqslant \delta$ *of the point* λ_0. *This implicit function is continuous.*

Consider as an example (10.141). We assume that the kernel $K(x, t, u; \lambda)$ and its derivative $K'_u(x, t, u; \lambda)$ are continuous with respect to all the variables.

Let $u_0(x)$ be the solution of (10.141) for $\lambda = \lambda_0$:

$$u_0(x) \equiv \int_\Omega K[x, t, u_0(t); \lambda_0]\, dt. \tag{10.145}$$

The equation (10.141) can be written in the form (10.142), where E and F are the space C of continuous functions on Ω and

$$f(u, \lambda) = u(x) - \int_\Omega K[x, t, u_0(t); \lambda]\, dt. \tag{10.146}$$

Obviously,

$$f'_u(u_0, \lambda_0)h = h(x) - \int_\Omega K'_u[x, t, u_0(t); \lambda_0]\, h(t)\, dt.$$

So, all the conditions of Theorem 3.1 are satisfied, if unity is not an eigenvalue of the kernel $K'_u[x, t, u_0(t); \lambda]$. From the fact that unity is not an eigenvalue of the mentioned kernel, it follows that (10.141) has a unique continuous solution $u(x, \lambda)(x \in \Omega)$, which is near to $u_0(x)$ for values of λ which are near to λ_0. The function $u(x, \lambda)$ is continuous with respect to all the variables.

Using more subtle continuity and differentiability conditions for the integral operators (see §1), we can obtain, with the help of Theorem 3.1, more general results about the solutions of (10.141).

We also note that an implicit function can be obtained as the limit of the successive approximations

$$u_{n+1}(\lambda) = u_n(\lambda) - \Gamma f[u_n(\lambda); \lambda] \qquad (n = 0, 1, 2 \ldots) \tag{10.147}$$

with the initial approximation

$$u_0(\lambda) \equiv u_0. \tag{10.148}$$

In a neighbourhood of the point λ_0, the approximations (10.147) converge uniformly with the rate of a geometric progression.

3.3 *Differential properties of an implicit function*

In the present subsection, we used the notion of the partial derivative of an operator which depends on two variables. In this subsection, as well as in some subsequent ones, we use higher order partial derivatives. We do not give the definitions of these derivatives here (see [8], [14]), for they can be introduced in a natural way. We shall assume below that all the partial derivatives under consideration depend continuously, in a neighbourhood of the point $\{u_0, \lambda_0\}$, on all the variables and parameters they contain.

THEOREM 3.2. *Let the conditions of Theorem* 3.1 *be satisfied. Let the operator* $\boldsymbol{f}(u, \lambda)$ *have all partial derivatives up to order k in a neighbourhood of a point* $\{u_0, \lambda_0\}$. *Then it follows from Theorem* 3.1 *that there exists an implicit function* $u(\lambda)$, *which has derivatives up to order k.*

We can calculate the derivatives by means of successive formal differentiation of

$$\boldsymbol{f}[u(\lambda); \lambda] \equiv 0. \tag{10.149}$$

The first differentiation gives

$$\boldsymbol{f}'_u[u(\lambda), \lambda]\, u'(\lambda) + \boldsymbol{f}'_\lambda[u(\lambda), \lambda] \equiv 0,$$

and hence

$$u'(\lambda) = -\Gamma(\lambda)\, \boldsymbol{f}'_\lambda[u(\lambda), \lambda], \tag{10.150}$$

where

$$\Gamma(\lambda) = \{\boldsymbol{f}'_u[u(\lambda), \lambda]\}^{-1}. \tag{10.151}$$

The existence of the operator (10.151) for values of λ which are near to λ_0 follows from the existence of the operator (10.144) and from the fact that $\boldsymbol{f}'_u[u(\lambda); \lambda]$ depends continuously on λ.

In an analogous way, we can obtain the formulas

$$\begin{aligned} u''(\lambda) = & -\Gamma(\lambda)\{\boldsymbol{f}''_{u^2}[u(\lambda), \lambda][u'(\lambda), u'(\lambda)] \\ & + 2\boldsymbol{f}''_{u\lambda}[u(\lambda), \lambda]\, u'(\lambda) + \boldsymbol{f}''_{\lambda^2}[u(\lambda), \lambda]\}, \end{aligned} \tag{10.152}$$

$$\begin{aligned} u'''(\lambda) = & -\Gamma(\lambda)\{\boldsymbol{f}'''_{u^3}[u(\lambda), \lambda][u'(\lambda), u'(\lambda), u'(\lambda)] \\ & + 3\boldsymbol{f}'''_{u^2\lambda}[u(\lambda), \lambda][u'(\lambda), u'(\lambda)] + 3\boldsymbol{f}'''_{u\lambda^2}[u(\lambda), \lambda]\, u'(\lambda) \\ & + \boldsymbol{f}'''_{\lambda^3}[u(\lambda), \lambda] + 3\boldsymbol{f}''_{u^2}[u(\lambda), \lambda][u'(\lambda), u''(\lambda)] \\ & + \boldsymbol{f}''_{u\lambda}[u(\lambda), \lambda]\, u''(\lambda)\}. \end{aligned} \tag{10.153}$$

The method often used for the calculation of derivatives (particularly in the case when derivatives at a fixed point are sought) is the method of undetermined coefficients. According to this method, we seek an implicit function $u(\lambda)$ in the form

$$u(\lambda) = u_0 + \boldsymbol{D}_1(\lambda - \lambda_0) + \ldots + \frac{1}{k}\boldsymbol{D}_k(\lambda - \lambda_0) + \omega(\lambda - \lambda_0), \quad (10.154)$$

where

$$\boldsymbol{D}_j(\mu) = \bar{\boldsymbol{D}}_j(\mu_1, \ldots, \mu_j) \quad (10.155)$$

and $\bar{\boldsymbol{D}}_j(\mu_1, \ldots, \mu_j)$ is an operator which is continuous with respect to all its variables and is linear with respect to any one of its variables, and

$$\omega(\lambda - \lambda_0) = o(\|\lambda - \lambda_0\|^k). \quad (10.156)$$

Substituting (10.154) into (10.149) and setting equal to zero all the members of order $(\lambda - \lambda_0)^j$ or less, we obtain k equations from which we determine the operators

$$\boldsymbol{D}_j(\lambda - \lambda_0) = u^j(\lambda_0)(\lambda - \lambda_0). \quad (10.157)$$

We note that when making actual calculations, it is convenient to substitute (10.154) not into (10.149) but into the equivalent identity

$$u - u_0 = \Gamma\{f(u; \lambda) - f'_u(u_0; \lambda)(u - u_0)\}. \quad (10.158)$$

Then the equations for the determination of the operators $\boldsymbol{D}_j$ reduce to a sequence of recurrence formulas. Besides, when calculating the derivatives up to order k, we can set $\omega(\lambda - \lambda_0) \equiv 0$ in (10.154).

As an example, consider again (10.141). For simplicity, we assume that λ is a scalar parameter. We also assume that the identity (10.145) holds and that unity is not an eigenvalue of the kernel $K'_u[x, t, u_0(t); \lambda_0]$. Let the function $K(x, t, u; \lambda)$ be k times continuously differentiable with respect to the variables u, λ. Then the following equality hold:

$$K[x, t, u_0(x) + v; \lambda_0 + \nu] - K[x, t, u_0(t); \lambda_0] = \sum_{m+l=1}^{k} K_{m,l}(x, t)\nu^m v^l + \omega(x, t, v; \nu),$$

where ω contains the small terms which can be neglected, and the kernels

$K_{m,l}(x, t)(x, t \in \Omega)$ are continuous with respect to all the variables. The equation (10.41) can be replaced by the equivalent formulation

$$v(x) = \sum_{m+l=1}^{k} (\lambda - \lambda_0)^m \int_\Omega K_{m,l}(x, t)[v(t)]^l \, dt + \omega_1, \qquad (10.159)$$

where ω_1 contains all the small terms which can be neglected.

The solution of

$$v(x) = \int_\Omega K'_u[x, t, u_0(t); \lambda_0] \, v(t) \, dt + h(x)$$

is determined by

$$v(x) = h(x) + \int_\Omega R(x, t) \, h(t) \, dt,$$

where $R(x, t)$ is a continuous kernel. Hence, (10.159) can be rewritten in the following equivalent form (the symbol $\sum_{m+l=1}^{k}{}'$ means that the sum does not contain the term corresponding to the indices $m = 0, l = 1$)

$$v(x) = \sum_{m+l=1}^{k}{}' (\lambda - \lambda_0)^m \int_\Omega \Pi_{m,l}(x, t)[v(t)]^l \, dt + \omega_2, \qquad (10.160)$$

where

$$\Pi_{m,l}(x, t) = K_{m,l}(x, t) + \int_\Omega R(x, \tau) \, K_{m,l}(\tau, t) \, d\tau, \qquad (10.161)$$

and ω_2 again contains the small terms which can be neglected. The equation (10.160) relates to (10.141) in the same way as (10.153) relates to (10.142).

We seek the solution of (10.141) (or, equivalently, (10.160)) in the form $u(x; \lambda) = u_0(x) + v(x; \lambda)$, where

$$v(x, \lambda) = (\lambda - \lambda_0) \, v_1(x) + \dots + (\lambda - \lambda_0)^k \, v_k(x) + o(|\lambda - \lambda_0|^k). \qquad (10.162)$$

We substitute the r.h.s. of (10.162) into (10.160) and set the coefficients of the same degrees of $\lambda - \lambda_0$ equal.

We obtain the recurrence relations

$$v_1(x) = \int_\Omega \Pi_{10}(x, t)\, dt, \tag{10.163}$$

$$v_2(x) = \int_\Omega \{\Pi_{02}(x, t)\, v_1^2(t) + \Pi_{11}(x, t)\, v_1(t) + \Pi_{20}(x, t)\}\, dt, \tag{10.164}$$

$$\begin{aligned} v_3(x) = \int_\Omega \{&2\Pi_{02}(x, t)\, v_1(t)\, v_2(t) + \Pi_{11}(x, t)\, v_2(t) \\ &+ \Pi_{03}(x, t)\, v_1^3(t) + \Pi_{12}(x, t)\, v_1^2(t) \\ &+ \Pi_{21}(x, t)\, v_1(t) + \Pi_{30}(x, t)\}\, dt, \end{aligned} \tag{10.165}$$

$$\begin{aligned} v_4(x) = \int_\Omega \{&\Pi_{02}(x, t)[v_2^2(t) + 2v_1(t)\, v_3(t)] + \Pi_{11}(x, t)\, v_3(t) \\ &+ 3\Pi_{03}(x, t)\, v_1^2(t)\, v_2(t) + 2\Pi_{12}(x, t)\, v_1(t)\, v_2(t) \\ &+ \Pi_{21}(x, t)\, v_2(t) + \Pi_{04}(x, t)\, v_1^4(t) + \Pi_{13}(x, t)\, v_1^3(t) \\ &+ \Pi_{22}(x, t)\, v_1^2(t) + \Pi_{31}(x, t)\, v_1(t) + \Pi_{40}(x, t)\}\, dt, \end{aligned} \tag{10.166}$$

. . . .

The complexity of the recurrence formulas increases rapidly.

3.4 *The analyticity of solutions*

We assume now that an operator $f(u; \lambda)$ is analytic with respect to all the variables. Then, under the conditions of Theorem 3.1, an implicit function $u(\lambda)$ is also analytic in a neighbourhood of the point λ_0. The determination of the coefficients in the decomposition of an implicit function into a series is equivalent to the determination of its successive derivatives. So, we can apply the method of undetermined coefficients of the previous subsection in order to reduce an analytic implicit function to its series representation.

Estimates of the radii ρ of the neighbourhoods $\|\lambda - \lambda_0\| \leqslant \rho$, in which the power series representation of an implicit function converges, play a basic role in some problems. For these estimates, see the papers [19], [20]. Probably, the most subtle estimates are due to Gel'man [3].

3.5 *The bifurcation equation*

We continue the study of the implicit functions defined by

$$f(u; \lambda) = 0. \tag{10.167}$$

As above, we assume that

$$f(u_0, \lambda_0) = 0. \tag{10.168}$$

In this and subsequent subsections, we consider *the degenerate* case, when the operator $f'_u(u_0, \lambda_0)$ has no continuous inverse.

The study of implicit functions in the degenerate cases is essentially a more complex problem. The general methods are due to Liapunov [15], Schmidt [22], Nekrasov [20], Nazarov [9] and others.

We limit consideration to (10.167) in the particular form

$$u = A(u; \lambda), \tag{10.169}$$

where the operator A is *completely continuous* in the sense that it is continuous with respect to all the variables in some neighbourhood of the point $\{u_0, \lambda_0\}$, and that its range is compact in this neighbourhood. We assume here that the domain and range of A belong to one and the same space E.

We can rewrite (10.168) as

$$u_0 = A(u_0; \lambda_0). \tag{10.170}$$

From the general theory of non-linear operators (see [9]), it follows that the derivative $A'_u(u_0, \lambda_0)$ of a non-linear completely continuous operator is a linear completely continuous operator. So the degeneracy of an implicit function problem, which is defined by (10.169), is equivalent to the condition that unity be an eigenvalue of the operator $A'_u(u_0, \lambda_0)$.

Every non-zero eigenvalue of a linear completely continuous operator has finite order. We denote by n the order of the eigenvalue 1 of the operator $B = A'_u(u_0, \lambda_0)$. Let $e_1, \ldots, e_n$ be a basis of the subspace of the corresponding eigenvectors. From the general theory of linear completely continuous operators, it follows that unity is also an eigenvalue of the same order n of the conjugate operator B^*. We denote by $f_1, \ldots, f_n$ the basis of the subspace of the corresponding eigenfunctions.

We choose a collection of functionals $g_1, \ldots, g_n$ in E^* such that

$$g_i(e_j) = \delta_{ij}, \tag{10.171}$$

and in E the set of elements $h_1, h_2, \ldots, h_n$ such that

$$f_i(h_j) = \delta_{ij}. \tag{10.172}$$

We introduce into the examination the operator

$$\boldsymbol{D}u = \sum_{i=1}^{n} g_i(u)h_i. \tag{10.173}$$

It is easy to see that (10.169) is equivalent to the system consisting of the vector equation

$$u = \boldsymbol{A}(u, \lambda) + \boldsymbol{D}(u - u_0) - \sum_{j=1}^{n} \xi_j h_j \tag{10.174}$$

and the n scalar equations

$$\xi_j = g_j(u - u_0) \qquad (j = 1, \ldots, n). \tag{10.175}$$

Here, $\xi_1, \ldots, \xi_n$ are additional scalar variables.

As we are interested in a solution u of (10.169), which is near to u_0, we seek only the solutions (10.174)–(10.175) which correspond to small $\xi_1, \ldots, \xi_n$.

We consider (10.174) in detail. This equation has the solution $u = u_0$ for $\lambda = \lambda_0$, $\xi_1 = \ldots = \xi_n = 0$. We examine the set $\mu = \{\lambda; \xi_1, \ldots, \xi_n\}$ as a vector parameter. Then (10.174) can be rewritten as

$$u = \boldsymbol{T}(u; \mu), \tag{10.176}$$

where $\boldsymbol{T}$ is a completely continuous operator. It is easy to see that

$$\boldsymbol{T}'_u(u_0; \mu_0) = \boldsymbol{A}'_u(u_0; \lambda_0) + \boldsymbol{D}, \tag{10.177}$$

where $\mu_0 = \{\lambda_0; 0, \ldots, 0\}$. A simple check shows that unity is not an eigenvalue of $\boldsymbol{T}'_u(u_0; \mu_0)$. Therefore, we can apply Theorem 3.1 to (10.176). In other words, we can determine a solution

$$u = u(\lambda; \xi_1, \ldots, \xi_n) \qquad (\|\lambda - \lambda_0\| \leqslant \rho, |\xi_j| \leqslant \rho) \tag{10.178}$$

from (10.174). So the system (10.174)–(10.175) can be replaced by (10.178) and by the n scalar equations

$$\xi_j = g_j[u(\lambda; \xi_1, \ldots, \xi_n) - u_0] \qquad (j = 1, \ldots, n). \tag{10.179}$$

The system (10.179) is called *a Liapunov-Schmidt bifurcation equation*. The determination of an implicit function in degenerate cases is equivalent to the determination of small solutions $\xi_1(\lambda), \ldots, \xi_n(\lambda)$ of the system (10.179) followed by their substitution in (10.178).

We now make some observations of a negative nature.

The system (10.179) has the obvious zero solution $\xi_1 = \xi_2 = \ldots = \xi_n = 0$ for $\lambda = \lambda_0$. It would be desirable to apply the fundamental Theorem 3.1 for this case. Unfortunately, this is impossible, for in this case the linearized system is the system of identities $0 = 0$. Consequently, the system (10.179) requires special study.

The second observation concerns the fact that a solution of (10.178) in explicit form can be found only in quite exceptional cases. Thus, as a rule, the equations of (10.179) cannot be written in explicit form.

Nevertheless, these disappointing observations do not prevent a successful study of the bifurcation equation. The facts are that the general methods for the study of non-linear systems of equations {the graphical method of Newton; see, for example, 'The Reference Book. Mathematical Analysis'. II, Ch. III, §2; and the elimination method; see, for example, Lefschetz' 'Geometric Theory of Differential Equations' and others} require in the general case either approximate formulas for the functions, which are contained in the equations, or knowledge about the first coefficients of the Taylor expansions for these functions. In order to give approximate expressions for the r.h.s. of (10.179), we can use approximations for the functions (10.178), obtained by means of the method of successive approximations (10.147). For the decomposition of the r.h.s. of (10.179) with respect to powers of $\lambda - \lambda_0, \xi_1, \ldots, \xi_n$, it is sufficient to find the decomposition of the functions (10.178) with respect to the same variables. Here, we can use the method of undetermined coefficients described in §3.3.

3.6 *The Nekrasov-Nazarov method*

As was shown in the preceding subsection, the study of the bifurcation equation (10.179) is a difficult and cumbersome problem. The basic problem is: to construct directly the solutions of (10.167), or, in particular (10.169), which are near to u_0. A method for this has been proposed by Nekrasov and Nazarov [19], [20].

We continue the study of (10.169). Let λ be a scalar parameter. An operator $A(u; \lambda)$ is considered to be analytic at a point $\{u_0; \lambda_0\}$ in the sense that

$$A(u; \lambda) - A(u_0; \lambda_0) = \sum_{m+l=1}^{\infty} (\lambda - \lambda_0)^m B_{ml}(u - u_0), \tag{10.180}$$

where

$$B_{ml}(u) = \bar{B}_{ml}(u_1, \ldots, u_l), \tag{10.181}$$

and $\bar{B}_{ml}(u_1, \ldots, u_l)$ is an operator, which is continuous with respect to all the variables and linear with respect to any one of them. Let the series (10.180) converge uniformly in a neighbourhood of the point $\{u_0, \lambda_0\}$. Under these assumptions, the r.h.s. of the bifurcation equation is an analytic function with respect to the variables $\lambda - \lambda_0, \xi_1, \ldots, \xi_n$ (both for complex and real valued spaces).

In the general case, the solutions $\xi_1(\lambda), \ldots, \xi_n(\lambda)$ of (10.179) can be expressed (see, for example, [1]) in the form of a power series with respect to integer or fractional powers of $\lambda - \lambda_0$. Let

$$\xi_j(\lambda) = \sum_{k=1}^{\infty} \eta_{jk}\mu^k \qquad (j = 1, 2, \ldots, n), \tag{10.182}$$

where

$$\mu = (\lambda - \lambda_0)^{1/p}, \tag{10.183}$$

and p be an integer. Then the solution $u(\lambda)$ of (10.169) can be expressed as the series

$$u(\lambda) = u_0 + \sum_{k=1}^{\infty} u_k \mu^k, \tag{10.184}$$

which converges uniformly for sufficiently small values of μ.

According to the Nekrasov-Nazarov Method, we use the method of undetermined coefficients for the determination of the coefficients u_k. We substitute the series (10.184) into (10.169) and decompose its r.h.s. into a power series with respect to μ. Setting the coefficients of the same degrees of μ on the r.h.s. and l.h.s. equal, we obtain an infinite system of equations which the coefficients u_k satisfy. A simple calculation shows that this system has the form

$$\begin{aligned} u_1 &= A'_u(u_0; \lambda_0)u_1 + h_0, \\ u_2 &= A'_u(u_0; \lambda_0)u_2 + h_1(u_1), \\ &\ldots\ldots\ldots\ldots\ldots \\ u_k &= A'_u(u_0; \lambda_0)u_k + h_{k-1}(u_1, \ldots, u_{k-1}), \\ &\ldots\ldots\ldots\ldots\ldots \end{aligned} \tag{10.185}$$

It is not known *a priori* what p must be taken, when we decompose $\mu(\lambda)$ of (10.169) into a power series with respect to fractional ($1/p$) or integer powers of $\lambda - \lambda_0$. It seems that the simplest way to find p is to use a trial-and-error method. Obviously, the systems (10.185) will be different for different p.

Suppose that we managed to find a solution $\{u_1, u_2, \ldots\}$ of the system (10.185). It does not follow that the series (10.184) with the calculated coefficients is a solution of (10.169). What is more, it is not clear, whether the series (10.184) converges for a non-zero μ. For these reasons, the series (10.184) is called *the formal solution* of (10.169).

It is easy to prove that a formal solution becomes a real solution, if (10.184) converges for small values of μ. In order to prove the convergence and, simultaneously, to find the range of convergence, the method of majorant series is used. We restrict attention to the papers of Nekrasov and Nazarov [19], [20], in which the analysis of integral equations is given, and to the later papers [4, 5].

Results of another type, concerning the convergence of formal solutions, are known (see [18]). We state one of them: *if u_0 is an isolated solution of* (10.169) *for $\lambda = \lambda_0$, then all formal solutions* (10.184) *are proper solutions.*

We now pass to the analysis of the system (10.185).

The first equation of this system is linear. Since unity is an eigenvalue of the linear compact operator $A'_u(u_0, \lambda_0)$, this equation does not have a solution for all h_0. If this equation does not have a solution, then the number p cannot be chosen correctly. We assume that the first equation has a solution. Then its solution has the form

$$u_1 = u_1^* + c_1^{(1)}e_1 + \ldots + c_n^{(1)}e_n, \tag{10.186}$$

where $e_1, \ldots, e_n$ is the basis for the eigenvectors of $A'_u(u_0, \lambda_0)$, which correspond to the eigenvalue 1, and $c_1^{(1)}, \ldots, c_n^{(1)}$ are arbitrary constants to be determined.

We now pass to the second equation of the system (10.185). Again this is a linear equation with $h_1(u_1)$ depending on $c_1^{(1)}, \ldots, c_n^{(1)}$. The solvability conditions for this equation yield a system of n scalar equations, the constants $c_1^{(1)}, \ldots, c_n^{(1)}$ of which can be determined. Thus, the solvability conditions for the second equation enables one to determine u_1.

We can determine the solution u_2 from the second equation with an accuracy up to the arbitrary constants $c_1^{(2)}, \ldots, c_n^{(2)}$. The values of these constants can be determined from the solvability conditions for the third equation.

This process continues inductively. In most cases, it enables one to determine successively all the coefficients u_k. As a rule, starting with any k, the equations for the determination of the constants $c_1^{(k)}, \ldots, c_n^{(k)}$, which are contained in the expression for u_k, can be derived from a system of algebraic equations. *In all such cases, formal solutions are proper solutions.*

Thus, the method of undetermined coefficients consists of the solution of linear integral equations and the successive solution of systems of scalar equations.

We also note that the determination of arbitrary constants in expressions for u_k can involve the solvability conditions for several successive equations of the system (10.185).

3.7 *The bifurcation points*

The study of the degenerate case can be simplified, if we know one solution of the given equation for all values of its parameter and not for one fixed value.

We continue the study of

$$u = \boldsymbol{A}(u, \lambda), \tag{10.187}$$

where λ is (for simplicity) a scalar parameter. We assume that we know a solution $u(\lambda)$ and seek the solutions which differ from $u(\lambda)$ in the form

$$u = u(\lambda) + v. \tag{10.188}$$

Then v will be determined from

$$v = \boldsymbol{T}(v, \lambda), \tag{10.189}$$

where

$$\boldsymbol{T}(v, \lambda) = \boldsymbol{A}[u(\lambda) + v; \lambda] - u(\lambda). \tag{10.190}$$

The equation (10.189) has a non-zero solution for all λ.

A number λ_0 is called a *bifurcation point* of the parameter λ, if for any $\varepsilon > 0$, there corresponds a $\lambda \in (\lambda_0 - \varepsilon, \lambda_0 + \varepsilon)$, for which the equation (10.189) has a non-zero solution $v(\lambda)$, satisfying the condition $\|v(\lambda)\| < \varepsilon$. We stress that the definition of bifurcation point tells nothing about the number of values λ for which (10.189) has small non-zero solutions. For example, every eigenvalue λ_0 of the equation $v = (1/\lambda)\boldsymbol{T}v$ with a linear operator $\boldsymbol{T}$ will be a bifurcation point. If λ_0 is an isolated eigenvalue, then for values of λ which are near to λ_0 and differ from λ_0, this equation has no non-zero solutions. In other problems, non-zero solutions exist only for values of λ which differ from λ_0.

Henceforth, in this subsection, we assume that the operators $\boldsymbol{A}$ and $\boldsymbol{T}$ are completely continuous and sufficiently smooth.

THEOREM 3.3. *Let λ_0 be a bifurcation point. Then unity is an eigenvalue of the linear operator $\boldsymbol{T}'_v(0, \lambda_0)$.*

Certainly, this theorem is only a necessary condition for λ_0 to be a bifurcation point. Theorem 3.3 enables one to select values of λ_0, which can be 'regarded as possible bifurcation points'. As a rule, in applications, unity is an eigenvalue of $\boldsymbol{T}'_v(0; \lambda)$ only for some isolated values of λ. If we managed to find these values of the parameter λ, then all of them must be subjected to special additional analysis.

As an example, consider (10.141) in the special form

$$v(x) = \lambda \int_{\Omega} K[x, t, v(t)]\, dt. \tag{10.191}$$

We assume that $K(x, t, 0) \equiv 0$: Then (10.191) has a non-zero solution for all values of λ. We also assume that the kernel $K(x, t, u)$ and its derivative $K'_u(x, t, u)$ are continuous with respect to the totality of variables $x, t \in \Omega$, $|u| < \rho$. Then the integral operator (10.191) can be considered as the operator equation (10.189) with a completely continuous operator

$$\boldsymbol{T}(v, \lambda) = \lambda \int_{\Omega} K[x, t, v(t)]\, dt \tag{10.192}$$

in the space C of continuous functions on Ω. Obviously,

$$\boldsymbol{T}'_v(0; \lambda)h = \lambda \int_{\Omega} K'_u(x, t, 0)\, h(t)\, dt, \tag{10.193}$$

and unity is an eigenvalue for not more than a countable number of values of the parameter λ, when $1/\lambda$ coincides with one of the eigenvalues of the kernel $K'_u(x, t, 0)$.

The following reasoning only applies to real valued spaces.

Let $\lambda = 1$ be an eigenvalue of the operator $\boldsymbol{T}'_v(0, \lambda_0)$ and let λ_0 be an isolated value of the parameter, which is thought to be a bifurcation point. Simple reasoning shows that values of λ, which are near to λ_0 and are greater than λ_0, are the same as $(-1)^{\beta(\lambda)}$, where $\beta(\lambda)$ is the sum of the multiplicities of the real eigenvalues of the operator $\boldsymbol{T}'_v(0, \lambda)$, which exceed unity. We denote $(-1)^{\beta(\lambda)}$ by $(-1)^{\beta(\lambda_0+0)}$. In an analogous way, we define the number $(-1)^{\beta(\lambda_0-0)}$.

THEOREM 3.4. *If*

$$(-1)^{\beta(\lambda_0-0)} \neq (-1)^{\beta(\lambda_0+0)}, \tag{10.194}$$

then λ_0 is a bifurcation point.

We stress that Theorem 3.4 uses only the properties of the equation which is linearized at zero.

THEOREM 3.5. *Let the derivative $\boldsymbol{T}'_v(0, \lambda)$ have the form*

$$\boldsymbol{T}'_v(0, \lambda) = \lambda \boldsymbol{B}, \tag{10.195}$$

where $\boldsymbol{B}$ is a linear compact operator. Let μ_0 be a non-zero eigenvalue of $\boldsymbol{B}$ of odd multiplicity. Then

$$\lambda_0 = 1/\mu_0 \tag{10.196}$$

is a bifurcation point.

Simple examples show that (10.196) can be a non-bifurcation point, if the eigenvalue μ_0 has an even multiplicity.

Theorem 3.5 can be applied directly to the study of (10.191). It follows from it that (10.196) is a bifurcation point for small non-zero solutions of (10.191), if μ_0 is an eigenvalue of the kernel $K'_u(x, t, 0)$ with an odd multiplicity.

As a second example, consider the integral equation

$$v(x) = \lambda\rho(x)\int_0^1 G(x, t)\, v(t)\, dt \left\{1 - \left[\int_0^1 G'_x(x, t)\, v(t)\, dt\right]^2\right\}^{1/2}, \tag{10.197}$$

where $G(x, t)$ is the Green function for the differential operator $-y''$ under the boundary conditions $y(0) = y(1) = 0$ (see §1, Chapter IV) and $\rho(x)$ is a positive continuous function. The solutions of this equation determine the loss of stability of a compressed rod, which has a rigidity $\rho(x)$ where the parameter λ is proportional to the load. The zero solution of (10.197) defines the rectilinear form of the rod. Bifurcation points are the critical loads. The equation (10.197) can be considered as an operator equation in the space C of functions which are continuous on $[0, 1]$. From Theorem 3.5, it follows that bifurcation points, i.e., the critical points of the parameter λ, coincide with the values of λ for which the linearized equation

$$v(x) = \lambda\rho(x)\int_0^1 G(x, t)\, v(t)\, dt \qquad (10.198)$$

has a non-zero solution, since all eigenvalues of the kernel $\rho(x)\, G(x, t)$ are simple. The formulated result is the basis of the well-known Euler method for the determination of critical loads in the problem of the loss of stability of a compressed rod.

3.8 *Supplementary notes about bifurcation points*

As we established in the previous subsection, the study of an isolated value λ_0 of the parameter λ, which is thought to be a bifurcation point, is a difficult problem, if the condition (10.194) is not satisfied. Then, we have to use either general methods (see §3.5–3.6) for the construction of implicit functions for the degenerate case, or some special tools.

One of the general theorems, concerning bifurcation points, is based on the notion of the index γ of a zero point of the completely continuous vector field $v - \boldsymbol{A}(v; \lambda_0)$. It turns out that *if among the three numbers*

$$(-1)^{\beta(\lambda_0 - 0)}, \qquad \gamma, \qquad (-1)^{\beta(\lambda_0 + 0)} \qquad (10.199)$$

at least two of them are different, then λ_0 is a bifurcation point. This statement certainly contains Theorem 3.4 as a special case. There exist effective methods for the calculation of the index γ (see, for example, [7], [9], [16]), but their description exceeds the limits of this reference book.

We recall that an operator $\boldsymbol{T}$ is called a *potential*, if it is the gradient of a functional. We assume now that a completely continuous operator $\boldsymbol{T}$ is the gradient of a uniformly differentiable and weakly continuous functional, acting in a real valued Hilbert space. As Lusternik has shown, the derivatives of the operator $\boldsymbol{T}$ are self-adjoint operators.

THEOREM 3.6. *Let $T(0) = 0$. Let $T'(0)$ be positive-definite. Then any non-zero eigenvalue λ_0 of the operator $T'(0)$ is a bifurcation point of the equation $v = (1/\lambda)T(v)$.*

We stress that the multiplicity of the eigenvalues has no importance in this theorem. The proof of Theorem 3.6 is based on topological tools.

As an example, consider the integral equation

$$v(x) = \frac{1}{\lambda} \int_{\Omega} K(x, t) f[t, v(t)] \, dt \tag{10.200}$$

with a continuous positive-definite kernel $K(x, t)$ and with a r.h.s. $f(t, u)$ which is continuous along with its derivative $f'_u(t, u)$. Suppose that $f(t, 0) \equiv 0$. It follows from Theorem 3.3 and 3.6 that the bifurcation values associated with small solutions of (10.200) coincide with the eigenvalues of the kernel $K(x, t) f'_u(t, 0)$.

More subtle applications of the Theorems 3.3–3.6 to non-linear integral equations can be obtained, if we apply stronger conditions of differentiability, potentiality and others to the integral operators acting in different functional spaces.

We cannot present a more complete analysis of bifurcation points because of lack of space, and therefore refer the reader to the monographs [9] and [10].

We conclude with a note about the case when (10.181) and (10.183) are considered in a complex valued space, λ is a complex valued parameter and the operator A is analytic. Then any isolated value λ_0, which is thought to be a bifurcation point, is actually a bifurcation point (see [17]).

REFERENCES FOR CHAPTERS I–III

[1] Akhiezer N. I., Shcherbina V. A., On the inversion of some singular integrals, *Zap. matem. otd. phys.-matem. f-ta khGU and Khark, matem. o-va 25* (1957) 191–198.

[2] Gol'dman M. A., Krachkovskiy S. N., Some properties of compact operators in a Hilbert space, *Izv. AN Latv. SSR 10 (39)* (1950) 93–106.

[3] Gol'dman M. A., Krachkovskiy S. N., On the principal part of the compact operator, *Dokl. Akad. Nauk SSSR 70* (1950) 945–948.

[4] Gol'dman M. A., Krachkovskiy S. N., On zero elements of the linear operators in their Fredholm range, *Dokl. Akad. Nauk SSSR 86* (1952) 15–17.

[5] Gokhberg I. Ts., Markus A. S., On one characteristic property of the kernel of the linear operator, *Dokl. Akad. Nauk SSSR 105* (1955) 893–896.

[6] Gokhberg I. Ts., Markus A. S., On some stable properties of normally solvable operators, *Matem. Sb. 40 (82)* (1956) 453–466.

[7] Gokhberg I. Ts., Krein M. G., Principal aspects concerning defect numbers, root numbers and indices of linear operators, *UMN XII, 2 (74)* (1957) 43–118.

[8] Goursat E., *Course on Mathematical Analysis*, vol. III., part II. GTTI, 1934. Translated from French: (Cours d'Analyse Mathematique, Tome III, Pt. II, Gauthier-Villars, Paris, 1923).

[9] Krachkovskiy S. N., On the kernel of smallest norm with right principal functions of a given square summable kernel, *Izv. Akad. Nauk Latv. SSR 3* (1947) 75–78.

[10] Krachkovskiy S. N., On the Lalesko condition, *Izv. Akad. Nauk. Latv. SSR 4* (1947) 101–108.

[11] Krachkovskiy S. N., Canonical representation of zero-elements of the linear operator in its Fredholm domain, *Dokl. Akad. Nauk SSSR 88* (1953) 201–204.

[12] Krachkovskiy S. N., On the properties of linear operators connected with their generalized Fredholm domain, *Dokl. Akad. Nauk SSSR 91* (1953) 1011–1013.

[13] Krachkovskiy S. N., On the extended domain of singularity of the operator $T_\lambda = I - \lambda A$, *Dokl. Akad. Nauk SSSR 95* (1954) 1101–1104.

[14] Lebedev N. N., Skal'skaya I. P., Uflyand T. S., *Collected Problems In Mathematical Physics*, Gostekhizdat, 1955.

[15] Lovitt W. V., *Linear integral equations*, Translated from English. Gostekhizdat, 1957. (McGraw-Hill Book Co. Inc., New York and London, 1924).

[16] Marcus A. S., On a theorem of F. Riesz, *Uch. Zap. Khishinevsk. un-ta 17* (1955) 73–75.

[17] Marcus A. S., On a property of the kernel of a linear operator, *Uch. Zap. Khishinevsk. un-ta 19* (1957) 25–27.

[18] Mikhlin S. G., Certain elementary boundary problems involving the wave equation, *Trud. Seism. Inst. Akad. Nauk SSSR 101* (1940).

[19] MIKHLIN S. G., On the convergence of Fredholm series, *Dokl. Akad. Nauk SSSR 42* (1944) 387–390.

[20] MIKHLIN S. G., *Lectures On Linear Integral Equations*, Physmatgiz, 1959. English translation by Pergamon Press, New York, 2nd edition, 1964.

[21] MIKHLIN S. G., SMOLITSKIY KH. T., *Approximate Methods For The Solution Of Differential and Integral Equations* (Moscow, Nauka, 1965), English translation by American Elsevier, New York, 1967.

[22] PETROVSKY I. G., *Lectures On The Theory Of Integral Equations*, GTTI, 2nd edition, 1951.

[23] PRIVALOV I. I., *Integral Equations*, Ot. Nauch. Tehn. Inf., 1935.

[24] RIESZ F., On linear functional equations (translated from German), *Usp. Mat. Nauk 1* (1936) 175–199.

[25] SMIRNOV V. I., *A Course In Higher Mathematics*, vol. IV (Moscow, 1951), English translation by Pergamon Press, Oxford, 1964.

[26] SNEDDON I., *The Fourier Transform* (translated from English), Inostr. Liter., 1955.

[27] TITCHMARSH E. C., *Introduction To The Theory Of Fourier Integrals* (translated from English), Gostekhizdat, 1948.

[28] TRICOMI F. G., *Integral Equations*, Inostr. Liter., 1960 (translated from English, Interscience Publishers, New York, London, 1957).

[29] CARLEMAN T., Zur Theorie der linearen Integralgleichungen, *Math. Zeitschrift 9* (1921) 196–217.

[30] SCHMEIDLER W., *Integralgleichungen mit Anwendungen in Physik und Technik. I. Lineare Integraleichungen*, Akad. Verlagsgesellsehaft, Leipzig, 1955.

REFERENCES FOR CHAPTER IV

[1] GANTMAKHER F. R., *The Theory Of Matrices* (3rd edition, 'Nauka', 1967), English translation by Interscience, J. Wiley, New York, 1959.

[2] GANTMAKHER F. R., KREIN M. G., *Oscillatory Matrices And Kernels, And Small Vibrations Of Mechanical Systems*, Gostekhizdat, 1950.

[3] KANTOROVICH L. V., VULIKH B. Z., PINSKER A. G., *Functional Analysis In Partly Ordered Spaces*, Gostekhizdat, 1950.

[4] KRASNOSEL'SKII M. A., *Positive Solutions Of Operator Equations*, Fizmatgiz, 1962.

[5] KREIN M. G., RUTMAN M. A., Linear operators, leaving the cone invariant in Banach space, *Usp. Mat. Nauk 3* (1948).

[6] MUKHTAROV S. N., STET'SENKO V. YA., On certain properties of equations with indecomposable linear operators, *Dokl. Acad. Nauk Tadzikskoi SSR 8* (1965).

[7] STET'SENKO V. YA., On an iterative method for the determination of spectral radius of linear positive operator, *Matem Sb. 67* (*109*) (1965).

[8] URISON P. S., Papers *On Topology and Other Fields Of Mathematics*, vol. I, Gostekhizdat, 1951, 45–77.

[9] ENTSCH R., Über Integralgleichungen mit positiven Kern, *J. für reine und angewandte Mathematik 141* (1912) 235–244.

[10] FRÉCHET M., Sur l'allure asymptotique des densités itérés dans les probabilités en chaîne, *Bull. de Soc. Mathem. de France 62* (1934) 68–83.

REFERENCES FOR CHAPTER V

[1] Banach S., *A Course In Functional Analysis* (Ukrainian translation), Kiev, 1948.

[2] Gokhberg I. Ts., Krein M. G., *Introduction To The Theory Of Linear Non-self-adjoint Operators*, Nauka, 1965.

[3] Gokhberg I. Ts., Krein M. G., *The Theory of Volterra Operators in Hilbert Spaces and Their Application*, Nauka, 1967.

[4] Goursat E., *Course On Mathematical Analysis*, vol. III, part I, Moscow, 1933. Translated from French (Cours d'Analyse Mathematique, Tome III, Pat II, Gauthier-Villars, Paris, 1923).

[5] Dunford N., Schwarz D. T., *Linear operations: General theory*, Inostr. Liter., 1963. Translated from English (Interscience, New York, 1958, 1963).

[6] Zabreyko P. P., Non-linear Integral Equations, *Trudy Seminara Po Funkzional Analizu* (Work of the Seminar on Functional Analysis) 8, Voronez, 1966.

[7] Zabreyko P. P., On the spectral radius of Volterra operators *Usp. Mat. Nauk 22* (1967) 167–168.

[8] Kantorovich L. V., On integral operators, *Usp. Mat. Nauk 11* (1956).

[9] Kantorovich L. V., Akilov G. P., *Functional Analysis In Normed Spaces* (Moscow, 1959), English translation by Pergamon Press, Oxford, 1964.

[10] Kantorovich L. V., Valikh B. Z., Pinsker A. G., *Functional Analysis In Partly Ordered Spaces*, Gostekhizdat, 1950.

[11] Krasnosel'skii M. A., *Positive Solutions Of Operator Equations*, Fizmatgiz, Moscow, 1962.

[12] Krasnosel'skii M. A., Zabreyko P. P., Pustyl'nik E. I., Sobolevskiy P. E., *Integral Operations In Space Of Summable Functions*, Nauka, Moscow, 1966.

[13] Krasnosel'skii M. A., Rutitskiy Ya. B., *Convex Functions and Orlicz Spaces*, Fizmatgiz, Moscow, 1959.

[14] Krein M. G., Rutman M. A., Linear operators, leaving the cone invariant in Banach space, *Usp. Mat. Nauk 3* (1948).

[15] Mikhlin S. G., *Lectures On Linear Integral Equations*, Fizmatgiz, 1959. English translation by Pergamon Press, New York, 2nd edition, 1964.

[16] Radon I., On linear functional transforms, *Usp. Mat. Nauk 1* (1936).

[17] Riesz F., Nagy B. Sz., *Lectures In Functional Analysis*, Moscow, 1954. English translation by Blackie, London, 1956.

[18] Smirnov V. I., *A Course In Higher Mathematics*, Vol. IV, 1957.

[19] Sobolev S. L., *Some Applications Of Functional Analysis To Mathematical Physics*, Novosibirsk, 1965.

[20] Stet'senko V. Ya., On an estimate of the spectrum of some classes of linear operators, *Dokl. Akad. Nauk 157* (1964).

[21] Tricomi F. G., *Integral Equations*, Inostr., Liter., 1960. English translation by Interscience Publishers, New York, London.

[22] HARDY G. G., LITTLEWOOD T. E., POLYA G., *Inequalities*, Inostr. Liter., 1948. (Cambridge University Press, 1934).
[23] ZAANEN A. C., *Linear Analysis*, New York–Amsterdam, 1953.

REFERENCES FOR CHAPTER VI

[1] ATKINSON F. V., Normal solvability of linear equations in normalized spaces, *Matem. Sbornik 28 (70)* (1951) 3–14.

[2] AKHIEZER N. I., On some inversion formulas for singular integrals, *Izv. Akad. Nauk SSSR 9* (1945) 275–290.

[3] BABAEV A. A., SALAEV V. V., On one analog of the Plemel-Privalov theorem for the case of non-smooth curves and its applications, *Dokl. Akad. Nauk 161* (1965) 267–269.

[3a] BABENKO K. I., On conjugate functions, *Dokl. Akad. Nauk SSSR 62* (1948) 157–160.

[4] VEKUA N. P., *Systems Of Singular Integral Equations And Some Boundary Problems*, GTTI, Moscow, 1950.

[5] GAKHOV F. D., *Boundary Value Problems*, 1st edition, 1958, 2nd edition, 1963, Fizmatgiz, Moscow.

[6] GEL'FAND I. M., SHILOV G. E., *Spaces Of Ordinary And Generalized Functions*, Fizmatgiz, Moscow, 1958.

[7] GEL'FAND I. M., SHILOV G. E., *Generalized Functions and Their Properties*, Fizmatgiz, Moscow, 1958.

[8] GERONIMUS YA. L., On certain integral equations, *Dokl. Akad. Nauk SSSR 98* (1954) 5–7.

[9] GOKHBERG I. Ts., On one application of the theory of normed rings to singular integral equations, *Usp. Matem. Nauk 7* (48) (1952) 149–156.

[10] GOKHBERG I. Ts., On systems of singular integral equations, *Uch. Zap. Khishinevsk, un-ta 11* (1954) 56–60.

[11] GOKHBERG I. Ts., On the number of solutions of the homogeneous, singular integral equation with continuous coefficients, *Dokl. Akad Nauk SSSR 122* (1958) 327–330.

[12] GOKHBERG I. Ts., On the range of applicability of Nöther-theorems, *Uch. Zap. Khishinevsk, un-ta 17* (1955) 35–44.

[13] ZYGMUND A., Trigonometric Series, vol. 1, Cambridge University Press, 1959.

[14] KARTSIVADZE N. I., On singular integral operators with discontinuous coefficients, *Dokl. Akad. Nauk SSSR 109* (1956) 450–452.

[15] KOSULIN A. E., One-dimensional singular equations in generalized functions, *Dokl. Akad. Nauk 163* (1965) 1054–1057.

[16] KREIN S. G. (editor), *Functional Analysis—A Reference Book*, first edition, Moscow, 1964. English translation Wolters–Noordhoff, Holland, 1972.

[17] MAGNARADZE L. G., On a generalization of Plemel–Privalov theorem, *Soobshch. Akad. Nauk Gruzinskoy SSR 8* (1947) 509–516.

[18] MAGNARADZE L. G., On a linear boundary problem of Riemann-Hilbert, *Soobshch. Akad. Nauk Gruzinskoy SSR 8* (1947) 585–590.

[19] MAGNARADZE L. G., On a generalization of the Privalov theorem and its application to certain boundary problems of the theory of functions and to singular integral equations, *Dokl. Akad. Nauk SSSR 68* (1949) 657–660.

[20] MARKUSHEVICH A. I., *The Theory of Analytic Functions*, Gostekhizdat, Moscow, 1950.

[21] MIKHLIN S. G., On the Tricomi integral equation, *Dokl. Akad. Nauk 59* (1948) 1053–1056.

[22] MIKHLIN S. G., Singular integral equations, *Usp. Matem. Nauk 3* (*25*) (1948) 29–112.

[23] MIKHLIN S. G., *Multidimensional Singular Integrals And Integral Equations* (Moscow, 1962), English translation by Pergamon, 1965.

[24] MIKHLIN S. G., On the calculation of the index of the system of one-dimensional singular equations, *Dokl. Akad. Nauk SSSR 168* (1966) 1248–1250.

[25] MUSKHELISHVILI N. I., *Singular Integral Equations*, Moscow, 1946—first edition, 1962—second edition; English translation by P. Noordhoff, Groningen-Holland, 1953.

[26] MUSKHELISHVILI N. I., VEKUA N. P., Riemann boundary problem for several unknown functions and its application to systems of singular integral equations, *Trudy Tbil. Mat. In-ta 12* (1943) 1–46.

[27] MUSKHELISHVILI N. I., KVESELAVA D. A., Singular integral equations with Cauchy-type kernels on unclosed boundaries, *Trudy Tbil. Mat. in-ta 11* (1942) 141–172.

[28] PRÖßDORF S., Operators admitting an unbounded regularization, *Vestn. Leningr. Gos. Univ., ser. matem., mekh. astr., 13* (*3*) (1965) 59–67.

[29] PRÖßDORF S., On the theory of singular integral systems with degenerate symbolic matrix, *Vestn. Leningr. Gos. Univ., ser. matem., mekh. astr., 19* (*4*) (1963) 58–73.

[30] PRIVALOV I. I., *The Cauchy Integral*, Saratovsk, un-t, 1919.

[31] PRIVALOV I. I., *Boundary Properties Of Analytic Functions*, Gostekhizdat, Moscow, 1950.

[32] ROGOZHIN V. S., Riemann boundary problem in the space of generalized functions and Faber polynomials, *Dokl. Akad. Nauk SSSR 152* (1963) 1308–1311.

[33] ROGOZHIN V. S., General theory for the solution of boundary problems in the space of generalized functions, *Dokl. Akad. Nauk SSSR 164* (1965) 277–280.

[34] SIMONENKO I. B., Riemann boundary problem with measurable coefficient, *Dokl. Akad. Nauk 135* (1960) 538–541.

[35] TRICOMI F. G., *On Linear Equations Of Mixed Type* (translated from Italian), Gostekhizdat, 1947.

[36] FREIDKIN S. A., The solution of one class of singular integral equations, *Uch. Zap. Khishinevsk. un-ta 11* (1954) 13–17.

[37] FREIDKIN S. A., The Riemann boundary problem and singular integral equations with piecewise constant coefficients in the case of countable set of intervals, *Uch. Zap. Khishinev. un-ta 70* (1964) 27–38.

[38] KHALILOV Z. I., Linear singular equations in the unitary ring, *Matem. Sbornik 25* (*67*) (1949) 169–188.

[39] KHARAZOV D. F., KHVEDELIDZE B. V., Some notes to the theory of singular integral equations with Cauchy type kernel, *Soobshch. Akad. Nauk Gruzinsk. SSR 28* (1962) 129–135.

[40] KHVEDELIDZE B. V., On the Poincare boundary problem in the theory of logarithmic potential for multi-connected region, *Soobshch Akad. Nauk Gruzinsk. SSR 2* (1941) 577–578, 865–872.

[41] KHVEDELIDZE B. V., Linear discontinuous boundary problems of the theory of functions, singular integral equations and some applications of them, *Trudy In-ta matem. Akad. Nauk Gruzinsk. SSR 23* (1956) 3–158.

[42] HÖRMANDER L., *Linear Differential Operators*, Springer, 1965.

[43] CHIKIN L. A., Particular cases of the Riemann boundary problems and singular integral equations, *Uch. Zap. Kazansk un-ta 113* (1953) 57–105.

[44] SHERMAN D. I., On some methods of solutions of certain singular integral equations, *Prikl. matem. i. mekh. XII* (1948) 423–452.

[45] SHERMAN D. I., On one regularization of singular equations, Nauka, *Prikl. matem. i. mekh. 15* (1952) 75–82.

[46] SHILOV G. E., *Mathematical Analysis. The Second Special Course*, Moscow, 1965.

[47] CARLEMAN T., Sur la résolution de certaines équations intégrales, *Arkiv för Mathematik, Astronomi och Physik 16 (26)* (1922).

[48] GELLERSTEDT S., Quelques problèmes mixtes pour l'équation $y^m z_{xx} + z_{yy} = 0$. *Arkiv för Mathematik, Astronomi och Physik 26 (3)* (1938).

[49] GIRAUD G., Sur une classe d'équations intégrales où figurent des valeurs principales d'intégrales simples, *Ann. Sci. de l'École Norm. Sup. 46* (1939) 119–172.

[50] HORVÀTH T., Sur l'itération de la transformée de Hilbert d'une distribution complexe, *C.R. Acad. Sci. 237* (1953) 1480–1482.

[51] NÖTHER F., Über eine Klasse singulärer Integralgleichungen, *Math. Ann. 82* (1921) 42–63.

[52] PRZEWORSKA-ROLEWICZ D., Équations avec opérations algébriques, *Studia Mathem. 22* (1963) 337–367.

[53] SCHWARZ L., Théorie des distributions, Tomes I, II. Paris, 1950.

[54] SOBOLEFF S. L., Méthode nouvelle à résoudre le problème de Cauchy pour les équations lineaires hyperboliques normales, *Matem. Sbornik 1 (43)* (1936) 39–72.

REFERENCES FOR CHAPTER VII

[1] BABICH V. M., et al., *The Linear Equations Of Mathematical Physics—A Reference Book*, Nauka, Moscow, 1964.

[2] VEKUA I. N., *New Methods For The Solution of Elliptic Equations*, Gostekhizdat, Moscow, 1948.

[3] GOURSAT É., *Course On Mathematical Analysis*, vol. III, p. 2., Moscow, 1934. Translated from French. (Cours d'analyse mathématique, tome III, pt. II, Gauthier-Villars, Paris, 1923).

[4] GUNTHER N. M., *The Theory of Potential and Its Application To Basic Problems Of Mathematical Physics*, Moscow, 1953. Translated from French: La théorie du potentiel et ses application aux problèmes fondamentaux de la physique mathématic, Paris, 1934.

[5] COURANT R., and HILBERT D., Methods of Mathematical physics, vol. I, Moscow, 1933—1st edition. English translation by Interscience Publishers, New York and London, 1953.

[6] MIKHLIN S. G., Integral Equations (2nd edition Moscow, 1949), English translation by Pergamon Press, New York, 2nd edition, 1964.

[7] MUSKHELISHVILI N. I., *Some Basic Problems of The Mathematical Theory of Elasticity*, Izdat. Akad. Nauk SSSR, 1949. English translation by Wolters-Noordhoff, Groningen, Holland, 1953.

[8] MUSKHELISHVILI N. I., *Singular Integral Equations*, 2nd edition Moscow, 1962; English translation by W. Noordhoff, Groningen, Holland, 1953.

[9] MÜNTZ G. U., Integral Equations, GTTI, Moscow, 1934.

[10] PRIVALOV I. I., Integral Equations, GTTI, Moscow, 1934.

[11] SMIRNOV V. I., *A Course in Higher Mathematics*, vol. IV (Moscow, 1951); English translation by Pergamon Press, Oxford, 1964.

[12] SOBOLEV S. L., *Equations of Mathematical Physics*, Gostekhizdat, Moscow, 1954.

[13] TIKHONOV A. N., SAMARSKIY A. A., *Equations of Mathematical Physics*, Moscow, 1966.

[14] TRICOMI, F., *Integral Equations*, Interscience Publishers, New York, London, 1957.

REFERENCES FOR CHAPTER VIII

[1] BERKOVICH F. D., On one integral equation on a half-axis, *Izvestia Vuzov, 'Matematika' 1* (1966) 13–14.

[2] VEKUA N. P., Systems of singular integral equations and some boundary problems, *Gostechizdat*, Moscow, 1950.

[3] PALEY R., WIENER N., *Fourier Transforms In The Complex Domain*. Published by the American Mathematical Society, New York, 1934.

[4] GANIN M. P., On a Fredholm equation with its kernel depending on the difference of arguments, *Izv. Vuzov, 'Matematika' 2* (1963) 31–34.

[5] GAKHOV F. D., Riemann boundary problem for the system of *n*-pairs of functions, *Usp. Matem. Nauk VII (4)* (1952) 3–54.

[6] GAKHOV F. D., *Boundary Problems*, Fizmatgiz, Moscow, 1958.

[7] GAKHOV F. D., SMAGINA V. I., Particular cases in integral equations of convolution type and equations of the first kind, *Izv. Akad. Nauk SSSR, ser. matem. 26* (1962) 361–390.

[8] GAKHOV F. D., CHERSKIY Y. I., Particular integral equations of convolution type, *Izv. Akad. Nauk SSSR 20* (1956) 32–52.

[9] GOKHBERG I. Ts., On the range of applicability of F. Nöther theorems, *Uch. Zap. Khishinevsk. un-ta XVII* (1965) 35–44.

[10] GOKHBERG I. Ts., KREIN M. G., Systems of integral equations on the half-line with kernels depending on the difference of arguments, *Usp. Matem. Nauk. XIII (2)* (1958) 3–72.

[11] GOKHBERG I. Ts., KREIN M. G., On a coupled integral equation and its transpose, *Teoreticheskaya i prikladnaya matematika, Izd. Lvovskogo un-ta 1* (1959) 58–78.

[12] GOKHBERG I. Ts., Problems of factorization in normalized rings of functions of isometric and symmetric operators and singular integral equations, *Usp. Matem. Nauk XIX (1)* (1964) 74–124.

[13] GOKHBERG I. Ts., FEL'DMAN I. A., On the approximate solution of some classes of linear equations, *Dokl. Akad. Nauk SSSR 160* (1965) 750–754.

[14] GOKHBERG I. Ts., FEL'DMAN I. A., On the method of reduction for systems of equations of Wiener-Hopf type, *Dokl. Akad. Nauk Nauk SSSR 165* (1965) 268–271.

[15] GRENANDER U., SZEGÖ G., *Toeplitz Forms And Their Applications*, University of California Press, Berkeley-Los Angeles, 1958.

[16] GRINBERG G. A., FOK V. A., *On The Theory Of Cliff-top Refraction Of Electromagnetic Waves*. Survey of radio wave propogation. Collected papers, vol. II, (edited by B. A. Vvedenskiy), Izdat. Akad. Nauk SSR, Moscow-Leningrad, 1948.

[17] DAVENPORT W. B., ROTT W. L., *An Introduction To The Theory Of Random Signals And Noise*, McGraw-Hill, New York, 1958.

[18] DOMBROVSKAYA I. N., On the approximate solution of a general singular integral equation, *Matem. zaniski Uralskogo un-ta IV* (1963) 38–45.

[19] DYBIN V. B., A particular case of integral equations of convolution type on the class of generalized functions, *Dokl. Akad. Nauk 161* (1964) 753–756.

[20] ZARIPOV R. H., Systems of particular integral equations of convolution type, *Izv. vuzov, 'Mathematika' 2* (1958) 93–105.

[21] IVANOV V. V., KARAGODOVA E. A., Approximate solution of convolution type equations by means of the Galerkin procedure, *Ukrainskiy Matem. Zhurnal 13* (1961) 28–38.

[22] KREIN M. G., Integral equations on the half-line with the kernel depending on the difference of arguments, *Usp Matem. Nauk XIII (5)* (1958) 3–103.

[23] LENING G. H., BETTIN R. G., *Random Processes In Problems Of Automatic Control*. Inostrannaya Literatura, Moscow, 1958.

[24] MIKHLIN S. G., Fourier integrals and multiple singular integrals, *Vestnic Leningr. Gos. Univ., ser Matem., mekh. i. astr. 7 (2)* (1957) 143–155.

[25] MORSE PH. M., FESHBACH H., *Methods Of Theoretical Physics*, McGraw-Hill book company, New York, 1953.

[26] *Some Mathematical Problems Of Neutron Physics; Collected Papers*. Izdat., Moscow Univ., 1960.

[27] NOBL B., *Methods Based On The Wiener-Hopf Technique For The Solution Of Partial Differential Equations*, Pergamon Press, New York, London, 1958.

[28] PISARENKO V. F., ROZANOV YU. A., On some problems connected with stationary processes involving integral equations similar to Wiener-Hopf type, *Problemi peredachi informatzii, 14*, 113–135. *Izd-vo Akad. Nauk SSSR*, 1963.

[29] RABOTNIKOV YU. N., Equilibrium of elastic media with after-effect, *Prikl. matem. i. mekh. XII* (1948) 58.

[30] RAKOVSHCHIK L. S., On the theory of convolution type equations, *Usp. Matem. Nauk 18* (1963) 171–177.

[31] RAKOVSHCHIK L. S., On the effective solution of some convolution type equations, *Trudy Lening. Inzh.-econ. in-ta, 63* (1966) 69–77.

[32] RAKOVSHCHIK L. C., Integral equations with almost convolution kernels, *Vestnic. Lening. Gos. Univ., ser. matem., mekh. i. astr. (3)* (1961) 52–72.

[33] RAKOVSHCHIK L. D., On the calculation of the index of almost convolution system of integral equations, *Sibir. matem. zhurn. 5* (1964) 904–909.

[34] RAPOPORT I. M., On some coupled integral and integrodifferential equations, *Sb. trudov In-ta matemat. Akad. Nauk SSSR 12* (1949) 102–118.

[35] SAMIKO S. G., General singular equation in a particular case, *Differen. uravnenya 1* (1965) 1108–1115.

[36] SIMONENKO I. B., On some integro-differential equation of convolution type, *Izv. vuzov, 'Matematika' 2* (1959) 213–226.

[37] SIMONENKO I. B., On some systems of convolution type integral equations, *Izv. vuzov, 'Matematika' 6* (1962) 119–130.

[38] SMAGINA V. I., Particular cases of convolution type equations and corresponding equations of the first kind in the class of functions of exponential growth; Equations of A-class, *Dokl. Akad. Nauk BSSR 7* (1963) 12–16.

[39] SMAGINA V. I., Particular cases of convolution type equations and corresponding

equations of the first kind in the class of functions of exponential growth; Equations of B-class, *Dokl. Akad. Nauk BSSR 6* (1963) 76–79.

[40] Smirnov V. I., *A Course In Higher Mathematics*, vol. IV, English translation by Pergamon Press, Oxford, 1964.

[41] Sobolev B. B., *Radiative Transfer In The Atmosphere Of Stars And Planets*, Gostechizdat, Moscow, 1956.

[42] Titchmarsh E. C., *Introduction To The Theory Of Fourier Integrals*, Gostekhizdat, Moscow-Leningrad, 1948.

[43] Fel'dman I. A., On the asymptotic solutions of systems of Wiener-Hopf type integral equations, *Sibir. matem. zhurn. 6* (1965) 596–615.

[44] Fok V. A., On some integral equations of mathematical physics, *Matem. sbornik 14* (1943) 3–48.

[45] Hörmander L., Estimates for translation invariant operators in L^p spaces, *Acta Mathematica 104* (1960) 93–140.

[46] Chandrasekhar S., *Radiative Transfer*, Oxford, 1950.

[47] Chebotarev G. N., On rings of functions integrable with a weight, *Izv. vuzov, 'Matematika' 5* (1963) 133–145.

[48] Cherskiy Yu. I., Convolution type equations, *Izv. Akad. Nauk SSSR, ser. matem. 22* (1958) 361–378.

[49] Cherskiy Yu. I., On the solution of mixed problems for partial differential equations, *Differential uravnenia 1* (1965) 647–662.

[50] Asplund E., The Wiener-Hopf equation in an algebra of Beurling, *Acta Math. 108* (1964) 89–111.

[51] Fuchs W. H. I., On the eigenvalues of an integral equation arising in the theory band limited signals, *J. Math. Analysis and appl. 9* (1964) 317–330.

[52] Kac M., Distribution of eigenvalues of certain integral equations, *Michigan Math. J. 3* (1955–1956) 141–148.

[53] London H. I., The eigenvalue behaviour of certain convolution equations, *Trans. Amer. Math. Soc. 115* (1965) 242–256.

[54] Rosenblatt M., Some results on the symptotic behaviour of eigenvalues for a class of integral equations with translation kernels, *Journ. of Math. and Mech. 12* (1963) 619–628.

[55] Slepian D., Pollak H. O., Prolate Spheroidal Wave Functions, Fourier analysis and uncertainty I, *Bell. Syst. tech. Journ. XL* (1961) 40–63.

[56] Volterra V., *Leçons sur les fonctions des lignes*, Paris, Gauthier-Villars, 1913.

[57] Wadom H., Equations of Wiener-Hopf type, *Illinois. Journ. of Mathem. 2* (1958) 261–270.

[58] Widom H., Extreme eigenvalues of translation kernels, *Trans. Amer. Math. Soc. 100* (1961) 252–276.

[59] Widom H., Asymptotic behaviour of the eigenvalues of certain integral equations, *Trans. Amer. Math. Soc. 109* (1963) 278–295.

[60] Widom H., Rapidly increasing Kernels, *Proc. Amer. Math. Soc. 14* (1963) 501–506.

[61] Widom H., Asymptotic behaviour of the eigenvalues of certain integral equations II, *Archiv. for rational Mech. and Analysis 17* (1964) 215–229.

[62] WIENER N., *The interpolation, extrapolation and smoothing of stationary time series*, New York, 1949.
[63] YOULA D., On the factorization of rational matrices, *IRE Trancs. inform. theory*, *IT-7* (1961) 172–188.

REFERENCES FOR CHAPTER IX

[1] Agranovich M. S., Elliptic singular integro-differential operators, *Usp. Matem. Nauk 20* (1965) 3–120.

[2] Berikashvili N. A., On the index of systems of singular integral equations on two-dimensional varieties, *Soobshch, Akad. Nauk Gruzin. SSR 34* (1964) 257–264.

[3] Vishik M. I., Eskin G. I., Convolution equations in a bounded domain, *Usp. Matem. Nauk 20* (1965) 89–152.

[4] Vol'pert A. I., On the index of two-dimensional singular integral equations, *Dokl. Akad. Nauk SSSR 142* (1962) 776–778.

[5] Vol'pert A. I., Elliptic equations on a sphere and two-dimensional singular integral equations, *Matem. Sbornik 59* (1962) 195–214.

[6] Gegelia T. G., On the boundedness of singular operators, *Soobshch. Akad. Nauk Gruzinsk, SSR XX* (1958) 517–523.

[7] Gokhberg I. Ts., On the theory of multi-dimensional singular integral equations, *Dokl. Akad. Nauk SSSR 132* (1952) 1279–1282.

[8] Gokhberg I. Ts., Certain questions of the theory of multidimensional singular integral equations, *Izv. Moldav, filiala Akad. Nauk SSSR 10* (*76*) (1960) 39–50.

[9] Dynin A. S., Singular operators of arbitrary order on a variety, *Dokl. Akad. Nauk SSSR 141* (1961) 285–287.

[10] Itskovich I. A., The inversion of Jiro's formulas, *Uch. zap. Khishinev. un-ta 11* (1954) 7–11.

[11] Krupnik N. Ya., On the continuity of multi-dimensional singular operator in the space of basic and generalized functions, *Usp. zap. Khishinev. un-ta 50* (1962) 111–117.

[12] Krupnik N. Ya., On the multi-dimensional singular operators in spaces of basic and generalized functions, *Dokl. Akad. Nauk SSSR 157* (1964).

[13] Krupnik N. Ya., On multi-dimensional singular operators in spaces of basic and generalized functions, *Uch. zap. Khishinov. un-ta 70* (1964) 39–48.

[14] Krupnik N. Ya., On multi-dimensional singular integral equations, *Usp. Matem. Nauk 20* (1965) 119–123.

[15] Kupradze V. D., *Methods Of Potential Theory In The Theory Of Elasticity*, Fizmatgiz, Moscow, 1963.

[16] Maz'ya V. G., Plamenevskiy B. A., (a) On singular equations with vanishing symbol, *Dokl. Akad. Nauk SSSR 160* (1965) 1250–1253; (b) On the Cauchy problem for hyperbolic singular integral equations of convolution type, *Vestnic. Lening. Univ. ser. matem. mekh. i. astron.* (*4*) (1965) 161–163.

[17] Mikhailova-Gubenko N. M., Singular integral equations in Lipschitz spaces; I. *Vestnic. Lening. Univ. 1* (1966) 51–63; II. *Vestnic. Lening. Univ. 7* (1966) 45–57.

[18] Mikhlin S. G., Multi-dimensional Singular Integrals and Integral Equations (Moscow, 1962); English translation by Pergamon, 1965.

[19] MIKHLIN S. G., On the index of the system of singular equations, *Dokl. Akad. Nauk SSSR 152* (1963) 555–558.

[20] MIKHLIN S. G., On the index of the system of singular integral equations, *Papers At the Joint Soviet-American Symposium on Partial Differential Equations*, Novosibirsk, 1963.

[21] PLAMENEVSKIY B. A., (a) Singular integral equations on the infinite cylinder, *Dokl. Akad. Nauk 158* (1963), 780–783; (b) On singular integral operators on a group, *Dokl. Akad. Nauk SSSR 164* (1965) 47–50.

[22] SIMONENKO I. B., Boundedness of singular operators in Orlicz space, *Sokl. Akad. Nauk SSSR 130* (1960) 984–987.

[23] SLOBODETSKIY L. N., Generalized Sobolev spaces and their applications to boundary problems of partial differential equations, *Uch. zap. Lening. pedag. in-ta 197* (1958) 54–112.

[24] SMIRNOV V. I., *A Course In Higher Mathematics*, Vol. V (Moscow, 1959); English translation by Pergamon Press, Oxford, 1964.

[25] SOBOLEV S. L., *Some Applications Of Functional Analysis In Mathematical Physics*, Izdat. Lening. Univ., 1950.

[26] ATIYAH M. F., SINGER I. M., The index of elliptic operators on compact manifolds, *Bull. Amer. Math. Soc. 69* (1963).

[27] BOJAESKI B., On the index problem for systems of singular integral equations, *Bull. Acad. Polon. Scienc. ser. math. astr. et. phys. 11* (1963).

[28] CALDERON A. P., Uniqueness in the Cauchy problem for partial differential equations, *Amer. Journ. of Mathem. 80* (1958) 16–36.

[29] CALDERON A. P., ZYGMUND A., On singular integrals, *Amer. Journ. of Mathem. 78* (1956) 289–309.

[30] CALDERON A. P., ZYGMUND A., Singular integral operators and differential equations, *Amer. Journ. of Mathem. 79* (1957) 901–921.

[31] GIRAUD G., Equations à intégrals principales, *Ann. Sc. de l'Ecole Norm. Sup. 51* (1934) 251–372.

[32] HORVÀTH J., Sur l'iteration de la transforme de Hilbert d'une distribution complexe, *C.R. Acad. Sci. 237* (1953) 1480–1482.

[33] HORVÀTH J., Transformadas de Hilbert de distribuciones, *2nd symposium sobre algunos problemas matematicas que se están estudiando en Latino America, Montevideo*, (1954) 61–70.

[34] HORVÀTH J., Hilbert transforms of distributions in R^n, *Proc. Internat. Congr. Math. 2*, Amsterdam (1954) 122–123.

[35] HORVÀTH J., Singular integrals and spherical harmonics, *Trans. Amer. Math. Soc. 82* (1956) 52–63.

[36] KOIZUMI SUMIYUKI, On singular integrals, I, II, III, *Proc. Japan. Acad. 34* (1958) 193–198, 235–240, 594–598.

[37] MALGRANGE BERNARD, (a) Opérateurs intégraux singuliers, théorèmes d'interpolation dans les espaces L^p; (b) Multiplicateurs de L^p; (c) Multiplicateurs de L^p; (d) Multiplicateurs de L^p; (e) Noyaux valeurs principales; (f) Noyaux valeurs principales; (g) Calcul symbolique approximatif. *Sémin. Schwarz. Fac. Sciences Paris*, 1959–1960, 4 année, Paris, 1960.

References for chapter IX

[38] SEELEY R. T., Singular integrals on compact manifolds, *Amer. Journ. of Mathem. 81* (1959) 658–690.

[39] SEELEY R. T., Integro-differential operators on vector bundles, *Trans. Amer. Math. Soc. 117* (1965) 167–204.

[40] SÉMINAIRE HENRI CARTAN, *Théorème d'Atiyah—Singer sur l'indice d'un operateur differential elliptique fasc. 1 et 2*, Paris (1965).

[41] STEIN E. M., Note on singular integrals, *Proc. Amer. Math. Soc. 8* (1957) 250–254.

[42] ERDÉLYI A., MAGNUS W., OBERHETTINGER F., TRICOMI F. G., *Higher transcendental functions*, Vol. 2, 1953.

[43] YAMAGUTI MÁSAYA, Sur l'inégalite d'energie pour le système hyperbolique, *Proc. Japan. Acad. 35* (1959) 37–41.

REFERENCES FOR CHAPTER X

[1] Bochner S., Martin W. T., *Functions of Several Complex Variables*, Princeton, 1948.
[2] Vainberg M. M., *Variational Methods In The Study Of Non-linear Operators*, Gostekhizdat, 1956; translation by Holden-Day, San Francisco, 1964.
[3] Gel'man A. E., Implicit abstract function theorems, *Dokl. Akad. Nauk SSSR 132* (1960) 501–503.
[4] Gel'man A. E., On analytic solutions of essentially non-linear equations, *Dokl. Akad. Nauk SSSR 144* (1962) 19–20.
[5] Gel'man A. E., On simple solutions of operator equations in the case of bifurcation, *Dokl. Akad. Nauk SSSR 152* (1963) 1142–1144.
[6] Zabreyko P. P., Non-linear integral operators, *Trudy seminara po functsional'momu analysu*, 8, Voronezh, 1966.
[7] Zabreyko P. P., Krasnosel'skii M. A., On the calculation of the index of the fixed point of completely continuous vector field, *Sibir. Matem. Zhurn. 5* (1964).
[8] Kantorovich L. V., Akilov G. P., *Functional Analysis In Normed Spaces* (Moscow, 1959); English translation by Pergamon Press, Oxford, 1964.
[9] Krasnosel'skii M. A., Topological Methods In The Theory of Non-linear Integral Equations (Moscow, 1956); English translation by Pergamon Press, Oxford, 1964.
[10] Krasnosel'skii M. A., *Positive Solutions of Operator Equations*, (Moscow. 1962); English translation by Wolters-Noordhoff, Groningen, Holland.
[11] Krasnosel'skii M. A., Zabreyko P. P., Pustil'nik E. I., Sobolevskiy P. E., *Integral Operators In Spaces Of Summable Functions*, Fizmatgiz, Moscow, 1966.
[12] Krasnosel'skii M. A., Rutitskiy Ya. B., *Convex Functions and Orlicz Spaces*, Fizmatgiz, Moscow, 1959.
[13] Krein M. G., Rutman M. A., Linear operators leaving the cone invariant in Banach space, *Usp. Matem. Nauk 3* (1948).
[14] Lusternic L. A., Sobolev V. I., *Elements of Functional Analysis*, Fitzmatgiz, Moscow, 1966.
[15] Liapunov A. M., Sur les figures d'équilibre peu différentes des ellipsiodes d'une masse liquide homogene douée d'un mouvement de rotation, *Zapiski Akad. Nauk, St.-Peterburg* (1906).
[16] Melamed V. B., On the calculation of rotation of the completely continuous vector field in the critical case, *Sibir. Matem. Zhurn 2* (1961).
[17] Melamed V. B., On bifurcation points of one class of equations, *Dokl. Akad. Nauk SSSR 152* (1963).
[18] Melamed V. B., On formal solutions of certain non-linear equations, *Sibir. Matem. Zhurn 6* (1965).

[19] NAZAROV N. N., Integral equations of Hammerstein type, *Trudy Sredne-Aziatskogo Gos. Univers. 33* Tashkent (1941).
[20] NEKRASOV A. I., Exact theory of steady waves on the surface of heavy liquid, *Izdat. Akad. Nauk SSSR* (1950).
[21] TRICOMI F. G., *Integral Equations*, Interscience Publishers, New York, London, 1957.
[22] SCHMIDT E., Zur Theorie der linearen und nichtlinearen Integralgleichungen, III Teil. Über die Auflösung der nichtlinearen Integralgleichungen und die Verzweigung ihrerer Lösungen, *Math. Ann. 65* (1908).

INDEX